Tourism, Hospitality & Event Management

This book series covers all topics relevant to the tourism, hospitality, and event industries. It includes destination management and related aspects of the travel and mobility industries as well as effects from developments in information and communication technologies. "Tourism, Hospitality & Event Management" embraces books both for professionals and scholars, and explicitly includes undergraduate and advanced texts for students. In this setting, the book series reflects the close connection between research, teaching, and practice in tourism research and tourism management and the related fields.

This series is indexed in SCOPUS.

Jorge Marques • Makhabbat Ramazanova
Helena Albuquerque
Editors

Tourism and Climate Change

Research on Global Challenges and Opportunities

Springer

Editors
Jorge Marques
REMIT—Research on Economics,
Management and Information Technologies
University Portucalense
Porto, Portugal

Makhabbat Ramazanova
REMIT—Research on Economics,
Management and Information Technologies
University Portucalense
Porto, Portugal

Helena Albuquerque
REMIT—Research on Economics,
Management and Information Technologies
University Portucalense
Porto, Portugal

ISSN 2510-4993 ISSN 2510-5000 (electronic)
Tourism, Hospitality & Event Management
ISBN 978-3-031-99127-1 ISBN 978-3-031-99128-8 (eBook)
https://doi.org/10.1007/978-3-031-99128-8

This Springer imprint is published by the registered company Springer Nature Switzerland AG
The registered company address is: Gewerbestrasse 11, 6330 Cham, Switzerland

Preface

Climate change is one of our most pressing global challenges, affecting multiple sectors of human activity. Tourism is particularly vulnerable to its impacts and a notable contributor to its causes. The book *Tourism and Climate Change: Challenges and Opportunities* brings together a diverse collection of chapters that explore the challenges and opportunities arising from the relationship between tourism and climate change from various perspectives and geographic contexts. This volume aims to provide an integrated and progressive reading experience, enabling readers to understand this complex phenomenon, reflect on its impacts, and engage with strategies for mitigation, adaptation, and sustainable transformation within the tourism sector.

The book is structured in three main parts, following a logical progression from broad theoretical and global perspectives to applied tools and strategies, culminating in case studies based on specific local realities. The first part, *Global and Theoretical Perspectives on Tourism and Climate Change*, gathers chapters that offer analytical and systemic views on the phenomenon. Makhabbat Ramazanova, Helena Albuquerque, and Jorge Marques open this section with *Climate Change and Tourism: A Global Challenge with Regional Impacts and Responses*, which examines the differentiated regional impacts of climate change on tourism and the corresponding adaptation and mitigation strategies. This is followed by *Tourism, Hospitality and Climate Changes: Insights in the Last Decade* by Ahmed H. Alsharif, Salmi Mohd Isa, and Lina Pilelienė, presenting a systematic literature review of research produced between 2012 and 2022, mapping key trends and gaps in the field. In *Sustainable Tourism and its Role in Economic Development*, Abhijit Chakraborty, Nisha Goyal, Abhijit Ranjan Das, and Priyanka Roy reflect on the contributions of sustainable tourism to economic growth while addressing the associated structural and environmental challenges. Micaela Pinho and Fátima Leal, in *The European Union's Potential in Mitigating Climate Change Caused by Tourism*, employ machine learning techniques to classify the pro-environmental behaviour of tourists from the 27 EU Member States, offering insights with strong policy relevance. Concluding this first section, Sanna-Mari Renfors, in *Educating for the Circular Economy: The Key Competencies of Future Tourism Professionals*, proposes a set of core competencies to be promoted in tourism higher education, aiming to support the transition to a circular economy within the sector.

The second part of the book, *Tools, Models and Strategies for Tourism Sustainability*, focuses on policy instruments, educational approaches, and conceptual frameworks applicable to sustainable tourism development. Marco Martins, Ricardo Jorge da Costa Guerra, Lara Santos, Luísa Lopes, and Ana Rita Conde contribute with *Enhancing Green Capability in the Hospitality Sector through Nudging: A Conceptual Framework*, where they propose a model linking nudging techniques to green capability building in the hospitality industry. Also addressing the regulatory dimension, Susana Cristina Rodrigues Aldeia, in *Taxation on Non-Recoverable Packaging as a Sustainability Measure in Tourist-Intensive Regions*, analyses Portuguese environmental tax policy on single-use packaging and its implications for sustainable practices in high-tourism regions. From a South American context, *Capacity Building Initiatives to Foster Sustainable Tourism in a Climate Change Context in Bolivia* by Georgina Chávez, Vidfa Garvizu, Freddy Soria, and Omar Salinas explores professional training initiatives as vital tools for building climate resilience in Bolivia's emerging tourism destinations. In the final chapter of this section, *Exploring the Ecomuseum Paradigm for Sustaining Siwa's Cultural Heritage Amid Climate Change*, Shaimaa Matar discusses how the ecomuseum model can serve as a sustainable, community-based tourism strategy in Egypt, combining cultural preservation with climate adaptation.

The third and final part, *Case Studies and Local Approaches to Sustainable Tourism*, presents context-specific examples of climate adaptation and sustainability in action. In *The Impact of Home-Based Food Enterprises for Sustainable Tourism in Bali*, I Ketut Sutapa, Pondang Polikarpus Nainggolan, I Nengah Subadra, and I Made Bayu Wisnawa examine the role of small culinary enterprises in fostering cultural preservation and sustainable tourism in Bali. Also based in Bali, L.K. Herindiyah Kartika Yuni and I Made Bayu Wisnawa present *Capturing Laduma Tourism Village: A Renewable Energy Based-Tourism Destination*, describing the development of a tourism model powered by renewable energy and structured using the Business Model Canvas. In Vietnam, Bui Thi Nga and Ho Ngoc Ninh, through *Climate Change Resilience in Tourism Destinations: An Exploratory Study of Tan Hoa Weather Adaptive Tourism Village in Vietnam*, explore how a rural community transformed climate adversity into a driver for innovative tourism development. In *Tourism and Agro-Silvopastoral Systems as Alternatives to Counteract Climate Change: Case of Cotopaxi, Ecuador*, Marco Antonio Toledo, José Luis Núñez, Katherine Joselyn Villarreal, and Nancy Carolina Montenegro propose an integrated model of agrotourism and agroforestry systems as a sustainable resource management strategy. Finally, *Sustainable Practices in Rural Tourism: Promoting Environmental Responsibility and Community Engagement in India* by Madhu Kumari, Sandeep Guleria, Suneel Kumar, and Nisha Devi discusses environmentally responsible tourism practices adopted in rural India, highlighting community engagement as a key pillar of sustainability.

By combining theoretical reflections, practical frameworks, and locally grounded examples, this book offers a comprehensive overview of the intersections between tourism and climate change. It is intended for researchers, students, professionals, and policymakers interested in understanding and promoting a more resilient, sustainable, and equitable future for tourism. We hope that the insights gathered in this volume will inspire further research, innovative practices, and meaningful policy actions in response to one of the most urgent challenges facing the tourism sector today.

Porto, Portugal Jorge Marques
Porto, Portugal Makhabbat Ramazanova
Porto, Portugal Helena Albuquerque

Contents

Global and Theoretical Perspectives on Tourism and Climate Change

Climate Change and Tourism: A Global Challenge with Regional Impacts and Responses

Makhabbat Ramazanova, Helena Albuquerque, and Jorge Marques

Abstract Climate change is a pressing global challenge that significantly impacts the tourism sector. At the same time, it is influenced by the sector's greenhouse gas emissions. This chapter examines the bidirectional relationship between tourism and climate change through a regional lens, analysing vulnerabilities, adaptation strategies, and mitigation efforts across Africa, the Americas, Asia-Pacific, Europe, and the Middle East. Findings highlight that tourism-dependent regions, mainly coastal and nature-based destinations, face heightened risks from rising temperatures, extreme weather events, and changing precipitation patterns. Concurrently, the tourism industry—primarily through air travel, accommodation, and activities—contributes substantially to global carbon emissions. Regional disparities in climate change impacts and in policy responses are evident, with some destinations implementing proactive adaptation measures while others remain highly vulnerable. This chapter underscores the necessity for collaborative governance, industry-wide carbon reduction commitments, and innovative destination management strategies to enhance tourism resilience. Future research should explore region-

M. Ramazanova
REMIT—Research on Economics, Management and Information Technologies, University Portucalense, Porto, Portugal
e-mail: ramazanova@upt.pt

H. Albuquerque
REMIT—Research on Economics, Management and Information Technologies, University Portucalense, Porto, Portugal

GOVCOPP—Research Unit on Governance, Competitiveness and Public Policies, University of Aveiro, Aveiro, Portugal
e-mail: helenaa@upt.pt

J. Marques (✉)
REMIT—Research on Economics, Management and Information Technologies, University Portucalense, Porto, Portugal

CEGOT—Centre of Studies in Geography and Spatial Planning, University of Coimbra, Coimbra, Portugal
e-mail: jorgem@upt.pt

J. Marques et al. (eds.), *Tourism and Climate Change*, Tourism, Hospitality & Event Management, https://doi.org/10.1007/978-3-031-99128-8_1

specific best practices and policy frameworks to ensure sustainable tourism development amid a changing climate.

1 Overview of Tourism and Climate Change

Climate change is now a global issue that is already happening all over the world. It significantly impacts the environment, society, and the global economy. It primarily affects human health, food, water resources, infrastructure, economy, and security (Mora et al., 2018).

According to the Intergovernmental Panel on Climate Change (IPCC), the Earth's climate has experienced major changes since the mid-twentieth century, resulting in higher temperatures, higher sea levels, less snow and ice cover, and increased atmospheric concentrations of greenhouse gases (IPCC, 2023). Global warming is one of the main drivers of climate change, primarily caused by anthropogenic factors such as burning fossil fuels, deforestation, industrial emissions, and the development of diverse economic sectors. In this context, tourism is also responsible for these changes, which cause global warming, and at the same time, its operation and activities are significantly affected by climate change's consequences. The UNWTO report (2023) indicates that in 2008, the tourism sector was responsible for about 5% of global greenhouse gas emissions, and in 2013, its contribution reached 8%. These figures highlight the growing environmental impact of tourism and related activities over time. Tourism and travel contribute to climate change through emissions of greenhouse gases, which are usually calculated for three subsectors: transportation, accommodation, and other tourism-related activities. According to a study by Sun et al. (2024), between 2009 and 2019, emissions from this sector increased from 3.7 Gt to 5.2 Gt, equivalent to the annual emissions of all Latin America and the Caribbean. Lenzen et al. (2018) analysed several subsectors and their contribution to CO-2 emissions and found that emissions from air travel account for 20% of tourism's global carbon footprint (Lenzen et al., 2018). Also relevant is the excessive energy consumption, which accounts for 21% of the total tourism sector's greenhouse gas emissions (Gössling & Lund-Durlacher, 2021).

The tourism industry is highly dependent on nature and the environment, and its interest is to contribute to climate change mitigation as it also suffers from the effects of climate change. Recognising the bidirectional and complex relationship between tourism and climate change, academics and researchers have, in turn, attempted to analyse this relationship by providing empirical evidence from the perspective of destinations, service providers, and consumers. It is clear that, each time, the challenges related to climate change, such as heatwaves, wildfires, and floods, are more evident in destinations (Lopes et al., 2024), for example rising sea levels and increased coastal erosion threaten beach tourism in destinations (Dube et al., 2024). Thus, weather is an important decision-making factor for tourists in nature-based destinations, leading them to choose another destination when weather conditions do not meet their expectations (Groulx et al., 2017). Climate change and sea level rise are significant concerns for geotourism destinations, mountainous and coastal areas where visitors are affected by temperature and rainfall variations (Gordon, 2023). Weather or climatic conditions also influence visitors' booking decisions (Dubois et al., 2016) because it makes it very difficult for tourists to stay in destinations with complicated weather conditions or extremely high temperatures, leading to discomfort (Chang et al., 2024). This means that when there is an increase in adverse weather conditions in tourist destinations, tourist activity will suffer. This context will be especially critical in tourist destinations that are overly dependent on tourism.

Generally, global data show that tourism has grown rapidly over the past several decades. According to UNWTO (2024), an estimated 1.3 billion international tourist arrivals were registered in 2023, with an increase of 34% compared to 2022, and thus reaching pre-pandemic levels (Fig. 1).

When regional performance is compared to pre-pandemic levels (Fig. 2), the Middle East remains the fastest-growing region, with 31% more international arrivals than in 2019. Others are almost back to pre-pandemic levels in 2023, indicating a strong recovery in tourism activity (UNWTO, 2024). Moreover, the sector is expected to grow, emphasising the need for urgent mitigation and adaptation measures to climate change.

In this context, tourism decision-makers, stakeholders, and international, national, and regional governmental institutions are challenged with solving the problems posed by climate change in destinations.

In 2015, the 195 countries in Paris at the United Nations Climate Conference (COP21) signed the first universal agreement to combat climate change and global warming. One of the main objectives is to limit global temperature increase in this century to below 2°C compared to preindustrial levels and, due to the serious risks,

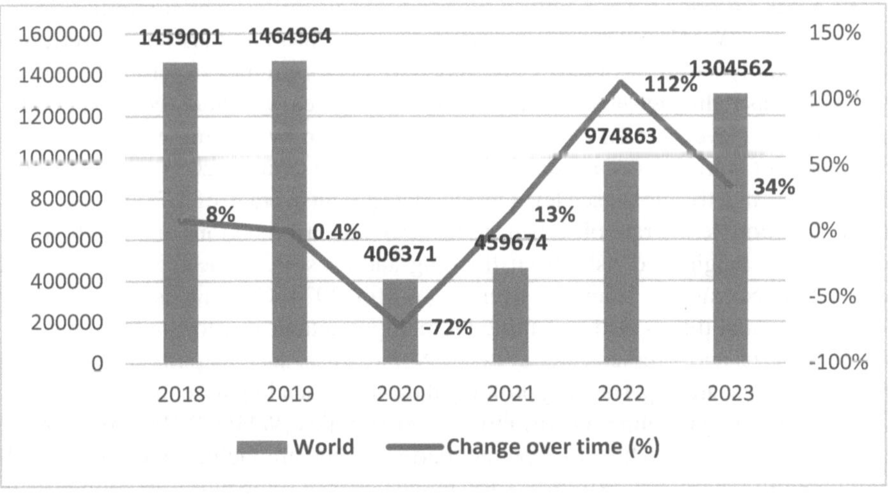

Fig. 1 International tourist arrivals. (Source: UNWTO (2024)

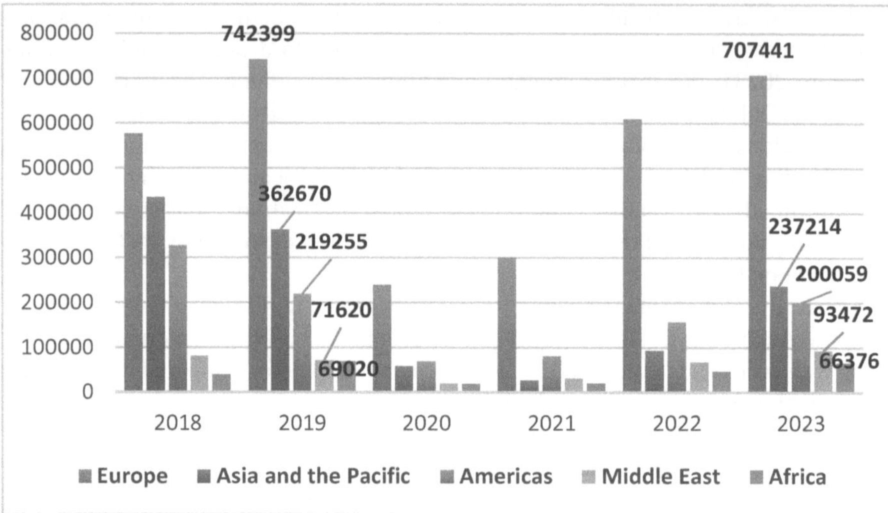

Fig. 2 International tourist arrival by regions. (Source: UNWTO (2024))

to strive for 1.5°C. This endeavour calls for an urgent need to address the dependence of economic growth on resource use and emissions to counter the effects of climate change. Thus, SDG 13 (Climate Action) in the global 2030 Agenda is critical, providing a roadmap to reduce emissions and enhance climate resilience.

However, global temperatures have reached unprecedented levels, with 2024 expected to be the warmest year on record, following closely on the heels of 2023 (WMO, 2024). The average temperature has increased since preindustrial times, as seen in Fig. 3, pushing closer to the critical 1.5°C threshold set by the Paris Agreement.

El Niño episodes have mainly influenced this warming, which amplifies temperature rises (WMO, 2024). These changes are not just numbers but are reflected in increasingly severe heatwaves and other weather extremes experienced worldwide, such as reducing snow cover and intensifying heavy rainfall (NOAA, 2024).

Rising sea levels are another pressing issue, driven by melting glaciers and the expansion of warming ocean waters. The State of the Climate 2024 report (WMO, 2024) notes that the global sea level is now rising at an average of 4.77 mm per year, more than double the rate observed in the 1990s. Coastal regions and small island nations face heightened risks from flooding and erosion, while the Antarctic and Arctic Sea ice extent is below the average (WMO, 2024; Copernicus, 2024).

Water availability is also becoming more unpredictable due to changing precipitation patterns and stressed river systems. Some areas, such as parts of Africa and South America, are experiencing prolonged droughts, leading to reduced river flows and impacts on agriculture and drinking water supplies (WMO, 2024). At the same time, other regions, such as Europe and parts of Asia, are facing excessive rainfall and flooding. These dual extremes—too much water in some places and too little in others—create global challenges for communities, economies, and ecosystems

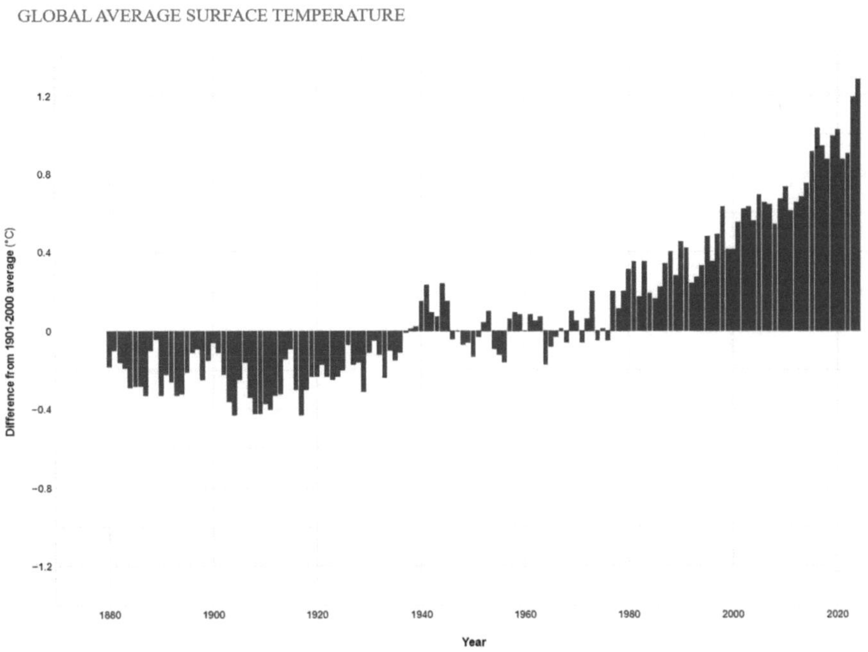

Fig. 3 Yearly surface temperature from 1880–2023 compared to the twentieth-century average (1901–2000). (Source: NOAA (2024))

(WMO, 2024). In 2022 and 2023, regional contributions to global temperature anomalies varied significantly across land and ocean areas, with notable patterns emerging (Fig. 4).

Land regions in the northern extratropic (20°N–90°N) played a dominant role in driving temperature anomalies, particularly during the second half of 2023, with pronounced contributions from regions experiencing warmer-than-average conditions. Similarly, tropical oceans (20°S–20°N) showed a marked shift from negative anomalies in 2022 to positive anomalies in 2023, mainly due to the transition from La Niña to El Niño and generally warmer conditions in the Atlantic and Indian Oceans. By the final four months of 2023, the combined effects of significant warming over both land and ocean, especially in the tropics and northern extratropic regions, contributed to record high global temperature anomalies (Copernicus, 2024).

The UNWTO, in collaboration with the United Nations Environment Program (UNEP), is taking the lead on a global initiative of Glasgow Declaration on Climate Action, aiming to engage tourism stakeholders in supporting the global goal of reducing emissions by half by 2030 (UNWTO, 2023). In turn, the World Travel and Tourism Council promoted a roadmap for tourism businesses, including accommodation, aviation, cruise, tour operators, and travel agents, with the aim to reach net zero emissions by 2050.

Undoubtedly, all regions are affected by the impact of climate change on tourism in different ways, depending on geographical location, climate zones, and the extent

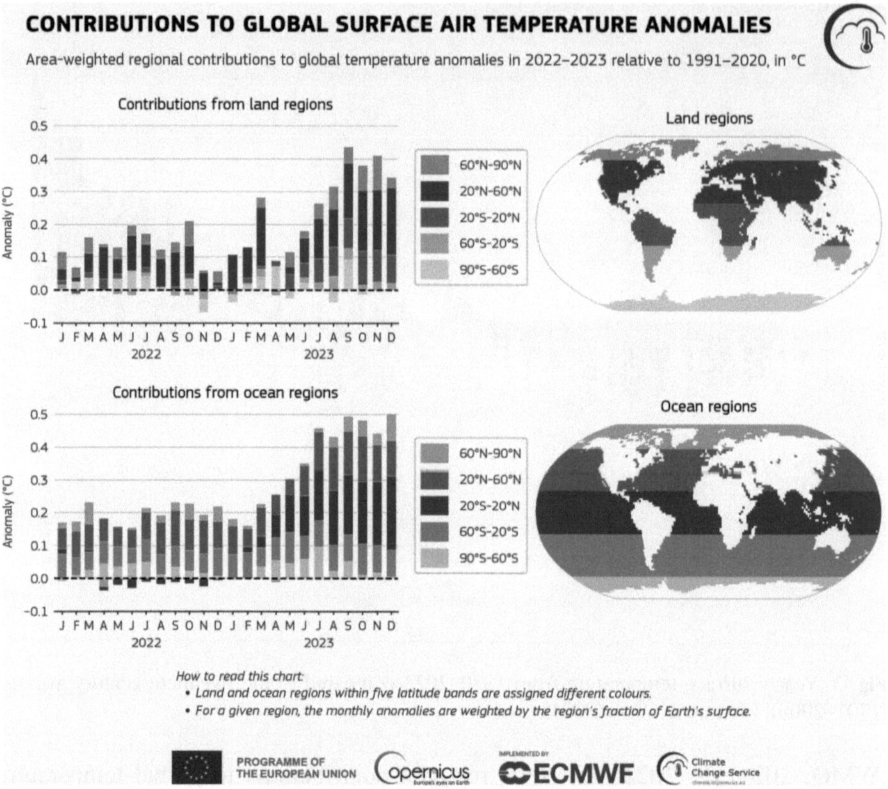

Fig. 4 Contributions to global surface air temperature anomalies. (Source: Copernicus, 2024)

to which policies and measures to address these issues have been implemented. The following section provides a general overview of the regions and discusses the impact of climate change on tourism in these regions based on recent studies and international reports.

The regions are presented based on the UNWTO classification system. The UNWTO typically divides the world into six main regions: Africa, the Americas, Asia and the Pacific, Europe, and the Middle East.

2 Regional Perspectives

2.1 *Africa*

Africa's tourism is highly vulnerable to climate change, given its dependence on nature-based tourism, extensive coastal zones, and fragile ecosystems. Climate variability has already disrupted tourism activities, with impacts varying across subregions (Dube et al., 2024; IPCC, 2023). Rising sea levels and increased coastal erosion threaten beach tourism in destinations like South Africa's Cape Town, Egypt's Red Sea resorts, and Mombasa in Kenya (Dube et al., 2024). The IPCC has highlighted that Africa's sea levels are rising above the global average, exacerbating coastal flooding and damaging tourism infrastructure (IPCC, 2023). Furthermore, coral reef degradation, caused by ocean acidification and increasing sea temperatures, negatively impacts marine tourism, particularly along the East African coastline (Dube et al., 2024). At the same time, extreme weather events, including cyclones and coastal storms, are becoming more frequent and severe, further threatening tourism-dependent communities and infrastructure (WMO, 2024).

Parallel to these coastal threats, severe droughts in various regions of the continent are worsening water scarcity, affecting tourism destinations and leading to conflicts over water resources. The World Wide Fund for Nature (WWF) (2022) reports that the reduction in water availability directly impacts major tourist areas, including Morocco, Egypt, and South Africa. The Day Zero phenomenon[1] crisis in Cape Town illustrated how water shortages can lead to a decline in tourist arrivals and substantial economic losses (Dube et al., 2024). Beyond urban areas, water scarcity is also threatening natural reserves such as Tanzania's Ngorongoro National Park and Botswana's Okavango Delta, where water availability is crucial for sustaining iconic wildlife species.

Changing migration patterns and habitat degradation are becoming increasingly evident, putting wildlife tourism at risk, one of the main economic pillars of several African countries (Musavengane & Musakwa, 2023).

Climate change also affects cultural and historical heritage. Dube et al. (2024) warn that extreme weather conditions are accelerating the deterioration of historic monuments, negatively impacting cultural tourism in several African countries. Rising temperatures also make visiting outdoor historical sites increasingly challenging, reducing tourist flows and affecting local economies (Dube et al., 2022). Furthermore, Chambwe and Shereni (2024) stress that local communities around protected areas are at the forefront of climate-induced challenges, where increased environmental pressures threaten conservation efforts and the livelihoods of those depending on tourism. Another primary concern is the growing frequency of extreme weather events, such as tropical storms and heavy rainfall, which have damaged infrastructure in key tourism areas. Flooding in national parks and coastal regions has disrupted roads, tourist accommodations, and transportation networks, making some destinations less accessible to visitors (Dube et al., 2024). At the same time, high temperatures are diminishing the appeal of outdoor activities such as hiking and wildlife safaris (Dube et al., 2022).

The tourism sector also contributes to climate change, primarily through carbon emissions associated with air travel and the high energy consumption of hotels and resorts. According to the World Travel & Tourism Council (WTTC) (2022), if

[1]According to Dube et al. (2022), the Day Zero phenomenon came after the 2015–2018 drought episodes on the tourism industry in the Western Cape as well as the industry's response. Day Zero was coined to indicate the day when taps would run dry.

mitigation efforts are not implemented, these emissions will continue to grow, exacerbating the climate crisis and putting more pressure on tourism destinations. Musavengane and Musakwa (2023) emphasise that sustainable tourism forms, such as ecotourism and climate-conscious destination management, are critical for reducing the tourism environmental footprint. In addition, the tourism industry must account for its carbon emissions, particularly in air transport, which represents a significant source of greenhouse gas emissions (WTTC, 2022). Faced with these challenges, the future of Africa's tourism remains uncertain, and the World Meteorological Organisation (WMO) (2024) emphasises that climate risks will continue to escalate unless coordinated efforts are made to develop resilience strategies for Africa's tourism sector. Therefore, the need for adaptation and a transition to a more sustainable model is a priority to ensure the sector's resilience.

2.2 Americas

The UNWTO classification system divides the Americas region into four subregions: North America, the Caribbean, Central America, and South America. The Americas offer a diverse array of tourism products, supported by the continent's wide range of climatic zones. This geographic diversity makes America a year-round destination for different tourist profiles and activities. For example, North America offers a diverse range of activities such as nature-based, cultural, historical, adventure, and wildlife during the whole year, thanks to the presence of all five climate groups (tropical, dry, temperate, continental, polar) (Rutty et al., 2024). This tourism potential has increased arrivals to the Americas, with arrivals in 2023 at 97% of 2019 levels (UNWTO, 2024). Other American subregions are known for their unique landscapes, coastal areas, natural wonders, and cultural heritage, which draw visitors worldwide. Most tourist arrivals are concentrated in North America, where coastal areas, mountain resorts, and national parks are particularly vulnerable to climate-related challenges. According to the Intergovernmental Panel on Climate Change, observed temperature change exceeds the global mean in North America, and future scenarios indicate continued warming, particularly in the winter season (IPCC, 2023).

A recent report of IPCC (2023) on climate change alerted on key risks in North America, such as risks of sea-level rise impacting coastal cities, settlements, and infrastructure, weather and climate extremes (floods, droughts, storms), human mortality and morbidity due to rising temperatures, degradation of marine, coastal, and terrestrial ecosystems, risk to freshwater resources, degraded water quality, risks to well-being, livelihoods, and economic activities. In the case of Central and South America, the main risks are related to water security, severe health effects due to increasing epidemics, disease in coral reef ecosystems degradation, risk to food security due to frequent droughts, damages to life and infrastructure due to floods, sea level rise, and storms (IPCC, 2023).

Recent studies have shown concerns about the future of winter sports tourism in North America, one of the segments most directly affected by warming temperatures (Malik et al., 2020; Rutty et al., 2024). Rising temperatures will lead to a later start date for the skiing season and reduce the number of skiable days, leading to a degrading number of arrivals. The evidence of the impact of climate change and global warming already exists in Florida, according to Tourism Economics (2018), where extreme storms such as Hurricane Irma led to an 11.6 % drop in hotel bookings, a 1.8 million decrease in tourism arrivals, and a $1.5 billion loss in tourism spending. Reyer et al. (2017) show a concern regarding the Latin America and Caribbean region, saying that climate change will severely affect them, even under lower levels of warming. They also stated that large businesses could suffer, which has implications for a large labour force, threatening less vulnerable populations. Rural and Indigenous communities in Latin America are particularly vulnerable to climate change due to their dependence on small-scale, rain-fed agriculture, natural resources, and traditional knowledge systems and cultural practices (Hoffman & Grigera, 2013; Reyer et al., 2017).

2.3 *Asia and the Pacific*

Climate change also impacts tourism in Asia and the Pacific, a region known for its diverse natural and cultural attractions. The Asia-Pacific region is experiencing some of the fastest sea-level rise rates globally (IPCC, 2023). Low-lying island nations like the Maldives and major coastal tourist destinations such as Vietnam and Thailand face risks of inundation that threaten critical tourism infrastructure, beaches, and natural ecosystems. Rising sea levels jeopardise the physical foundations of coastal resorts and attractions, directly undermining the tourism-dependent economies of these countries. Compounding this is the increased frequency and intensity of extreme weather events, as highlighted by the WMO (2024a). Cyclones, typhoons, and heatwaves disrupt tourism operations and the safety of visitors and local populations.

Coral reefs, vital for marine tourism in the Pacific, are undergoing rapid degradation due to rising sea temperatures and ocean acidification (IPCC, 2023). Iconic ecosystems such as the Great Barrier Reef and smaller reef systems across Southeast Asia have experienced mass bleaching events, diminishing their ecological and aesthetic value for visitors. Meanwhile, changing weather patterns have altered the seasonality of tourism.

While climate change poses a significant threat to the tourism sector, the industry itself is also a considerable contributor to greenhouse gas emissions. Air travel, identified by the UNWTO (2023) as the fastest-growing mode of transport in Asia-Pacific, creates most tourism-related emissions. Without mitigation, emissions from this sector are projected increase substantially. Additionally, the energy demands of accommodation, particularly for cooling systems in tropical destinations, further exacerbate emissions (UNWTO & ITF, 2019).

2.4 Europe

In Europe, the year 2023 was considered until now to be the hottest year on record. The summer of 2023 registered the highest temperatures since 1940, with extreme heatwaves, wildfires, and floods that severely impacted tourism destinations (ETC, 2023; C3S, 2023; IPCC, 2023). Mediterranean regions such as Greece, Southern Italy, Malta, and Eastern Spain experienced exceptional temperatures, posing increased risks to human health and altering tourist flows (ETC, 2023; WMO, 2024).

The tourism sector is a major contributor to these impacts, mainly through carbon emissions from air travel and energy-intensive accommodations (Sesana et al., 2018; Scott & Gössling, 2022; Cholakova & Dogramadjieva, 2023; Steiger & Scott, 2024). According to the European Environment Agency, emissions from tourism-related transport and infrastructure continue to rise, highlighting the urgent need to accelerate decarbonisation efforts (EEA, 2024).

As the region faces an increase in extreme weather events, including wildfires, floods, and marine heatwaves affecting several ecosystems (C3S, 2023; IPCC, 2023; FAO, 2024), these events are causing damage to tourism infrastructure, cultural heritage, and the overall travel experience (Sesana et al., 2018; ETC, 2023; EEA, 2024). Climate projections suggest that extreme weather patterns will persist, making tourism seasonality less predictable and potentially shifting peak seasons in some regions while extending them in others (IPCC, 2023; WMO, 2024; Scott & Gössling, 2022).

Coastal tourism, representing a substantial portion of European tourism revenue, is increasingly threatened by sea level rise, coastal erosion, and extreme storm events (IPCC, 2023; Scott & Gössling, 2022; WMO, 2024). Beach destinations along the Mediterranean and Atlantic coasts risk habitat loss, infrastructure damage, and reduced attractiveness due to changing environmental conditions (European Environment Agency, 2024; Amelung & Moreno, 2012; FAO, 2024). The impact of climate change on European beaches may lead to shifts in demand towards northern coastal areas, which are projected to experience milder summers (Scott & Gössling, 2022; Amelung & Moreno, 2012; IPCC, 2023).

Ski tourism is among the most vulnerable sectors in Europe. Rising temperatures and decreasing snowfall put at risk ski resorts in the Alps, Pyrenees, and other mountainous regions (Cholakova & Dogramadjieva, 2023). Many destinations increasingly rely on artificial snow production, but this solution is costly, energy-intensive, and unsustainable in the long term (Cholakova & Dogramadjieva, 2023).

As climate risks increase, European tourism must adapt to ensure long-term resilience. Strategies such as climate-smart destination planning, diversification of tourism products, and sustainability measures are crucial to mitigating the effects of climate change while maintaining economic stability (IPCC, 2023; WMO, 2024b; Scott & Gössling, 2022).

2.5 *Middle East*

The Middle East is also climate-vulnerable, experiencing rising temperatures, extreme heat waves, water scarcity, and intensified extreme weather events. This region is warming with temperature increases of approximately 0.45°C per decade (Zittis et al., 2022). Heatwaves are becoming more frequent and prolonged, with some regions expected to exceed 56°C by the end of the twenty-first century, posing severe risks to human health and economic sectors, including tourism (Francis & Fonseca, 2024).

Tourism in this region faces multiple climate threats. Extreme heat reduces outdoor tourism appeal, particularly in the UNESCO World Heritage sites (Fitchett & Roshan, 2020), where visitors are exposed to high temperatures (World Bank, 2024). Another example is in Lebanon, where snow days are expected to decline significantly by 2050, jeopardising winter tourism and ski resorts (World Bank, 2024).

In coastal areas, infrastructures are at risk due to sea level rise and increased coastal erosion (Bigano et al., 2008), and the frequency of heatwaves and wildfires threatens these areas (World Bank, 2024). Water scarcity is another major issue, where decreasing rainfall and groundwater depletion threaten the sustainability of ecotourism and rural tourism (Zittis et al., 2022; World Bank, 2024).

The economic implications of climate change for tourism are severe. In the Middle East, climate change is expected to reduce international arrivals in different countries, impacting employment and revenue streams in major tourism hubs (World Bank, 2024). Without proactive adaptation, climate change will continue to reshape tourism in the Middle East, threatening economic stability and long-term sustainability

3 Conclusions

This chapter has provided some regional perspectives on the relationship between tourism and climate change, discussing the impacts of climate change on tourism sector, which are becoming increasingly evident. It has also emphasised the importance of the tourism sector in contributing to climate change mitigation, given its dependence on environmental sustainability. Concerning the African region, tourism is highly vulnerable to climate change, which already led to the disruption of some tourism activities, with impacts varying across subregions (IPCC, 2023). Adaptation efforts are emerging (UNWTO, 2023), and there is also increasing consensus that protected areas and conservation initiatives must integrate climate adaptation strategies, ensuring that communities are directly involved in managing these areas (Chambwe & Shereni, 2024). In the case of the Americas, there is a serious concern about sea level rise impacting coastal cities, settlements, infrastructure, weather and climate extremes, risks to water security, and disease in coral reef ecosystems degradation, all of which may negatively affect tourism demand in the region (IPCC, 2023). The Asia-Pacific is also experiencing some of the fastest rates of sea level rise globally (IPCC, 2023). Europe is facing significant impacts of climate change, especially in the Mediterranean regions, where exceptional temperatures are occurring, creating increased health risks and altering tourism flows (WMO, 2024). The Middle East is facing water scarcity, where decreasing rainfall and groundwater depletion threaten the sustainability of ecotourism and rural tourism (World Bank, 2024).

The tourism sector recognises its role in contributing to climate change but also recognises the need to contribute to climate change mitigation as it also suffers from its impacts, given it is an activity that is highly dependent on nature and the environment. Its interests are closely tied to the health of these systems. Collaboration between governments, the private sector, and local communities will strengthen tourism resilience and ensure it remains an economic and social development driver.

Climate change is a long-term challenge, but the need for urgent action now is clear. In particular, the tourism sector needs to address this problem seriously, as its emissions have continued to increase over the past few decades (Gössling et al., 2023).

The analysis and discussion presented in this chapter are mainly exploratory and are intended to broaden the examination of the relationship between climate change and tourism. Therefore, a more detailed analysis requires a thorough understanding of the impact of climate change on tourism across these regions, as well as specific mitigation actions being implemented by the tourism sector in each region. It is essential to assess these regional efforts and explore the potential for regions to learn from each other. This collaborative approach represents an important direction for future research.

References

Amelung, B., & Moreno, A. (2012). Climate change and coastal tourism: Impacts and adaptation in Europe. *Tourism Management, 33*(5), 941–951. https://www.sciencedirect.com/science/article/pii/S0261517712000241

Bigano, A., Bosello, F., Roson, R., & Tol, R. S. J. (2008). Economy-wide impacts of climate change: A joint analysis for sea level rise and tourism. *Mitigation and Adaptation Strategies for Global Change, 13*, 765–791. https://doi.org/10.1007/s11027-007-9139-9

C3S. (2023). *European state of the climate 2023 Summary*. Copernicus Climate Change Service. https://climate.copernicus.eu/esotc/2023

Chambwe, M., & Shereni, N. C. (2024). Navigating the Nexus: Climate change, protected areas, and local tourism communities in Sub-Saharan Africa. *African Journal of Hospitality, Tourism and Leisure, 13*(4), 693–698. https://doi.org/10.46222/ajhtl.19770720.554

Chang, D., Bu, N., Zhang, N., & Xiao, H. (2024). Climate change and tour-ism demand: Risks for extreme heat? *Heliyon, 10*(17), e37186.

Cholakova, S., & Dogramadjieva, E. (2023). Climate change adaptation in the ski industry: Stakeholders' perceptions regarding a mountain resort in Southeastern Europe. *Journal of Outdoor Recreation and Tourism, 42*(2023), 100611. https://doi.org/10.1016/j.jort.2023.100611

Copernicus. (2024). The 2023 annual climate summary: Global climate highlights 2023, Retrieved January 7, 2025, from https://climate.copernicus.eu/global-climate-highlights-2023

Dube, K., Nhamo, G., & Chikodzi, D. (2022). Climate change-induced droughts and tourism: Impacts and responses of Western Cape province, South Africa. *Journal of Outdoor Recreation and Tourism, 39*, 100319. https://doi.org/10.1016/j.jort.2020.100319

Dube, K., Nhamo, G., Kilungu, H., Hambira, W. L., El-Masry, E. A., Chikodzi, D., Chapungu, L., & Molua, E. L. (2024). Tourism and climate change in Africa: Informing sector responses. *Journal of Sustainable Tourism, 32*(9), 1811–1831. https://doi.org/10.1080/09669582.2023.2193355

Dubois, G., Ceron, J. P., Gössling, S., & Hall, C. M. (2016). Weather preferences of French tourists: lessons for climate change impact assessment. *Climatic Change, 136*(2), 339–351.

EEA. (2024). European climate risk assessment. EEA Report 01/2024. European Environment Agency

ETC. (2023). *European tourism: Trends & prospects (Q3/2023)*. European Travel Commission. https://etc-corporate.org/reports/

FAO. (2024). Temperature change statistics 1961–2023 – Global, regional and country trends. FAOSTAT Analytical Briefs, No. 84. Rome. https://doi.org/10.4060/cd0082en

Fitchett, J., & Roshan, G. (2020). Climate change threats to cultural and heritage tourism in Iran, Chapter 14. In S. Seyfi & C. M. Hall (Eds.), *Cultural and heritage tourism in the Middle East and North Africa: Complexities, management and practices*. Routledge.

Francis, D., & Fonseca, R. (2024). Recent and projected changes in climate patterns in the Middle East and North Africa (MENA) region. *Scientific Reports, 14*, 10279. https://doi.org/10.1038/s41598-024-60976-w

Gordon, J. E. (2023). Climate change and geotourism: Impacts, challenges, and opportunities. *Tourism and Hospitality, 4*(4), 514–538.

Gössling, S., Balas, M., Mayer, M., & Sun, Y. Y. (2023). A review of tourism and climate change mitigation: The scales, scopes, stakeholders and strategies of carbon management. *Tourism Management, 95*, 104681.

Gössling, S., & Lund-Durlacher, D. (2021). Tourist accommodation, climate change and mitigation: An assessment for Austria. *Journal of Outdoor Recreation and Tourism, 34*, 100367.

Groulx, M., Lemieux, C. J., Lewis, J. L., & Brown, S. (2017). Understanding consumer behaviour and adaptation planning responses to climate-driven environmental change in Canada's parks and protected areas: a climate futurescapes approach. *Journal of Environmental Planning and Management, 60*(6), 1016–1035.

IPCC. (2023). Climate Change 2023: Synthesis Report. In H. Lee and J. Romero, Core Writing Team, (eds.), *Contribution of Working Groups I, II and III to the Sixth Assessment Report of the Intergovernmental Panel on Climate Change* (pp. 35–115). Geneva, Switzerland: IPCC. https://doi.org/10.59327/IPCC/AR6-9789291691647

Lenzen, M., Sun, Y. Y., Faturay, F., Ting, Y. P., Geschke, A., & Malik, A. (2018). The carbon footprint of global tourism. *Nature Climate Change, 8*(6), 522–528.

Lopes, H. S., Remoaldo, P. C., Ribeiro, I., Ribeiro, V., & Alves, J. A. (2024). Global warming and its influence on leisure activities, events, and sports management. *Events, and Sports Management*, 1–9.

Malik, K., McLeman, R., Robertson, C., & Lawrence, H. (2020). Reconstruction of past backyard skating seasons in the original six NHL cities from citizen science data. *The Canadian Geographer/Le Geographe Canadien, 64*(4), 564–575. https://doi.org/10.1111/cag.12640

Mora, C., Spirandelli, D., Franklin, E. C., Lynham, J., Kantar, M. B., Miles, W., et al. (2018). Broad threat to humanity from cumulative climate hazards intensified by greenhouse gas emissions. *Nature Climate Change, 8*(12), 1062–1071.

Musavengane, R., & Musakwa, W. (2023). Climate change impacts on wildlife protected areas in Southern Africa: A bibliometric analysis. *Studia Periegetica, 4*(44), 73–90. https://doi.org/10.58683/sp.620

National Oceanic and Atmospheric Administration/NOAA. (2024). Climate change: Global temperature. Retrieved January 7, 2025, from https://www.climate.gov/news-features/understanding-climate/climate-change-global-temperature

Reyer, C. P., Adams, S., Albrecht, T., Baarsch, F., Boit, A., Canales Trujillo, N., et al. (2017). Climate change impacts in Latin America and the Caribbean and their implications for development. *Regional Environmental Change, 17*, 1601–1621.

Rutty, M., Hewer, M., Knowles, N., & Ma, S. (2024). Tourism & climate change in North America: Regional state of knowledge. *Journal of Sustainable Tourism, 32*(9), 1924–1947.

Scott, D., & Gössling, S. (2022). Climate and carbon risk of tourism in Europe. *Journal of Sustainable Tourism, 30*(4), 1893–1923. https://doi.org/10.1080/09669582.2022.2163653

Sesana, E., Gagnon, A. S., Bertolin, C., & Hughes, J. (2018). Adapting cultural heritage to climate change risks: Perspectives of cultural heritage experts in Europe. *Geosciences, 8*, 305. https://doi.org/10.3390/geosciences8080305

Steiger, R., & Scott, D. (2024). Climate change adaptation in the Ski industry: Challenges and responses in Europe. *Journal of Sustainable Tourism, 32*(9), 1893–1923. https://doi.org/10.1080/09669582.2022.2163653

Sun, Y. Y., Faturay, F., Lenzen, M., Gössling, S., & Higham, J. (2024). Drivers of global tourism carbon emissions. *Nature Communications, 15*(1), 1–10.

Tourism Economics. (2018). The impact of Hurricane Irma on the Florida tourism economy. https://www.visitflorida.org/media/46680/fl-hurricane-impacts.pdf

UNWTO. (2023), Climate action in tourism – An overview of methodologies and tools to measure greenhouse gas emissions, World Tourism Organization, Madrid.. https://doi.org/10.18111/9789284423927

UNWTO. (2024). UN tourism barometer. https://www.unwto.org/un-tourism-world-tourism-barometer-data

UNWTO & ITF. (2019). Transport-related CO2 emissions of the tourism sector – Modelling results, UNWTO, World Tourism Organization and International Transport Forum, Madrid. 10.18111/9789284416660

WMO. (2024). State of the climate in Africa 2023. World Meteorological Organization. https://wmo.int/publication-series/state-of-climate-africa-2023

WMO. (2024a). State of the climate 2024 - Update for COP29, Climate Statement, Geneva, 12p. World Meteorological Organization

World Bank (2024). Report "Rising to the Challenge: Success Stories and Strategies for Achieving Climate Adaptation and Resilience". Available at https://www.worldbank.org/en/publication/rising-to-the-challenge-climate-adaptation-resilience

WTTC. (2022). African travel & tourism sector's climate footprint revealed. World Travel & Tourism Council. Retrieved March 12, 2025, from https://wttc.org/news-article/african-travel-and-tourism-sectors-climate-footprint-revealed

WWF. (2022). UNFCCC COP27: Para uma África Resiliente ao Clima. World Wide Fund for Nature. https://wwfafrica.awsassets.panda.org/downloads/wwf_cop27_expectations_portuguese.pdf

Zittis, G., Almazroui, M., Alpert, P., Ciais, P., Cramer, W., Dahdal, Y., et al. (2022). Climate change and weather extremes in the Eastern Mediterranean and Middle East. *Reviews of Geophysics, 60*, e2021RG000762. https://doi.org/10.1029/2021RG000762

Tourism, Hospitality, and Climate Changes: Insights in the Last Decade

Ahmed H. Alsharif, Salmi Mohd Isa, and Lina Pilelienė

Abstract Tourism and hospitality have been widely studied. However, no previous research was performed to conduct a descriptive and systematic analysis of tourism and hospitality. This study conducts a comprehensive descriptive and systematic analysis to fill the gap in the literature. This study is designed to follow the Preferred Reporting Items for Systematic Reviews and Meta-Analyses (PRISMA) protocol to select and extract the relevant articles from the Scopus database. We analyzed 1247 documents that were published between 2012 and 2022. We systematically analyzed the selected documents relevant to these authors' keywords. In addition, results indicate a growing number of publications over the years. Furthermore, keywords related to tourism and hospitality include "CSR" and "hotels and hotel industry." As far as we know, this is the first descriptive and systematic analysis that provides insights into the current state of tourism and hospitality services, tourism and hospitality industry, tourism and climate change, online reviews, hotels and hotels industry, and CSR and offers directions for future research.

1 Introduction

In the early months of 2020, a distressing and unfortunate event unfolded with the emergence of a novel disease called "Covid-19" (Davies & Wenham, 2020). This illness, similar to acute pneumonia, swiftly spread across the globe, impacting countries such as the USA, China, Italy, and others (Ren, 2020). The consequences were dire, resulting in a significant loss of lives and severe economic repercussions

A. H. Alsharif (✉) · S. M. Isa
Graduate School of Business, University of Science Malaysia (USM), Pulau Pinang, Malaysia
e-mail: ahmedalsharif07@gmail.com; salmi.mohd.isa@usm.my

L. Pilelienė
Vytautas Magnus University, Kaunas, Lithuania
e-mail: lina.pileliene@vdu.lt

J. Marques et al. (eds.), *Tourism and Climate Change*, Tourism, Hospitality & Event Management, https://doi.org/10.1007/978-3-031-99128-8_2

that affected the global tourism industry for an extended period because the countries locked down their borders (Kaushal & Srivastava, 2021). Therefore, this lockdown led to the collapsing economy of some countries, such as but not limited to Sri Lanka, which is highly dependent on tourism to get foreign currency (Fatoye et al., 2021). Furthermore, European nations, particularly Spain (−76.1%), Greece (−76.3%), Malta (−78.1%), Cyprus (−82%), and others, experienced drastic reductions in tourism activity, as evidenced by the lowest values recorded on the Perkal index, which measures tourism intensity (Korinth & Wendt, 2021). Furthermore, the overall decline in international tourism was staggering, with reports indicating a 69% decrease in international tourist arrivals across Europe in 2020 (Melnychenko et al., 2021). The pandemic's repercussions were not limited to Europe; countries like Indonesia faced substantial economic losses, with projections estimating a USD ($) 350 million loss due to a steep decline in international tourist arrivals (Atmojo & Fridayani, 2021). The situation has been further exacerbated in regions like French Polynesia (Bora-Bora), where tourism constitutes a significant portion of the economy, leading to a critical need for sustainable management strategies to mitigate the impacts of such crises (Lecchini et al., 2021; Pilelienė et al., 2024). Additionally, studies from China highlighted a significant shift in tourist destination preferences due to health concerns, further exacerbating the decline in tourism (Li et al., 2021), where domestic tourism revenue plummeted by 69% (Rudyanto et al., 2021).

Tourism is a multifaceted domain with a crucial economic role, which encompasses a broad range of activities, for example, travel, accommodation, food service, transportation, beverage services, and entertainment, all aimed at satisfying the leisure, business, or personal needs of individuals or groups (Buhalis et al., 2019; Davahli et al., 2020). As a critical sector, tourism involves various stakeholders, including governments, businesses, and local communities, who collaborate to plan, develop, market, and manage tourist destinations and services (Koseoglu et al., 2016). Its contributions to employment, foreign exchange earnings, and economic growth are substantial for numerous countries (World Tourism Organization, 2020). Furthermore, tourism possesses the potential to facilitate cultural exchange, environmental preservation, and sustainable development (Chang et al., 2020). Nevertheless, the industry grapples with overcrowding, environmental degradation, and socio-cultural disruptions. To overcome these obstacles and ensure the long-term sustainability of tourism, a comprehensive comprehension of its intricacies, impacts, and effective management strategies is of utmost importance (Davahli et al., 2020; Kaushal & Srivastava, 2021).

According to Ahmed et al. (2023a); Kaushal & Srivastava (2021), hospitality is a fundamental component of the tourism sector, including services and facilities for guests or tourists. Davahli et al. (2020) mention that hospitality focuses on delivering personalized and memorable experiences, exceeding guest expectations, and ensuring their comfort and well-being. Successful hospitality relies on excellent service quality, effective communication, attention to detail, and a deep understanding of guest needs and preferences (World Tourism Organization, 2020). There are several systematic literature reviews (SLR) that investigate hospitality. For instance,

Yu et al. (2020) have examined abusive supervision in hospitality and tourism, defined as the perceptions of engaging supervisors in sustained displays of hostile verbal and nonverbal behavior that are salient for employees and have long-term negative outcomes. Furthermore, Górska-Warsewicz and Kulykovets (2020) have extracted and analyzed twenty-six documents to explore brand loyalty toward hotels. Similarly, Sharma et al. (2020) have selected and analyzed four-hundred and three documents to investigate the application of green hospitality.

Aware of the growing interest in tourism and hospitality, several studies analyzing scientific output in these fields have already been published (Górska-Warsewicz & Kulykovets, 2020). However, no previous research was performed to map the "(tourism OR tourism industry) AND ((hospitality OR hospitality industry) OR (hotel OR hotel industry) OR (tourism and hospitality))" research production in the Scopus (SC) database. Thus, this research distinguishes itself from other review articles, primarily focusing on global academic research trends within the SC database from 2012 to 2022. The period (2012–2022) was chosen to highlight the most recent decade of research activity, focusing on current trends and recent advancements. Analyzing this period allows researchers to assess the field's current state and understand the factors driving recent research momentum. This review paper aims to conduct extensive descriptive and systematic analysis, encompassing crucial elements such as identifying the most prominent authors' keyword occurrences. The contributions and steps of this descriptive and systematic analysis study are summarized as follows:

- To evaluate the annual growth trends of scientific publications in the field, providing insights into the evolving research activity over time.
- To determine the most interesting and prevalent topics within the selected articles, offering a thematic overview of the current research landscape in the field.
- To provide systematic analysis relevant to this study's prominent authors' keyword occurrences.

This research follows a well-organized structure, which can be outlined as follows: Section 2 presents a detailed explanation of the methodology adopted for this study. In Sect. 3, a comprehensive descriptive and systematic literature review analysis. Section 4 offers a concise and conclusive conclusion from the study's results.

2 Methodology

There are various types of review articles. These encompass bibliometric analysis, as demonstrated in studies by Nusair et al. (2019); Pilelienė et al. (2022). Another approach is a systematic review and meta-analysis (Ahmed et al., 2023b; Alsharif et al., 2024). Theoretical and conceptual review (Devkota et al., 2023), literature review (content) analysis (Ahmed et al., 2022b; Alsharif et al., 2024). Bibliometric reviews are divided into two main categories: domain-based (Ahmed et al., 2021; Khraiwish & Alsharif, 2024; Pimonenko et al., 2021) or journal-based Paul & Bhukya (2021).

This study rigorously followed the instructions of the Preferred Reporting Items for Systematic Reviews and Meta-Analyses (PRISMA) protocol, ensuring a meticulous identification of relevant scholarly papers (Moher et al., 2015). PRISMA Protocol has been widely used in several fields to extract data from the Scopus or Web of Science databases (Ahmed et al., 2024). This study adopted a descriptive and systematic analysis method to provide a concise and comprehensive overview of the global academic trends in (tourism OR tourism industry) AND ((hospitality OR hospitality industry) OR (hotel OR hotel industry) OR (tourism and hospitality)). This comprehensive investigation encompassed the analysis of various facets, including the highly prominent countries, authors, institutions, journals/sources, highly cited papers, and hot keywords, all serving as critical indicators for assessing the progress in scholarly publications within this domain.

To achieve this, three research questions were formulated to guide the structure of the analysis and gain a thorough understanding of the existing scientific research in the analyzed domain. These research questions were thoughtfully designed to shed light on key areas of interest and contribute to the improvement of knowledge in the relevant areas, as follows:

- Is there a yearly increase in scientific publications in the field, and if so, what is it?
- What are the most common authors' keywords in the articles chosen?
- What pertinent information can be found in the articles chosen?

This study starts by extracting English articles and review articles from the SC database in March 2023. The following query was applied to the title, abstract, and keywords: "TITLE-ABS-KEY ((tourism OR tourism AND industry) AND ((hospitality OR hospitality industry) OR (hotel OR hotel AND industry) OR (tourism AND hospitality)) AND (LIMIT-TO (OA, "all")) AND (LIMIT-TO (PUBYEAR, 2022) OR LIMIT-TO (PUBYEAR, 2021) OR LIMIT-TO (PUBYEAR, 2020) OR LIMIT-TO (PUBYEAR, 2019) OR LIMIT-TO (PUBYEAR, 2018) OR LIMIT-TO (PUBYEAR, 2017) OR LIMIT-TO (PUBYEAR, 2016) OR LIMIT-TO (PUBYEAR, 2015) OR LIMIT-TO (PUBYEAR, 2014) OR LIMIT-TO (PUBYEAR, 2013) OR LIMIT-TO (PUBYEAR, 2012)) AND (LIMIT-TO (LANGUAGE, "English")) AND (LIMIT-TO (DOCTYPE, "ar") OR LIMIT-TO (DOCTYPE, "re")) AND (LIMIT-TO (PUBSTAGE, "final"))".

The initial search parameters were set from 1973 to 2023, which is likely to capture a comprehensive, long-term view of the topic's research trajectory, allowing scholars to understand its foundational developments and how interest has evolved over time. This broader range can reveal historical trends, early studies, and shifts in research focus that may not be visible if limited to 2012–2022. By including earlier years, researchers gain insight into the topic's full academic context and can better analyze factors contributing to the recent surge in publications. Figure 1 demonstrates the selection process of relevant documents.

Through this process, a total of 1247 articles and review articles were identified, all of which were published between 2012 and 2022.

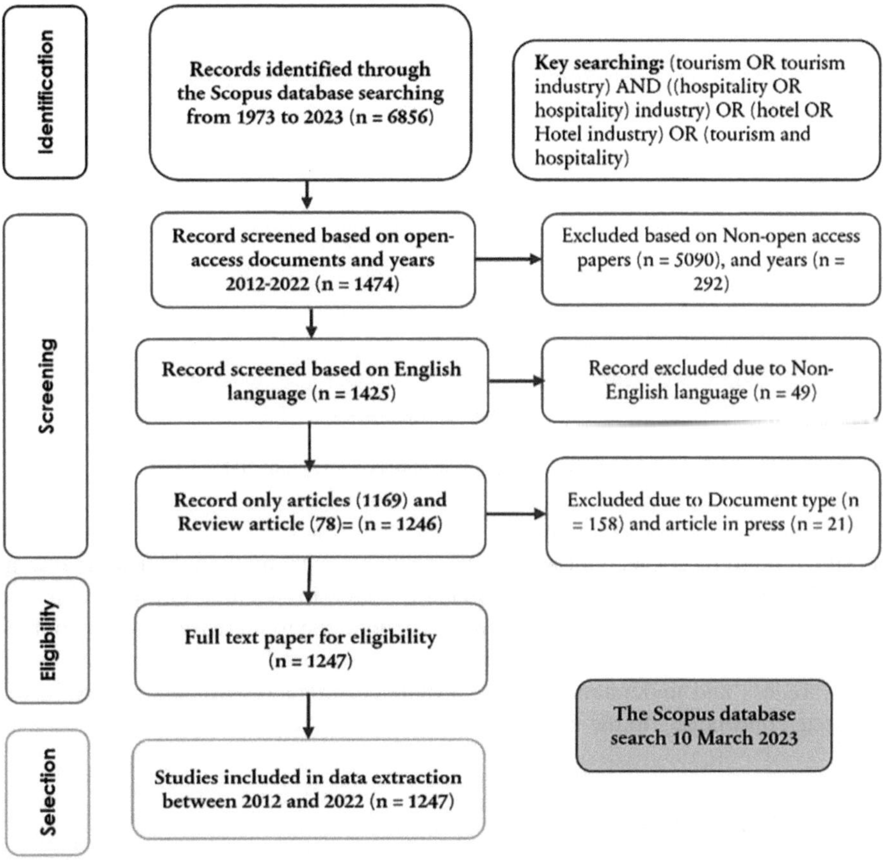

Fig. 1 PRISMA process for extracting documents

3 Results and Discussions

3.1 A Descriptive Analysis

3.1.1 Annual and Accumulative Growth

As a result of the procedure, 1247 articles and review articles in the relevant field were identified. The analysis revealed a significant publication growth, with over 60% of the papers (797 documents) published between 2020 and 2022. Figure 2 illustrates the annual publications published between 2012 and 2022, with 21 documents published in 2012 and 324 documents in 2022. The increasing interest among scholars in the relevant fields has led to increased publications. The increase in publications reflects growing recognition of the topic's importance and relevance,

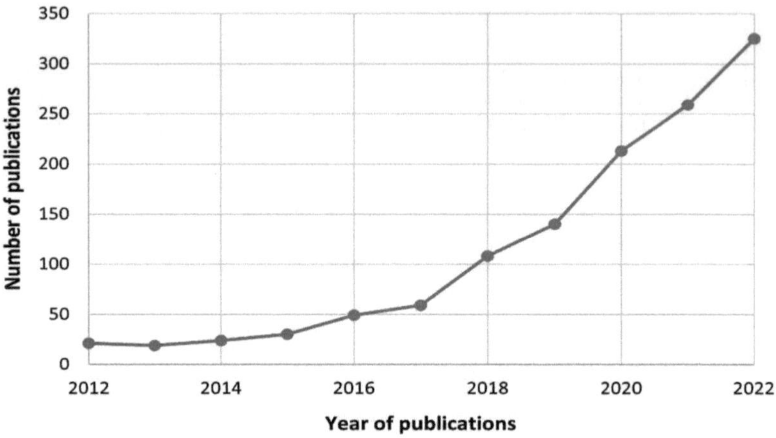

Fig. 2 The annual and accumulative publications between 2012 and 2022

possibly due to recent technological advancements, new methodologies, or emerging real-world applications that have made research in this area more feasible and impactful. Scholars may also be drawn to this field due to increased funding opportunities, interdisciplinary interest, or pressing societal needs that the topic addresses. As awareness and understanding have expanded, more researchers are exploring this area, contributing to the notable rise in scholarly output.

3.1.2 Keywords Analysis

Keywords are a key metric in understanding the strength of associations between paired keywords. A higher frequency indicates stronger connections (Ahmed & Salmi, 2024; Alsharif et al., 2023; Halsharif et al., 2020; Pilelienė & Jucevičius, 2023). The most popular authors' terms in the dataset with at least five occurrences each are summarized in Table 1. With a total link strength (TLS) of 377 and 251 occurrences "Tourism" is the most frequently appearing. Next on the list are the terms "Hospitality" (133 occurrences) and "Covid-19 and Covid-19 pandemic" (146 occurrences) which are accompanied by TLS values of 215 and 239 respectively. Along with "corporate social responsibility (CSR)," "online reviews," and "climate changes," the following terms are often used: tourism hospitality tourism industry hotel(s) and hotel industry. Among the specific terms "hotel(s)" appears 91 times while "hotel industry" appears 63 times indicating a TLS of 159 and 96 respectively. With 30 mentions the term "tourism industry" has a total phrase length score (TLS) of 38. These keywords do a great job of capturing the essence of the topics: the hotel sector tourism and climate change internet reviews corporate social responsibility and the tourist and hospitality industry

Table 1 Top twelve authors' keywords with five occurrences of at least

#	Keyword	Occurrences	TLS
1	Tourism	251	377
2	Covid-19/covid-19 pandemic	146	215
3	Hospitality	133	239
4	Hotel(s)	91	159
5	Hotel industry	63	96
6	Hospitality industry	59	66
7	Corporate social responsibility or CSR	37	85
8	Tourism and hospitality	36	41
9	Tourism industry	30	38
10	Tourism and hospitality services	27	29
11	Online reviews	17	25
12	Climate change	7	12

TLS total link strength

3.2 A Systematic Analysis and Thematic Evolution

According to the twelve authors' keywords, as tabulated in Table 1, the authors will discuss these core topics and combine topics that have the same ideas or structure, such as but not limited to hotels and the hotel industry, in the following sections.

3.2.1 Tourism and Hospitality Industry

Sustainable tourism practices and climate change: The tourism industry grapples with adopting sustainable practices, particularly in the face of climate change's adverse impact (Scott, 2021). Climate-related effects significantly affect tourism, emphasizing the urgency for sustainable policies. For instance, the impacts of climate change on tourism are multifaceted, including the dangers of species extinction, decreased freshwater, increased wildfire accidents, heat waves, and illnesses, leading to changes in visitor preferences and behaviors (Shariff, 2022). Climate change not only affects tourism demand but also influences the economic sensitivity of the sector, particularly in regions dependent on winter sports tourism (Tranos & Davoudi, 2014). The economic dimension of climate change impacts on tourism can be seen in studies assessing variations in tourism flows and their economic consequences for Mediterranean countries (Roson & Sartori, 2014). Research explores the relationship between sustainable tourism policies, destination management, and development, aiming to balance economic growth and environmental considerations (Pahrudin et al., 2022).

Tourism's environmental impact, especially regarding carbon emissions, has gained prominence. Recognizing the interplay between tourism and carbon emissions is crucial for promoting sustainable practices and minimizing the industry's environmental footprint (Chen et al., 2023). Moreover, understanding the spatial and temporal characteristics of tourism urbanization informs urban planning strategies, ensuring the creation of sustainable and appealing tourism destinations (Xie et al., 2021). Resident perceptions and attitudes toward tourism, particularly in the context of environmental impacts, emphasize the importance of engaging communities in decision-making processes for sustainable tourism practices (Medeiros et al., 2021).

Economic uncertainty and socio-economic impacts: Economic uncertainty emerges as a pivotal factor influencing tourism consumption patterns, with its effects varying based on the composition of a country's informal and official sectors (Nguyen et al., 2022). Economic uncertainty is influenced by geopolitical risks (e.g., military threats, political instability, political violence, wars, ethnic tensions) (Kumar, 2022), economic policy uncertainties (e.g., fluctuations in government policies and regulations) (Işık et al., 2020), and external shocks (e.g., COVID-19 pandemic, inflation, exchange rate fluctuations) (Khan et al., 2022). Thus, understanding the link between economic uncertainty and tourism consumption is vital for devising strategies to mitigate negative impacts and capitalize on opportunities. The socio-economic impacts of tourism, examined in specific regions, provide nuanced insights into the challenges and benefits associated with tourism development (Halsharif et al., 2022; Kozhokulov et al., 2019).

The impact of COVID-19 on the hospitality industry: The tourism and hospitality industry have undergone significant challenges and transformations, particularly during the COVID-19 pandemic (Zopiatis et al., 2021). A study by Hemmington & Neill (2022) explored the impact of the pandemic on the hospitality industry in New Zealand, emphasizing the necessity of distinguishing the economic role of the hospitality sector from the broader tourism sector. The study suggested a comprehensive evaluation of the hospitality sector to gain a more holistic perspective.

Initially, the pandemic instigated a dramatic decline in travel demand, with estimates indicating a loss of $2.1 trillion in global tourism revenue in 2020 alone, alongside the jeopardization of approximately $75 million jobs in the sector (Candia & Pirlone, 2021). The long-term impacts of COVID-19 are expected to include a reconfiguration of tourism supply chains and a heightened focus on sustainability and local engagement, as evidenced by the rise of local volunteer tourism initiatives (Abbas et al., 2021). Furthermore, the pandemic has accelerated the adoption of digital technologies, such as virtual reality (VR), facilitating remote and virtual tourism experiences that may redefine the industry's landscape in a post-pandemic world (Zheng et al., 2022). As the sector navigates these challenges, government support and strategic planning will be crucial in ensuring a sustainable recovery and the long-term viability of tourism and hospitality businesses (Chernikov & Thi, 2023).

Resilience and adaptation in tourism-dependent businesses: Ntounis et al. (2022) surveyed tourism-dependent businesses in English towns to explore their perceptions of resilience during the crisis (e.g., the COVID-19 pandemic). The study provides evidence of the challenges the tourism and hospitality sector faces and the need for resilience and adaptation in response to the pandemic. It underscores the importance of understanding the experiences of businesses directly impacted by the crisis to inform effective recovery strategies.

Challenges in the tourism and hospitality industry: The impact of the COVID-19 pandemic on the tourism and hospitality industry has been studied extensively, including its effects on medical tourism (Ghasemi et al., 2021). Ghasemi et al. (2021) highlighted the need for standardized procedures and measurement indices in medical tourism, emphasizing the importance of adapting and improving operational performance in the face of unprecedented challenges. The "Great Resignation" phenomenon observed in various sectors, including hospitality and tourism, has led to significant shifts in the workforce landscape, necessitating a re-evaluation of recruitment and retention strategies within the industry (Liu-Lastres et al., 2023). In addition, the wine tourist industry has also felt the effects. According to research, there was a change in consumer motives and intentions toward local experiences and quality above quantity in wine tourism after the pandemic. Additionally, the pandemic has made people rethink food hygiene and safety procedures, affecting their dining habits and tastes (Abbas et al., 2021). This has far-reaching consequences for gastro-tourism. The adventure tourism industry has also had difficulties due to the travel and outdoor activity limitations that have limited the choices available to adventure seekers. As a result, there has been a reevaluation of how these experiences may be safely provided in the post-pandemic world (Huang & Wang, 2023).

As the epidemic exposed the cruise industry's weaknesses and dependence on customer trust, it encountered unprecedented difficulties. Due to the lack of preparation for this health catastrophe, numerous cruise lines had to suspend operations starting in March 2020, which resulted in substantial financial losses and operational problems (Radic et al., 2020). In addition, health and safety policies have been reevaluated due to the pandemic, focusing on strict testing and social distancing techniques to regain travelers' trust (Chowell et al., 2021). Like other traditional practices, the pandemic has impacted heritage and cultural tourism. Local economies rely significantly on cultural tourism, which has been hit hard by the closure of cultural sites and the cancellation of events, reducing the number of tourists (Chi et al., 2022). On the other hand, new developments have emerged due to the epidemic, with digital platforms being used to keep audiences engaged and promote virtual experiences (Li et al., 2022). The pandemic has also had a major impact on religious tourism since pilgrimages and other religious events have declined due to travel and gathering restrictions (Prasetio et al., 2022).

Challenges in the hospitality industry: Despite its resilience and adaptability, the hospitality industry faces challenges, particularly due to the profound impact of the COVID-19 pandemic, leading to disruptions in travel and tourism. For instance, the World Travel and Tourism Council reported that the hospitality industry faced unprecedented financial losses, with some companies experiencing a market value drop of up to 80% due to the uncertainty surrounding the pandemic's duration and effects (Cepni et al., 2023). This immediate fallout necessitated drastic measures, including layoffs and business closures, which have had lasting implications for the workforce and the overall economic landscape of the hospitality sector (Shereni & Rogerson, 2023). Research indicates that the crisis has catalyzed a shift toward more sustainable tourism consumption, with businesses exploring new ways (e.g., investments in green projects and digitalization) to offer value that aligns with consumer expectations for sustainability. This transition is not merely a response to the pandemic. Still, it may represent a long-term shift in how the hospitality sector operates, potentially leading to a more equitable and sustainable industry in the future (Shereni & Rogerson, 2023).

Moreover, the long-term health implications for employees in the hospitality sector cannot be overlooked. Many workers who contracted COVID-19 may experience persistent health issues, such as fatigue and respiratory complications, which could affect their ability to work effectively (Townsend et al., 2021). The need for ongoing support and rehabilitation for these individuals is critical, as their health challenges could further strain the already beleaguered workforce in hospitality. Therefore, integrating sustainability into business models is not merely a trend but a strategic necessity for hospitality businesses aiming to thrive in the post-pandemic landscape (Halpin et al., 2021).

The role of online reviews and digitalization in hospitality: With the rise of technology and increasing online interactions, there has been a notable shift in how consumers interact with the tourism and hospitality industry (Chakraborty & Biswal 2020). Several studies (Chakraborty & Biswal 2020) investigated the impact of online reviews on hotel booking intentions, emphasizing the significance of online reputation management and the role of customer reviews in shaping consumer decision-making. Moreover, the COVID-19 pandemic has accelerated technology adoption and digitalization within the industry, leading to the increased use of digital platforms for booking accommodations, contactless check-ins, and virtual experiences, which is a long-term shift in consumer behavior due to COVID-19.

Digital transformation and survival strategies for hotels: The COVID-19 pandemic has prompted accelerated digitalization within the tourism and hospitality industry. The use of digital platforms for booking accommodations (Deale & Crawford, 2018), contactless check-ins (Hadjielias et al., 2022), and virtual experiences (Wong et al., 2023) has become increasingly prevalent. The need for enhanced safety measures and improved customer experiences has driven this shift toward digitalization. Chiang (2020) explored the impact of digitalization on the hotel industry during the pandemic. The study highlighted the importance of digital transformation in ensuring the survival and competitiveness of hotels in the face of unprecedented challenges. The adoption of AI and robotics in hotel management has been explored, providing insights into the role of digitalization in the industry during the pandemic. It emphasized the need for hotels to adapt to changing consumer behaviors and preferences by embracing technology and leveraging digital platforms. Navío-Marco et al. (2018) discuss the evolution of eTourism and the use of digital platforms for booking in the tourism industry. They highlight the significant impact of the Internet on the tourism sector and the widespread adoption of digital platforms for various tourism-related activities, including booking accommodations.

Importance of AI in tourism and hospitality: Jiang and Wen (2020) found that AI can open up research streams at the intersection of health crises and hotel management, considering the Covid-19 pandemic, providing insights into the role of digitalization, including AI and robotics, in the hotel industry during the pandemic. The investigation by Hao et al. (2020) proposed a framework to manage a disaster that addresses the anti-pandemic phases, principles, and strategies; additionally, discussed post-pandemic strategies, including multi-business and multi-channels, product design and investment preference, digital and intelligent transformation, and market reshuffle.

3.2.2 Tourism and Hospitality Services

Dynamics of tourism and hospitality services: Tourism and hospitality services play a pivotal role in the travel industry, offering experiences and accommodations (Abbas et al., 2021). These industries have undergone significant transformations in recent years, driven by technological advancements, changing consumer preferences, and global events like the COVID-19 pandemic (Liang et al., 2017). The pandemic prompted a considerable shift toward virtual experiences and contactless services as travelers became more cautious about health and safety (Modgil et al., 2022). Additionally, the economic recession has compelled businesses to innovate and adapt to new market realities, fostering a culture of resilience and flexibility within the industry (Li & Liu, 2022). As a result, tourism enterprises are increasingly leveraging data analytics to understand consumer behavior and preferences, enabling them to tailor their offerings and improve customer satisfaction (Oduro et al., 2022; Yallop & Seraphin, 2020). Researchers have explored various facets of tourism and hospitality services, delving into service quality, innovation, technology, religious impacts, economic and social impacts, environmental impacts, and factors influencing customer satisfaction.

Navigating customer satisfaction in tourism and hospitality: Customer satisfaction (CS) is a crucial aspect of the tourism and hospitality industry (Koc, 2020). Numerous studies have examined performance levels, contributing factors, and the interplay of elements influencing customer satisfaction levels (Susilo, 2020). These factors include the design of service structures, location, facilities, and the alignment of customer expectations and experiences (Susilo, 2020). Understanding customer satisfaction is essential for businesses to attract and retain customers and foster loyalty (Susilo, 2020). Understanding customer preferences and delivering superior value and satisfaction are essential for maintaining competitiveness in the industry (Suchánek & Králová, 2019).

Ethical considerations and religious impacts: Ethical considerations are pivotal in tourism and hospitality. Transparency, informed consent, and adherence to ethical guidelines in data collection and utilization are crucial for maintaining trust and respecting privacy (Abbasi et al., 2023). Religion is one area of research that focuses on the role of religiosity in shaping tourist satisfaction and behavior. Studies have explored the influence of Islamic religiosity on the relationship between perceived value and tourist satisfaction (Eid & El-Gohary, 2015). Additionally, the role of religiosity in word-of-mouth communication and Muslim-friendly tourism marketing has been investigated (Abror et al., 2021). For instance, Islamic religious locations like Mecca, the birthplace of the Prophet Muhammad and the site of the Kaaba, attract millions of Muslim pilgrims annually, particularly during the Hajj season. The Hajj pilgrimage is one of the largest religious gatherings in the world, highlighting the city's importance as a religious tourism destination (Tabash et al., 2023). The motivations for religious tourism to Mecca are multifaceted. Pilgrims are driven by spiritual fulfillment, the desire to connect with their faith, and the opportunity to engage in communal worship. The experience of performing rituals such as Tawaf (circumambulating the Kaaba) and Sa'i (walking between the hills of Safa and Marwah) is central to the pilgrimage experience, which enhances the spiritual and emotional connection of the pilgrims to their faith (Hassan et al., 2022).

On the other hand, Christian religious locations like Lourdes, a prominent pilgrimage site, attract millions of visitors annually, wherein pilgrims seek spiritual renewal, healing, and a connection to the divine, central to their experiences (Klimiuk & Moriarty, 2021). The significance of Lourdes as a pilgrimage destination is underscored by its historical and spiritual importance within the Catholic faith, drawing individuals seeking both religious fulfillment and personal healing experiences (Griffiths, 2011). The motivations for visiting such sites are deeply intertwined with the spiritual and emotional needs of the pilgrims, as many individuals report transformative experiences during their visits, which often include prayer, reflection, and participation in rituals (Thomas et al., 2018). Furthermore, studies have shown that the motivations of pilgrims often intertwine with cultural and heritage tourism, as visitors engage with the historical and spiritual significance of the site (Amaro et al., 2018). This blend of motivation necessitates a nuanced approach to hospitality services, where providers must cater to both the spiritual needs of pilgrims and the expectations of tourists seeking cultural experiences (Bond et al., 2015). These studies highlight the importance of understanding tourists' religious beliefs and preferences in designing and delivering tailored tourism experiences.

Knowledge management and service innovation: Knowledge management has emerged as a critical factor in the tourism and hospitality industry, offering solutions to complex problems (e.g., fluctuating customer preferences, seasonal demand variations, and the need for sustainable practices) and fostering innovation and competitiveness (Mohamad et al., 2022). For example, integrating information and communication technologies (ICTs) has transformed service delivery in the hospitality sector, leading to innovations such as online booking systems and personalized marketing strategies that enhance customer engagement (Gonzalez et al., 2022). Applying knowledge management frameworks has also been associated with developing new service offerings and operational efficiencies, demonstrating its impact on innovation (Okumus, 2013). Organizations can differentiate themselves by providing knowledge-based services and enhancing customer experiences and satisfaction (Mohamad et al., 2022). Service quality (SQ) and service innovation (SI) are also key research areas, providing insights into the current state of research and identifying potential areas for future exploration. Service quality (SQ) and service innovation (SI) are other key research areas in tourism and hospitality. The findings of Shin and Perdue (2022) provide insights into the current state of research and identify potential areas for future exploration, contributing to the advancement of service innovation in the industry.

Technology, pandemic impact, and sustainability: The impact of social media and digital platforms on tourism decision-making has also been extensively studied (Pencarelli, 2020). Research has investigated the influence of social media images, consumer engagement, and involvement in electronic word-of-mouth (eWOM) in tourism destinations (Abbasi et al., 2023). Understanding the value of social media content and its impact on consumer behavior can help tourism businesses develop effective marketing strategies and engage with their target audience.

The COVID-19 pandemic profoundly impacted the tourism and hospitality industry, leading to the total disappearance of sit-in guests and travelers. Researchers have examined the economic and social consequences of the pandemic, highlighting the need for government interventions to support the recovery of these industries (Habiba et al., 2022). The pandemic has also accelerated technology adoption in the tourism and hospitality sector, with studies exploring customers' behavioral intentions toward technology adoption (Iskender et al., 2022). The COVID-19 pandemic has also highlighted the importance of health-related crises in tourism destination management (Vong et al., 2021). For instance, the findings of Vong et al. (2021) provided valuable insights into crisis management strategies and the resilience of these industries. The pandemic underscored the need for sustainable practices and the integration of health and safety measures in tourism and hospitality services. In addition, efficiency evaluation and analysis have been studied to understand the performance of low-carbon tourism and ecological construction (Kuzior et al., 2021). Guo and He (2022) examined the factors that affect the efficiency of tourism operations, including ecological factors, economic development levels, and policy and regional development strategies. The findings contribute to developing sustainable tourism practices and optimizing resource allocation.

Sustainability is a key concern in the tourism and hospitality industry, focusing on minimizing negative environmental and social impacts (Molina-Collado et al., 2022). The quality of hospitality services, including accommodation and food and beverage, plays a vital role in sustainable tourism; providing high-quality services can contribute to the overall sustainability of destinations and enhance the satisfaction of tourists (Pavić et al., 2019). Community involvement and power dynamics are important considerations in developing tourism destinations. The engagement of local communities and their influence on decision-making processes can lead to more sustainable and inclusive tourism development. Balancing stakeholder interests and power dynamics is crucial for tourism destinations' long-term success and sustainability (Dong & Nguyen, 2023).

Emerging technologies and research methodologies: The technological revolution, combined with the intensive use of data, information, and knowledge, has transformed the markets' dynamics, structures, rules, and productive sectors. AI and virtual reality (VR) impact customer experiences and service delivery (Daqar & Smoudy, 2019). Additionally, integrating sustainability practices and responsible tourism principles can contribute to the long-term viability and success of the industry (Mihalic et al., 2021). Big data offers organizations opportunities to gain insights into consumer behavior, personalize experiences, and improve operational efficiency (Yichuan et al., 2018). However, it also raises concerns about data governance and privacy, highlighting the need for effective and ethical data management practices in these industries (Yallop & Seraphin, 2020). The impact of technology on the tourism and hospitality industry has been a subject of interest, with studies examining the adoption and acceptance of technology by customers and organizations. For example, neuromarketing technology can be used to conduct high-quality experiments in research to study consumer unconscious behavior (Ahmed et al., 2023c, 2022a; Alsharif & Isa, 2024) and create more effective advertising (Ahmed, 2023; Alsharif & Khraiwish, 2024).

Advancements in technology have also transformed the landscape of tourism and hospitality services, leading to the emergence of smart services (Kabadayi et al., 2019). These services leverage technologies like AI and automation to provide personalized and seamless experiences for travelers (Kabadayi et al., 2019). The accelerated growth of smart tourism induces fundamental changes in the tourism sector (Kumar et al., 2023). Integrating smart services can revolutionize how tourists interact with destinations, accommodations, and other service providers (Kabadayi et al., 2019). Using robotics in hospitality and tourism services is another area of interest, wherein service robots have the potential to provide predictable and efficient services, enhancing customer experiences (Xu et al., 2020). However, introducing robots also raises questions about the impact on the human touch and the overall hospitality experience (Xu et al., 2020). Balancing automation with the preservation of personalized and empathetic service is a key consideration for the future of tourism and hospitality services.

There has been a growing interest in research methodologies that apply quantitative and qualitative approaches to tourism and hospitality studies (Provenzano & Baggio, 2020). Integrating these approaches (e.g., mixed-methods approach) allows for a more comprehensive understanding of the complex dynamics within the industry (Provenzano & Baggio, 2020).

3.2.3 Tourism and Climate Changes

Impacts of climate change on tourism resources: Tourism is significantly influenced by climate change, a topic extensively studied in recent years (Zhong et al., 2019). This impact extends to key tourism resources like landscapes, biodiversity, and ecosystems, with alterations in their quality and composition affecting the sustainable development of tourist destinations (Jenkins & Nicholls, 2010). The tourism industry's heavy dependence on the natural environment and climate renders it particularly vulnerable to climatic variability and change. Projections also indicate shifts in tourism trends toward higher elevations and latitudes due to climate change, underscoring the dynamic relationship between climate shifts and tourism patterns (Wu et al., 2023). Research highlights the multifaceted impacts of climate change on tourism resources, emphasizing its potential to significantly alter their distribution and availability. While rising temperatures initially benefit certain tourism activities like trekking, long-term projections reveal negative effects such as water scarcity and climate-induced disasters (Ritika et al., 2021). By 2050, substantial climate change impacts are expected to lead to fluctuations in tourist numbers and economic slowdowns in specific destinations (Franzoni & Pelizzari, 2016; Hoogendoorn et al., 2016).

Accordingly, climate change may open up new opportunities for tourism and hospitality in areas previously considered too cold or inaccessible, such as Northern Europe, Canada, and Russia. In Northern Europe, projections suggest climate change may increase tourism potential as warmer temperatures make these regions more hospitable for visitors (Barrios & Ibañez, 2015). Furthermore, reducing sea ice in the Arctic facilitates access to previously remote locations, extending the tourist season and creating new opportunities for adventure tourism and eco-tourism in these regions (Gawor & Dolnicki, 2018). Similarly, Jacobsen et al. (2011) suggest that global climate change could shift tourism demand toward higher latitudes, creating additional opportunities for tourism in Northern Europe. In Canada, Becken (2013) indicates that warmer temperatures could lead to significant increases in visitation to national parks, with projections estimating a 6% to 10% increase in the 2020s and a 10% to 36% increase by the 2050s. This trend suggests climate change may enhance the attractiveness of previously less-visited cold regions, thereby stimulating local economies reliant on tourism. Furthermore, Scott et al. emphasize that the tourism sector must adapt to these climatic shifts, which could transform destination competitiveness and sustainability (Scott et al., 2012). Weather influences tourist behavior, and understanding these dynamics is crucial in mitigating climate change's adverse effects on tourism activities (Caldeira & Kastenholz, 2018).

In addition, climate change is anticipated to reshape the spatial and temporal distribution of climate resources, potentially altering domestic and international tourism flows (Kovács & Király, 2021). For instance, Northern and cooler regions are likely to benefit as warm temperatures make them more attractive for summer tourism. For instance, Nicholls and Amelung (2015) highlight that while winter tourism has traditionally been a significant draw for the Nordic region, the future summer tourism prospects are promising due to improved climatic conditions compared to the hotter Mediterranean regions. Pala (2020) also suggests that Northern European countries are likely to gain from warmer temperatures, enhancing their attractiveness as summer tourist destinations, while Southern Europe may face challenges such as droughts. This is further supported by Gawor and Dolnicki (2018), who indicate that global climate change may shift tourism demand to higher latitudes, thereby creating additional opportunities for tourism in northern regions. Conversely, research by Amengual et al. (2014) corroborates this by projecting a notable improvement in climatic conditions for northern Europe, contrasting with the deteriorating conditions in the Mediterranean, which could lead to a shift in tourist preferences toward cooler destinations. Additionally, low-lying island nations and coastal cities, such as the Maldives and parts of the Caribbean, face rising sea levels and extreme weather threats, deterring visitors (Tanoamchard & Limjirakan, 2012). The Maldives, in particular, is recognized as one of the most vulnerable regions to climate change, with rising sea levels posing existential threats to its population and infrastructure (Beke, 2022). Studies indicate that sea-level rise is expected to exacerbate coastal flooding and erosion, impacting the attractiveness of these tourist destinations (Amores et al., 2021). These climatic changes deter visitors, as the risks associated with extreme weather and rising sea levels create an unappealing environment for tourism.

The anticipated shifts in tourism patterns toward higher altitudes and latitudes could impact destination attractiveness and the profitability of tourism-based businesses. These shifts are expected to alter the spatial and temporal distribution of climate resources, potentially changing international or domestic tourism flows (Kovács & Király, 2021). Consequently, the tourism industry must adapt and develop strategies to mitigate climate change's adverse effects on tourism resources (Kilungu & Munishi, 2024). Understanding these dynamics is vital for ensuring the long-term sustainability and resilience of the tourism sector in the face of climate change challenges.

Tourist perceptions and industry concerns regarding climate change: Climate change has emerged as a significant concern for the tourism industry, as various studies highlight its impacts on tourist perceptions, travel destinations, and the sector overall. Acknowledgment of the industry's contribution to climate change, particularly through activities like air travel, has been underscored (Becken, 2007). Additionally, the potential consequences of climate change on migration, travel, and tourism have been identified, emphasizing the need for health systems to prepare for population health implications (Semenza & Ebi, 2019). A nuanced analysis of tourists' perceptions of climate change and its mitigation behavior indicates limited awareness of tourism's contribution to climate change, which can hinder the adoption of energy-saving practices (Qiao & Gao, 2017). Climate change affects tourism attractions and the psychology of tourists and influences the tourism industry as a whole (Zhong et al., 2019). Furthermore, the climate-dependent nature of tourism and the implications of climate change on tourism flows and seasonality highlight the industry's vulnerability to environmental shifts (Amelung et al., 2007). As a result, the tourism sector faces challenges adapting to climate change due to perceived uncertainties, necessitating the development of climate services to support sustainable tourism and adaptation efforts (Scott et al., 2011).

Concerns and adaptation measures within the tourism industry: Concerns and adaptation measures within the tourism industry have become increasingly critical due to the impacts of climate change. The tourism industry is particularly vulnerable to climate change, with various studies emphasizing the need for both adaptation and mitigation strategies to address this issue (Mahadew & Appadoo, 2019). Adaptation measures such as snowmaking and other technical strategies have been identified as crucial for the ski industry and winter tourism in regions affected by climate change (Steiger, 2010). Spandre et al. (2019) highlighted that whether snowmaking is a relevant adaptation measure under future climate change is a widely debated issue in mountainous regions, with major implications on the supply side of this tourism industry.

The integration of adaptation and mitigation within a unified framework for sustainable development has been highlighted in the literature. Grafakos et al. (2019) stressed the importance of a common policy or regulatory framework to enhance the integration of adaptation and mitigation. Furthermore, Ajani et al. (2013) pointed out that indigenous knowledge is often overlooked when designing and implementing modern mitigation and adaptation strategies. The tourism industry underscores the importance of reducing carbon emissions and implementing various adaptation measures to minimize climate change-related damages (Bhattarai, 2015). Additionally, the involvement of local stakeholders and residents in planning adaptive measures has been recognized as essential for the sustainable development of tourism in the face of climate change (Hambira et al., 2021). The tourism industry needs to address concerns related to climate change and implement effective adaptation measures to ensure its long-term sustainability and resilience. Direct and indirect consequences of climate change, such as flooding, wildfires, and severe storms, have been observed to have lasting effects on tourism sites and visitor patterns (Carroll Ph D et al., 2019). These findings underscore the importance of understanding and addressing the impact of climate change on tourism, as well as the need for adaptation measures within the industry.

***Climate-resilient tourism strategies*:** The development of climate-resilient tourism strategies is increasingly recognized as essential for the long-term sustainability and success of the tourism industry, particularly in light of escalating climate-related challenges. Collaboration among government, local communities, and tourism operators is essential for effective climate change mitigation and adaptation strategies, considering the unique vulnerabilities of each destination (Abdelwahab & El-saghier, 2023). Moreover, the integration of resilience and sustainability concepts is vital, as they complement each other in enhancing the adaptive capacity of tourism destinations. This synergy helps prepare for climate-related disruptions and promotes sustainable practices that can mitigate the industry's environmental impact (Espiner et al., 2017). Moreover, the transition toward a more sustainable tourism industry could reveal a more resilient tourism system, especially in the face of climate change and the obligations contained in international agreements such as the Paris Agreement (Antonsen et al., 2022). It is evident that managing risk, reducing vulnerability, and enhancing resilience are crucial components of a successful strategy that allows tourism destination systems to adapt to climate risks and enhance destination well-being proactively (Loehr et al., 2022). Therefore, tourism stakeholders must seek tailored climate services to enhance the sustainability and resilience of the tourism sector to adverse weather and climate impacts (Nalau et al., 2017).

The literature on tourism resilience highlights the importance of understanding the multifaceted nature of resilience, which encompasses individual, organizational, and community dimensions (Hu & Xu, 2022). For instance, the resilience of small and medium-sized enterprises (SMEs) in tourism can catalyze broader community resilience, thereby enhancing the overall capacity of tourism destinations to withstand and recover from climate-related shocks (Badoc-Gonzales et al., 2022). Furthermore, empirical studies have shown that the financial security of tourism businesses is integral to their resilience, as it influences their ability to adapt to climate-related events (Pathak et al., 2021). This financial aspect underscores the necessity for robust organizational strategies that focus on immediate recovery and long-term sustainability and adaptation (Jiang et al., 2019).

3.2.4 Hotels and Hotels Industry

Sustainability challenges and adaptive practices in the hotel industry: The hotel industry is facing evident transformations caused by changes in consumer behavior, technological evolution, and the constant evolution of hoteliers (Marques & Marques, 2023). It has undergone substantial challenges and transformations between 2020 and 2023, primarily attributed to the profound impact of the Covid-19 pandemic (Anguera-Torrell et al., 2021). Sustainability within the hospitality sector faced a critical juncture, prompting hotels to adapt and reconsider their practices (Jones & Comfort, 2020). The study by Jones and Comfort (2020) highlighted the intricate relationship between sustainability and the hospitality industry during the Covid-19 crisis. The findings underscored the crucial role of creativity and innovation in developing sustainable solutions to the unprecedented challenges faced by the industry, emphasizing the need for forward-thinking strategies in the face of adversity.

Short-term COVID-19 pandemic impact on the hotel industry: The COVID-19 pandemic has had significant repercussions on hotel workers, manifesting as job insecurity and distributive injustice (Alyahya et al., 2021). The study conducted by Alyahya et al. (2021) delved into the psychological influences of Covid-19 on hotel workers, particularly focusing on job insecurity and turnover intention. The findings shed light on the severe psychological and psychosocial impacts experienced by hotel workers, who found themselves among the most affected due to the nature of their work. Mass layoffs and lockdowns contributed to heightened job insecurity and social loafing behavior, emphasizing the urgent need for interventions and support systems to address the psychological well-being of hotel workers during and post-pandemic.

Long-term COVID-19 pandemic impacts the hotel industry: Beyond the immediate impacts of the pandemic, the hotel industry has grappled with adapting to evolving consumer preferences and expectations. The pandemic acted as a catalyst, accelerating technology adoption and hotel digitalization (Morrone et al., 2021). Contactless check-in and check-out processes, mobile apps for room service, and stringent cleaning protocols have become integral to ensuring the safety and well-being of guests (Bharwani & Mathews, 2021). A study by Li et al. (2022) affirmed the positive impact of technology adoption on guest satisfaction and loyalty, underscoring the pivotal role of embracing digital solutions in the post-pandemic era. Additionally, the pandemic emphasized the critical need for hotels to prioritize health and safety measures, with a study by Gursoy and Chi (2021) revealing the significance of effective communication and transparency in conveying these measures to guests. Trust emerged as a key factor influencing guests' intentions to stay at hotels during the ongoing pandemic, reinforcing the importance of robust health and safety protocols in instilling confidence among guests and ensuring their overall well-being.

3.2.5 Corporate Social Responsibility

The evolving landscape of corporate social responsibility: Corporate social responsibility (CSR) has become a focal point for businesses, reflecting a paradigm shift where companies are expected to contribute to societal and environmental well-being beyond profit-making endeavors (Latapí Agudelo et al., 2019). Expectations surrounding CSR significantly differ across generational lines. For instance, research indicates that younger individuals, particularly Millennials and Generation Z, prioritize ethical business practices and are more likely to support companies that are committed to social and environmental issues (Titko et al., 2021). Conversely, older generations, while still valuing CSR, may place greater emphasis on traditional business metrics such as profitability and stability, leading to a more cautious approach toward CSR initiatives (Rosati et al., 2018). This concept encompasses voluntary actions taken by companies to address social and environmental issues, emphasizing their responsibilities to stakeholders (Manuel & Herron, 2020). CSR activities span a wide range, including charitable initiatives, environmental sustainability efforts, ethical business practices, and community engagement (Sroka & Szántó, 2018).

CSR in the context of the covid-19 pandemic: The Covid-19 pandemic has underscored the critical role of CSR, compelling businesses to respond to the needs of both internal and external stakeholders. Many companies have initiated charitable actions and programs to support communities and tackle the challenges posed by the pandemic (Manuel & Herron, 2020). In the short term, these companies faced immediate operational disruptions due to government-imposed lockdowns and travel restrictions, leading to drastic revenue losses and a need for rapid adaptation to new health protocols (Yeon et al., 2021). The pandemic heightened the importance of CSR as businesses were compelled to demonstrate their commitment to employee welfare and community support amidst widespread economic distress (Popkova et al., 2021). For instance, many firms-initiated CSR activities focused on health and safety measures, such as implementing stringent sanitation protocols and supporting frontline workers, which were crucial for rebuilding consumer trust (Zhong et al., 2022). In the long term, the pandemic has catalyzed a transformation in consumer behavior and expectations, necessitating a reevaluation of CSR strategies within the tourism and hospitality sectors (Madanaguli et al., 2022). Companies are now expected to integrate sustainability and ethical practices into their core operations, reflecting a shift toward more responsible tourism that prioritizes environmental and social governance (García et al., 2024). The crisis has also underscored the need for stakeholder collaboration to foster resilience and adaptability in the face of future disruptions (Zhong et al., 2022). Moreover, the financial strain experienced during the pandemic has led to a re-consideration of CSR investments, with some studies suggesting that firms engaging in proactive CSR initiatives may better withstand economic shocks and enhance their long-term viability (Deng & Zhou, 2022). Motivated by utilitarian principles to maximize overall welfare and moral considerations to fulfill obligations, these actions highlight the ethical foundations that guide CSR decisions and emphasize the pivotal role of businesses in addressing societal needs during times of crisis (Manuel & Herron, 2020) and non-crisis contexts (Macassa et al., 2017).

Marketing, globalization, and the CSR-firm performance nexus: The practice of CSR is significantly influenced by marketing strategies and globalization trends. In the competitive landscape of today's markets, companies recognize CSR as a tool to build trust, enhance reputation, and ensure long-term sustainability (Shen et al., 2021). While some view CSR as a burden, others perceive it as an opportunity to differentiate themselves and contribute positively to society (Shen et al., 2021). The relationship between CSR and firm performance is a subject of research interest, exploring its impact on financial performance, risk management, and competitiveness (Javeed & Lefen, 2019). Studies suggest that CSR can positively influence financial performance, especially when coupled with responsible corporate governance practices and engagement in green innovation (Padilla-Lozano & Collazzo, 2021). However, this relationship is intricate and influenced by factors like CEO power, ownership structure, and industry context (Javeed & Lefen, 2019).

Ethical leadership and corporate governance in fostering CSR: Ethical leadership and responsible corporate governance play pivotal roles in promoting CSR within organizations. Ethical leaders set the tone for ethical behavior and cultivate a culture valuing social responsibility (Saha et al., 2020). Transparent disclosure of CSR activities and adopting good corporate governance practices enhance accountability, transparency, and stakeholder trust (Abdelfattah & Aboud, 2020). Despite the positive impact of CSR, challenges exist, including the practicality of working ethically in every business aspect and concerns about greenwashing, where companies engage in superficial CSR practices to enhance their image without making substantial changes (Padilla-Lozano & Collazzo, 2021). Authentic and effective CSR initiatives require transparency, measurement, and robust reporting mechanisms (Halkos & Nomikos, 2021).

3.2.6 Online Reviews

The impact of online reviews on consumer behavior: Online reviews have become an integral part of the consumer decision-making process in the digital age (Zhang et al., 2014). These reviews offer valuable information and insights about products, services, and businesses, profoundly influencing consumers' purchasing (Ma et al., 2022). Research has consistently shown that online reviews significantly shape consumers' perceptions (Filieri et al., 2018), attitudes (Siddiqi et al., 2020), and purchase intentions (Fernandes et al., 2022). Dissatisfied customers are less likely to recommend a hotel than satisfied customers are. Notably, a specific aspect of online reviews that has attracted attention is their impact on impulse buying behavior (Chakraborty & Biswal, 2020). Zhang et al. (2018) conducted research to study shopping websites in China (ju.taobao.com, dianping.com, and meituan.com). The findings reveal that online reviews increase consumers' impulsive behavior. Impulse buying affects products and services differently, with products—especially low-cost, appealing items—being more prone to spontaneous purchases (Atulkar & Kesari, 2018). Services are generally less impulsive but can still attract last-minute decisions, such as hotel upgrades or excursions (Sharma et al., 2013). In the tourism and hospitality industry, impulse buying is common as travelers often seek to maximize their experiences (Szymkowiak et al., 2021). Limited-time offers and exclusive deals on services encourage tourists to make unplanned purchases, enhancing their trip (Kwon & Ahn, 2021). This highlights the importance of understanding how online reviews trigger impulsive tendencies and influence consumers' purchasing decisions.

Online influencers play a significant role in the tourism and hospitality industries, leveraging their large followings to shape travel trends, destinations, and consumer choices (Varga & Gabor, 2021; (Stoldt et al., 2019). Influencers often showcase unique experiences, destinations, and accommodations on social media, which can inspire their followers to visit these places or book similar experiences (Razak & Mansor, 2021). Their content, which combines personal recommendations with visually appealing imagery, often creates a sense of authenticity and aspirational value, encouraging followers to make travel decisions based on these endorsements (Wang et al., 2022). Moreover, influencers can quickly spread awareness of new destinations, hidden gems, or special promotions, which can drive interest and bookings (Stoldt et al., 2019). The impact of influencer marketing is gaining attention. It could be valuable for future research, as it has the potential to further transform consumer behavior in tourism and hospitality by blending social influence with digital marketing strategies.

The influence of online reviews on reviewers' behavior: Online reviews also play a significant role in shaping the behavior of those who write them (Chen & Huang, 2013). Review disconfirmation, where a consumer's expectations are not met based on the online reviews they read (Li et al., 2020), is critical. Li et al. (2020) conducted a mixed-method investigation, revealing that review disconfirmation can lead to various types of customer online review behavior, including writing negative reviews, seeking compensation, or engaging in retaliatory behaviors. This underscores the importance of managing customer expectations and addressing discrepancies between online reviews and actual experiences to mitigate negative review behavior.

Manipulation of online reviews: The manipulation of online reviews is prevalent in the tourism and hospitality industry, as businesses recognize reviews' strong influence on consumer decisions (Hu et al., 2012). Many travelers rely heavily on platforms such as TripAdvisor, Google Reviews, and Yelp to assess accommodations, restaurants, and attractions, making positive reviews essential for attracting customers (Luca & Zervas, 2016). This dependency has led some businesses to engage in unethical practices, such as posting fake positive reviews or incentivizing guests to leave favorable feedback to improve their ratings and gain a competitive edge (Choi et al., 2017). Conversely, competitors have also left false negative reviews to harm a business's reputation (Anagnostopoulou et al., 2019). Although major review platforms have implemented measures to detect and reduce review manipulation, it remains challenging due to the high volume of reviews and the sophistication of some deceptive practices (Chen et al., 2019). As a result, this manipulation can create a misleading picture of a business's quality and negatively impact consumers' trust in online reviews (Mayzlin et al., 2014). Thus, manipulation makes it challenging for consumers to make informed decisions and stresses the need for platforms and regulators to implement measures to detect and prevent fraudulent review activities. Ensuring the integrity of online review systems is crucial for maintaining trust between consumers and the information provided by these platforms.

4 Conclusions

4.1 General Conclusions

The present study embarked on a remarkable journey, guided by the PRISMA framework, to identify pertinent articles and review articles within the realm of (tourism OR tourism industry) AND ((hospitality OR hospitality) industry) OR (hotel OR hotel industry) OR (tourism and hospitality) research. With meticulous precision, 1247 documents were selected from the Scopus database for this descriptive analysis. With the descriptive analysis, the researchers embarked on a grand expedition into global academic research. Here, interesting themes and topics were brought to light, promising to revolutionize future studies while saving valuable time for newbie scholars.

Tourism involves traveling for leisure or exploration, often encompassing visits to various destinations and immersion in different cultures. On the other hand, hospitality creates a warm and welcoming experience for guests, typically in settings like hotels, restaurants, or entertainment venues. Tourism and hospitality services refer to the range of offerings provided to travelers, including accommodation, dining, entertainment, and transportation. The tourism and hospitality industry comprises a diverse array of sectors, such as hotels, airlines, and tourism agencies, dedicated to meeting the needs and desires of guests. Customer satisfaction is a crucial metric in this industry, gauging how content and happy customers are with the services they receive. Online reviews, posted on websites and social media platforms, play a significant role in shaping perceptions of businesses and experiences, as they reflect the opinions and feedback of users. Within this industry, the hotel sector focuses specifically on providing lodging and accommodation options to travelers. In today's digital age, social media platforms are powerful communication and community engagement tools, influencing how businesses connect with their customers. CSR, or Corporate Social Responsibility, represents a company's commitment to ethical and responsible business practices. These practices extend beyond profit generation to benefit society, the environment, and various stakeholders.

The hospitality and tourist sector faces threats and opportunities from climate change, necessitating flexible responses to both immediate and distant dangers. As a result of climate change, increasing sea levels, and other natural disasters, many well-known tourist spots may soon be in jeopardy. However, other areas may soon present themselves as appealing alternatives, changing the face of tourism around the world. The sector needs to take the initiative to deal with these changes by funding sustainable practices, encouraging environmentally conscious tourism, and building infrastructure that can withstand the effects of climate change. As the number of eco-conscious vacationers and those looking for "last-chance" adventures in endangered areas continues to rise, companies will have to adapt to meet the demands of this demographic. The future of sustainability hinges on the tourist and hospitality industry's ability to address climate change in a way that promotes both economic growth and environmental stewardship.

The descriptive analysis highlights key findings in tourism and hospitality research. Frequent words in publications include "tourism," "hospitality," "tourism and hospitality services," "tourism and hospitality industry," "customer satisfaction," "online reviews," "hotels industry," "social media," and "CSR." The tourism and hospitality industry revolves around providing exceptional experiences to travelers, with customer satisfaction, online reviews, social media, and CSR all playing pivotal roles in shaping its landscape.

4.2 Theoretical and Social Implications

This study provides a comprehensive systematic analysis of the keywords and yields profound theoretical and social implications. Theoretically, this research offers vital insights for academic fields in the tourism and hospitality sectors. It provides insights into climate-resilient tourism strategies for the long-term sustainability and success of the tourism industry, especially in the face of increasing climate-related challenges. Moreover, understanding the dynamics of online reviews and leveraging social media platforms can bolster a brand's reputation and reach a wider audience. Simultaneously, the social ramifications are equally significant. A thriving tourism and hospitality industry generates economic growth and job opportunities, preserves cultural heritage, and fosters community development. Furthermore, an emphasis on customer satisfaction and corporate social responsibility (CSR) practices contributes to socially responsible tourism, benefiting both travelers and local communities. As this analysis offers valuable directions for future research, it serves as a cornerstone for ongoing efforts to enhance the sustainability and societal impact of the tourism and hospitality sectors.

4.2.1 Limitations of This Study

The objective of this paper was to minimize methodological constraints in the study; however, despite efforts, some limitations were encountered, prompting recommendations for future research. The study focused exclusively on articles and review articles published in English-language journals between 2012 and 2022, specifically those indexed in the Scopus database. This narrow scope omitted other documents, such as conference papers and editorials, potentially introducing bias into the study. While the study acknowledges its methodological restrictions, it is a valuable resource for understanding the landscape of the relevant research during the specified period.

4.2.2 Future Directions of This Study

Future research directions in tourism, hospitality, and related areas encompass a wide spectrum of critical themes. Firstly, the ongoing digital transformation within the industry calls for exploration into the role of emerging technologies, such as AI and blockchain, in revolutionizing customer experiences and ensuring data security. Sustainable tourism development is another crucial avenue, focusing on eco-friendly practices, responsible tourism, and community engagement to mitigate environmental impact and ensure long-term sustainability. The digital age demands an investigation into the influence of VR and chatbots on customer satisfaction and brand loyalty. Managing online reviews effectively, employing sentiment analysis, automated response systems, and reputation management is a key research area.

Furthermore, the evolving landscape of destination marketing via social media platforms and the impact of influencer marketing and user-generated content require in-depth analysis. CSR's evolving role, particularly in environmental conservation and cultural preservation, necessitates examination. Crisis management and resilience strategies, cross-sector collaborations, data privacy, security, and education and training in the industry constitute additional areas of study. These future directions underscore the industry's need for continual research and innovation to navigate dynamic challenges and seize emerging opportunities.

References

Abbas, J., Mubeen, R., Iorember, P. T., Raza, S., & Mamirkulova, G. (2021). Exploring the impact of COVID-19 on tourism: Transformational potential and implications for a sustainable recovery of the travel and leisure industry. *Current Research in Behavioral Sciences, 2*, 100033. https://doi.org/10.1016/j.crbeha.2021.100033

Abbasi, A. Z., Tsiotsou, R. H., Hussain, K., Rather, R. A., & Ting, D. H. (2023). Investigating the impact of social media images' value, consumer engagement, and involvement on eWOM of a tourism destination: A transmittal mediation approach. *Journal of Retailing and Consumer Services, 71*, 103231. https://doi.org/10.1016/j.jretconser.2022.103231

Abdelfattah, T., & Aboud, A. (2020). Tax avoidance, corporate governance, and corporate social responsibility: The case of the Egyptian capital market. *Journal of International Accounting, Auditing and Taxation, 38*, 100304. https://doi.org/10.1016/j.intaccaudtax.2020.100304

Abdelwahab, M. A., & El-saghier, E. (2023). Negative consequences of climate change on marine and coastal tourism activities in Ras Sudr: Evidence from stakeholders. *Journal of Association of Arab Universities for Tourism and Hospitality, 24*(2), 709–731. https://doi.org/10.21608/jaauth.2023.224420.1487

Abror, A., Patrisia, D., Trinanda, O., Omar, M. W., & Wardi, Y. (2021). Antecedents of word of mouth in Muslim-friendly tourism marketing: The role of religiosity. *Journal of Islamic Marketing, 12*(4), 882–899. https://doi.org/10.1108/jima-01-2020-0006

Ahmed, H. A. (2023). *The enhancing Islamic advertising effectiveness through emotional processes and consumer-centric elements*. 2023 International Conference on Sustainable Islamic Business and Finance (SIBF), Bahrain. 5–11. https://doi.org/10.1109/SIBF60067.2023.10379973

Ahmed, H. A., NorZafir, M. S., Lina, P., & Shaymah Ahmed, A.-Z. (2023a). Exploring the tourism, neuro-tourism, and hospitality Nexus: A comprehensive bibliometric analysis. *Journal of Tourism and Services, 27*(14), 10.29036/jots.v14i27.606.

Ahmed, H. A., NorZafir, M. S., Mazilah, A., Ahmad, K., & Azmirul, A. (2023b). Neuromarketing tools used in the marketing mix: A systematic literature and future research agenda. *SAGE Open, 13*(1), 1–23. https://doi.org/10.1177/21582440231156563

Ahmed, H. A., NorZafir, M. S., Rami Hashem, E. A., Ahmad, K., Lennora, P., & Lily, S. M. A. (2023c). Exploring factors influencing neuromarketing implementation in malaysian universities: Barriers and enablers. *Sustainability, 15*(5), 4603–4632. https://doi.org/10.3390/su15054603

Ahmed, H. A., NorZafir, M. S., Rohaizat, B., & Alharthi, R. H. E. (2021). Neuromarketing research in the last five years: A bibliometric analysis. *Cogent Business & Management, 8*(1), 1–26. https://doi.org/10.1080/23311975.2021.1978620

Ahmed, H. A., NorZafir, M. S., Rohaizat, B., Hassan, A., & Rami Hashem, E. A. (2022a). A global research trends of neuromarketing: 2015–2020. *Revista de. Comunicación, 21*(1), 15–32. https://doi.org/10.26441/rc21.1-2022-a1

Ahmed, H. A., NorZafir, M. S., Shaymah Ahmed, A.-Z., & Ahmad, K. (2022b). Consumer behaviour to be considered in advertising: A systematic analysis and future agenda. *Behavioral Sciences, 12*(12), 472–493. https://doi.org/10.3390/bs12120472

Ahmed, H. A., & Salmi, M. I. (2024). Revolutionizing consumer insights: The impact of fMRI in neuromarketing research. *Future Business Journal, 10*(1), 79–105. https://doi.org/10.1186/s43093-024-00371-z

Ahmed, H. A., Salmi, M. I., & Mohammad, N. A. (2024). Smart tourism, hospitality, and destination: A systematic review and future directions. *Journal of Tourism and Services, 15*(29), 72–110. https://doi.org/10.29036/jots.v15i29.746

Ajani, E., Mgbenka, R., & Okeke, M. (2013). Use of indigenous knowledge as a strategy for climate change adaptation among farmers in sub Saharan Africa. Implications for policy. *Asian Journal of Agricultural Extension Economics & Sociology, 2*(1), 23–40. https://doi.org/10.9734/ajaees/2013/1856

Alsharif, A. H., & Isa, S. M. (2024). Electroencephalography studies on marketing stimuli: A literature review and future research agenda. *International Journal of Consumer Studies, 49*(1), e70015. https://doi.org/10.1111/ijcs.70015

Alsharif, A. H., & Khraiwish, A. (2024). Tools in marketing research: Exploring emotional responses to Stimuli. *Scientific Annals of Economics and Business, 71*(2), 173–192. https://doi.org/10.47743/saeb-2024-0009

Alsharif, A. H., Salleh, N. Z. M., Alrawad, M., & Lutfi, A. (2024). Exploring global trends and future directions in advertising research: A focus on consumer behavior. *Current Psychology, 1–24*. https://doi.org/10.1007/s12144-023-04812-w

Alsharif, A. H., Salleh, N. Z. M., & Lina, P. (2023). A comprehensive bibliometric analysis of fNIRS and fMRI technology in neuromarketing. *Scientific Annals of Economics and Business, 70*(3), 1–14. https://doi.org/10.47743/saeb-2023-0031

Alyahya, M. A., Elshaer, I. A., & Sobaih, A. E. E. (2021). The impact of job insecurity and distributive injustice post COVID-19 on social loafing behavior among hotel workers: Mediating role of turnover intention. *International Journal of Environmental Research and Public Health, 19*(1), 411. https://doi.org/10.3390/ijerph19010411

Amaro, S., Antunes, A., & Henriques, C. (2018). A closer look at Santiago de Compostela's pilgrims through the lens of motivations. *Tourism Management, 64*, 271–280. https://doi.org/10.1016/j.tourman.2017.09.007

Amelung, B., Nicholls, S., & Viner, D. (2007). Implications of global climate change for tourism flows and seasonality. *Journal of Travel Research, 45*(3), 285–296. https://doi.org/10.1177/0047287506295937

Amengual, A., Homar, V., Romero, R., Ramis, C., & Alonso, S. (2014). Projections for the 21st century of the climate potential for beach-based tourism in the Mediterranean. *International Journal of Climatology, 34*(13), 3481–3498. https://doi.org/10.1002/joc.3922

Amores, A., Marcos, M., Pedreros, R., Le Cozannet, G., Lecacheux, S., Rohmer, J., Hinkel, J., Gussmann, G., van der Pol, T., & Shareef, A. (2021). Coastal flooding in the Maldives induced by mean sea-level rise and wind-waves: From global to local coastal modelling. *Frontiers in Marine Science, 8*, 665672. https://doi.org/10.3389/fmars.2021.665672

Anagnostopoulou, S. C., Buhalis, D., Kountouri, I. L., Manousakis, E. G., & Tsekrekos, A. E. (2019). The impact of online reputation on hotel profitability. *International Journal of Contemporary Hospitality Management, 32*(1), 20–39. https://doi.org/10.1108/ijchm-03-2019-0247

Anguera-Torrell, O., Aznar-Alarcón, J. P., & Vives-Perez, J. (2021). COVID-19: Hotel industry response to the pandemic evolution and to the public sector economic measures. *Tourism Recreation Research, 46*(2), 148–157. https://doi.org/10.1080/02508281.2020.1826225

Antonsen, K. M., Dale, B., & Mayer, S. (2022). "It's in our DNA": Climate change and perceived resilience and adaptive capacity in nature-based tourism in Lofoten, Norway. *Weather, Climate, and Society, 14*(1), 177–190. https://doi.org/10.1175/wcas-d-21-0026.1

Atmojo, M. E., & Fridayani, H. D. (2021). An assessment of covid-19 pandemic impact on Indonesian tourism sector. *Journal of Governance and Public. Policy, 8*(1), 1–9. https://doi.org/10.18196/jgpp.811338

Atulkar, S., & Kesari, B. (2018). Role of consumer traits and situational factors on impulse buying: Does gender matter? *International Journal of Retail & Distribution Management, 46*(4), 386–405. https://doi.org/10.1108/ijrdm-12-2016-0239

Badoc-Gonzales, B. P., Mandigma, M. B. S., & Tan, J. J. (2022). SME resilience as a catalyst for tourism destinations: A literature review. *Journal of Global Entrepreneurship Research, 12*(1), 23–44. https://doi.org/10.1007/s40497-022-00309-1

Barrios, S., & Ibañez, J. N. (2015). Time is of the essence: Adaptation of tourism demand to climate change in Europe. *Climatic Change, 132*(4), 645–660. https://doi.org/10.1007/s10584-015-1431-1

Becken, S. (2007). Tourists' perception of international air travel's impact on the global climate and potential climate change policies. *Journal of Sustainable Tourism, 15*(4), 351–368. https://doi.org/10.2167/jost710.0

Becken, S. (2013). Measuring the effect of weather on tourism: A destination-and activity-based analysis. *Journal of Travel Research, 52*(2), 156–167. https://doi.org/10.1177/0047287512461569

Beke, R. (2022). Global warming and displacement: Challenges and solutions in the Maldives. *Revista Tecnológica-ESPOL, 34*(1), 75–91. https://doi.org/10.37815/rte.v34n1.899

Bharwani, S., & Mathews, D. (2021). Post-pandemic pressures to pivot: Tech transformations in luxury hotels. *Worldwide Hospitality and Tourism Themes, 13*(5), 569–583. https://doi.org/10.1108/WHATT-05-2021-0072

Bhattarai, U. (2015). Tourism and climate change: Socioeconomic implications, mitigation and adaptation measures. *International Journal of Environment, 4*(2), 355–373. https://doi.org/10.3126/ije.v4i2.12664

Bond, N., Packer, J., & Ballantyne, R. (2015). Exploring visitor experiences, activities and benefits at three religious tourism sites. *International Journal of Tourism Research, 17*(5), 471–481. https://doi.org/10.1002/jtr.2014

Buhalis, D., Harwood, T., Bogicevic, V., Viglia, G., Beldona, S., & Hofacker, C. (2019). Technological disruptions in services: Lessons from tourism and hospitality. *Journal of Service Management, 30*(4), 484–506. https://doi.org/10.1108/JOSM-12-2018-0398

Caldeira, A. M., & Kastenholz, E. (2018). It's so hot: Predicting climate change effects on urban tourists' time–space experience. *Journal of Sustainable Tourism, 26*(9), 1516–1542. https://doi.org/10.1080/09669582.2018.1478840

Candia, S., & Pirlone, F. (2021). Tourism environmental impacts assessment to guide public authorities towards sustainable choices for the post-COVID era. *Sustainability, 14*(1), 18. https://doi.org/10.3390/su14010018

Carroll, J., Brennan, A., Huff, A., & Thornburg, M. K. (2019). Effects of climate change on tourism in the Mid-Atlantic. *Journal of Tourism Insights, 9*(1), 1–22. https://doi.org/10.9707/2328-0824.1095

Cepni, O., Dogru, T., & Ozdemir, O. (2023). The contagion effect of COVID-19-induced uncertainty on US tourism sector: Evidence from time-varying granger causality test. *Tourism Economics, 29*(4), 906–928. https://doi.org/10.1177/13548166221077633

Chakraborty, U., & Biswal, S. K. (2020). Impact of online reviews on consumer's hotel booking intentions: Does brand image mediate? *Journal of Promotion Management, 26*(7), 943–963. https://doi.org/10.1080/10496491.2020.1746465

Chang, C.-L., McAleer, M., & Ramos, V. (2020). A charter for sustainable tourism after COVID-19. *Sustainability, 12*(9), 3671. https://doi.org/10.3390/su12093671

Chen, H.-N., & Huang, C.-Y. (2013). An investigation into online reviewers' behavior. *European Journal of Marketing, 47*(10), 1758–1773. https://doi.org/10.1108/EJM-11-2011-0625

Chen, L., Li, W., Chen, H., & Geng, S. (2019). Detection of fake reviews: Analysis of sellers' manipulation behavior. *Sustainability, 11*(17), 4802. https://doi.org/10.3390/su11174802

Chen, S., Tan, Z., Chen, Y., & Han, J. (2023). Research hotspots, future trends and influencing factors of tourism carbon footprint: A bibliometric analysis. *Journal of Travel & Tourism Marketing, 40*(2), 131–150. https://doi.org/10.1080/10548408.2023.2227852

Chernikov, S., & Thi, T. H. P. (2023). Challenges of post-Covid Vietnam tourist brand recovery. *International Trade and Trade Policy, 9*(3), 134–154. https://doi.org/10.21686/2410-7395-2023-3-134-154

Chi, Y., Fang, Y., & Liu, J. (2022). Spatial–temporal evolution characteristics and economic effects of China's cultural and tourism industries' collaborative agglomeration. *Sustainability, 14*(22), 15119. https://doi.org/10.3390/su142215119

Chiang, C.-T. (2020). Developing an eMarketing model for tourism and hospitality: A keyword analysis. *International journal of contemporary hospitality management, 32*(10), 3091–3114. https://doi.org/10.1108/ijchm-03-2020-0230

Choi, S., Mattila, A. S., Van Hoof, H. B., & Quadri-Felitti, D. (2017). The role of power and incentives in inducing fake reviews in the tourism industry. *Journal of Travel Research, 56*(8), 975–987. https://doi.org/10.1177/0047287516677168

Chowell, G., Dahal, S., Bono, R., & Mizumoto, K. (2021). Harnessing testing strategies and public health measures to avert COVID-19 outbreaks during ocean cruises. *Scientific Reports, 11*(1), 15482. https://doi.org/10.1101/2021.01.24.21250408

Daqar, M. A. A., & Smoudy, A. K. (2019). The role of artificial intelligence on enhancing customer experience. *International Review of Management and Marketing, 9*(4), 22. 10.32479/irmm.8166.

Davahli, M. R., Karwowski, W., Sonmez, S., & Apostolopoulos, Y. (2020). The hospitality industry in the face of the COVID-19 pandemic: Current topics and research methods. *International Journal of Environmental Research and Public Health, 17*(20), 7366. https://doi.org/10.3390/ijerph17207366

Davies, S. E., & Wenham, C. (2020). Why the COVID-19 response needs international relations. *International Affairs, 96*(5), 1227–1251. https://doi.org/10.1093/ia/iiaa135

Deale, C. S., & Crawford, A. (2018). Providers' perceptions of the online community marketplace for lodging accommodations. *Tourism and Hospitality Research, 18*(4), 470–477. https://doi.org/10.1177/1467358416682067

Deng, F., & Zhou, C. (2022). Sustainable development of corporate governance in the hospitality and tourism industry: The evolution and the future. *Sustainability, 14*(7), 4286. https://doi.org/10.3390/su14074286

Devkota, N., Gajdka, K., Siwakoti, R., Klímová, M., & Dhakal, K. (2023). Promoting sustainable tourist behavior through promotional marketing. *Journal of Tourism and Services, 14*(26), 219–241. https://doi.org/10.29036/jots.v14i26.512

Dong, X. D., & Nguyen, T. Q. T. (2023). Power, community involvement, and sustainability of tourism destinations. *Tourist Studies, 23*(1), 62–79. https://doi.org/10.1177/14687976221144335

Eid, R., & El-Gohary, H. (2015). The role of Islamic religiosity on the relationship between perceived value and tourist satisfaction. *Tourism Management, 46*, 477–488. https://doi.org/10.1016/j.tourman.2014.08.003

Espiner, S., Orchiston, C., & Higham, J. (2017). Resilience and sustainability: A complementary relationship? Towards a practical conceptual model for the sustainability–resilience nexus in tourism. *Journal of Sustainable Tourism, 25*(10), 1385–1400. https://doi.org/10.1080/09669582.2017.1281929

Fatoye, F., Gebrye, T., Arije, O., Fatoye, C. T., Onigbinde, O., & Mbada, C. E. (2021). Economic Impact of COVID-19 lockdown on households. *Pan African Medical Journal, 40*(1). https://doi.org/10.11604/pamj.2021.40.225.27446

Fernandes, S., Panda, R., Venkatesh, V., Swar, B. N., & Shi, Y. (2022). Measuring the impact of online reviews on consumer purchase decisions–A scale development study. *Journal of Retailing and Consumer Services, 68*(3), 103066. https://doi.org/10.1016/j.jretconser.2022.103066

Filieri, R., McLeay, F., Tsui, B., & Lin, Z. (2018). Consumer perceptions of information helpfulness and determinants of purchase intention in online consumer reviews of services. *Information & Management, 55*(8), 956–970. https://doi.org/10.1016/j.im.2018.04.010

Franzoni, S., & Pelizzari, C. (2016). Weather risk management in tourism industry. *Symphonya Emerging Issues in Management, 1*(1), 45–55. https://doi.org/10.4468/2016.1.05franzoni.pelizzari

García, C. J., Herrero, B., & Morillas-Jurado, F. (2024). The impact of the environmental, social and governance dimensions of sustainability on firm risk in the hospitality and tourism industry. *Corporate Social Responsibility and Environmental Management*. https://doi.org/10.1002/csr.2715

Gawor, Ł., & Dolnicki, P. (2018). Arctic tourism development with regard to legal regulations and environmental protection. *Studies of the Industrial Geography Commission of the Polish Geographical Society, 32*(2), 289–298. https://doi.org/10.24917/20801653.322.20

Ghasemi, M., Nejad, M. G., & Aghaei, I. (2021). Knowledge management orientation and operational performance relationship in medical tourism (overview of the model performance in the COVID-19 pandemic and post-pandemic era). *Health Services Management Research, 34*(4), 208–222. https://doi.org/10.1177/0951484820971438

Gonzalez, R., Gasco, J., & Llopis, J. (2022). Information and communication technologies in food services and restaurants: A systematic review. *International journal of Contemporary Hospitality Management, 34*(4), 1423–1447. https://doi.org/10.1108/ijchm-05-2021-0624

Górska-Warsewicz, H., & Kulykovets, O. (2020). Hotel brand loyalty—A systematic literature review. *Sustainability, 12*(12), 4810. https://doi.org/10.3390/su12124810

Grafakos, S., Trigg, K., Landauer, M., Chelleri, L., & Dhakal, S. (2019). Analytical framework to evaluate the level of integration of climate adaptation and mitigation in cities. *Climatic Change, 154*(1–2), 87–106. https://doi.org/10.1007/s10584-019-02394-w

Griffiths, M. (2011). Those who come to pray and those who come to look: Interactions between visitors and congregations. *Journal of Heritage Tourism, 6*(1), 63–72. https://doi.org/10.1080/1743873x.2010.536234

Guo, E., & He, J. (2022). Efficiency evaluation and analysis of low-carbon tourism and ecological construction based on DEA model. *Mathematical Problems in Engineering, 2022*. https://doi.org/10.1155/2022/1218900

Gursoy, D., & Chi, C. G. (2021). Celebrating 30 years of excellence amid the COVID-19 pandemic—An update on the effects of COVID-19 pandemic and COVID-19 vaccines on hospitality industry: Overview of the current situation and a research agenda. *Journal of Hospitality Marketing & Management, 30*(3), 277–281. https://doi.org/10.1080/19368623.2021.1902052

Habiba, M. S. A. M., Preeti, B., & Abdullah, K. H. A. (2022). The economic and social impact of COVID-19 on tourism and hospitality industry: A case study from Oman. *Journal of Public Affairs, 22*, e2786. https://doi.org/10.1002/pa.2786

Hadjielias, E., Christofi, M., Christou, P., & Drotarova, M. H. (2022). Digitalization, agility, and customer value in tourism. *Technological Forecasting and Social Change, 175*, 121334. https://doi.org/10.1016/j.techfore.2021.121334

Halkos, G., & Nomikos, S. (2021). Corporate social responsibility: Trends in global reporting initiative standards. *Economic Analysis and Policy, 69*, 106–117. https://doi.org/10.1016/j.eap.2020.11.008

Halpin, S. J., McIvor, C., Whyatt, G., Adams, A., Harvey, O., McLean, L., Walshaw, C., Kemp, S., Corrado, J., & Singh, R. (2021). Postdischarge symptoms and rehabilitation needs in survivors of COVID-19 infection: A cross-sectional evaluation. *Journal of Medical Virology, 93*(2), 1013–1022. https://doi.org/10.1002/jmv.26368

Halsharif, A., Salleh, N., & Md, & Baharun, R. (2020). Research trends of neuromarketing: A bibliometric analysis. *Journal of Theoretical and Applied Information Technology, 98*(15), 2948–2962.

Halsharif, A., Salleh, N. Z. M., Baharun, R., Abuhassna, H., & Alsharif, Y. H (2022). *Neuromarketing in Malaysia: Challenges, limitations, and solutions*. International Conference on Decision Aid Sciences and Applications (DASA), 2022, Chiangrai, Thailand. 740–745. https://doi.org/10.1109/dasa54658.2022.9765010

Hambira, W. L., Saarinen, J., Atlhopheng, J. R., & Manwa, H. (2021). Climate change, tourism, and community development: Perceptions of Maun residents. *Botswana. Tourism Review International, 25*(2–3), 105–117. https://doi.org/10.3727/154427220x16059054538773

Hao, F., Xiao, Q., & Chon, K. (2020). COVID-19 and China's hotel industry: Impacts, a disaster management framework, and post-pandemic agenda. *International Journal of Hospitality Management, 90*, 102636. https://doi.org/10.1016/j.ijhm.2020.102636

Hassan, T., Carvache-Franco, M., Carvache-Franco, W., & Carvache-Franco, O. (2022). Motivations as predictors of religious tourism: The Muslim pilgrimage to the city of Mecca. *Journal of Cultural Heritage Management and Sustainable Development, 14*(3), 419–435. https://doi.org/10.1108/jchmsd-01-2022-0005

Hemmington, N., & Neill, L. (2022). Hospitality business longevity under COVID-19: The impact of COVID-19 on New Zealand's hospitality industry. *Tourism and Hospitality Research, 22*(1), 102–114. https://doi.org/10.1177/1467358421993875

Hoogendoorn, G., Grant, B., & Fitchett, J. M. (2016). Disjunct perceptions? Climate change threats in two-low lying South African coastal towns. *Bulletin of Geography. Socio-economic Series, 31*(31), 59–71. https://doi.org/10.1515/bog-2016-0005

Hu, H., & Xu, K. (2022). Visualizing the development of research on tourism resilience with mixed methods. *SAGE Open, 12*(3), 21582440221126684. https://doi.org/10.1177/21582440221126684

Hu, N., Bose, I., Koh, N. S., & Liu, L. (2012). Manipulation of online reviews: An analysis of ratings, readability, and sentiments. *Decision Support Systems, 52*(3), 674–684. https://doi.org/10.1016/j.dss.2011.11.002

Huang, S., & Wang, X. (2023). COVID-19 two years on: A review of COVID-19-related empirical research in major tourism and hospitality journals. *International Journal of Contemporary Hospitality Management, 35*(2), 743–764. https://doi.org/10.1108/ijchm-03-2022-0393

Işık, C., Sirakaya-Turk, E., & Ongan, S. (2020). Testing the efficacy of the economic policy uncertainty index on tourism demand in USMCA: Theory and evidence. *Tourism Economics, 26*(8), 1344–1357. https://doi.org/10.1177/1354816619888346

Iskender, A., Sirakaya-Turk, E., Cardenas, D., & Harrill, R. (2022). COVID or VOID: A systematic literature review of technology adoption and acceptance in hospitality and tourism since the breakout of COVID-19. *Tourism and Hospitality Research, 14673584221133667*. https://doi.org/10.1177/14673584221133667

Jacobsen, J. K. S., Denstadli, J. M., Lohmann, M., & Førland, E. J. (2011). Tourist weather preferences in Europe's Arctic. *Climate Research, 50*(1), 31–42. https://doi.org/10.3354/cr01033

Javeed, S. A., & Lefen, L. (2019). An analysis of corporate social responsibility and firm performance with moderating effects of CEO power and ownership structure: A case study of the manufacturing sector of Pakistan. *Sustainability, 11*(1), 248. https://doi.org/10.3390/su11010248

Jenkins, K., & Nicholls, S. (2010). The impacts of climate variability and potential climate change on tourism business in Torbay, England and implications for adaptation. *Tourism Analysis, 15*(1), 17–30. https://doi.org/10.3727/108354210x12724734223513

Jiang, Y., Ritchie, B. W., & Verreynne, M. L. (2019). Building tourism organizational resilience to crises and disasters: A dynamic capabilities view. *International Journal of Tourism Research, 21*(6), 882–900. https://doi.org/10.1002/jtr.2312

Jiang, Y., & Wen, J. (2020). Effects of COVID-19 on hotel marketing and management: A perspective article. *International Journal of Contemporary Hospitality Management, 32*(8), 2563–2573. https://doi.org/10.1108/ijchm-03-2020-0237

Jones, P., & Comfort, D. (2020). The COVID-19 crisis and sustainability in the hospitality industry. *International journal of Contemporary Hospitality Management, 32*(10), 3037–3050. https://doi.org/10.1108/ijchm-04-2020-0357

Kabadayi, S., Ali, F., Choi, H., Joosten, H., & Lu, C. (2019). Smart service experience in hospitality and tourism services: A conceptualization and future research agenda. *Journal of Service Management, 30*(3), 326–348. https://doi.org/10.1108/josm-11-2018-0377

Kaushal, V., & Srivastava, S. (2021). Hospitality and tourism industry amid COVID-19 pandemic: Perspectives on challenges and learnings from India. *International Journal of Hospitality Management, 92*, 102707. https://doi.org/10.1016/j.ijhm.2020.102707

Khan, N. U., Alim, W., Begum, A., Han, H., & Mohamed, A. (2022). Examining factors that influence the international tourism in pakistan and its nexus with economic growth: Evidence from ARDL approach. *Sustainability, 14*(15), 9763. https://doi.org/10.3390/su14159763

Khraiwish, A., & Alsharif, A. H. (2024). Insights and influencers: A decade of social media marketing research revealed through bibliometrics. *Scientific Annals of Economics and Business, 71*(4), 1–22. https://doi.org/10.47743/saeb-2024-0023

Kilungu, H., & Munishi, P. K. (2024). Eco-parcel: An approach to identify and describe attractions to support adapting nature-based tourism destinations to climate-change impacts. *African Journal of Ecology, 62*(1), e13211. https://doi.org/10.1111/aje.13211

Klimiuk, J., & Moriarty, K. J. (2021). The Lourdes pilgrimage and the impact on pilgrim quality of life. *Journal of Religion and Health, 60*(6), 3775–3787. https://doi.org/10.1108/tr-08-2013-0046

Koc, E. (2020). Do women make better in tourism and hospitality? A conceptual review from a customer satisfaction and service quality perspective. *Journal of Quality Assurance in Hospitality & Tourism, 21*(4), 402–429. https://doi.org/10.1080/1528008X.2019.1672234

Korinth, B., & Wendt, J. A. (2021). The impact of COVID-19 pandemic on foreign tourism in European countries. *Studies of the Industrial Geography Commission of the Polish Geographical Society, 35*(3), 186–204. https://doi.org/10.24917/20801653.353.11

Koseoglu, M. A., Rahimi, R., Okumus, F., & Liu, J. (2016). Bibliometric studies in tourism. *Annals of Tourism Research, 61*, 180–198. https://doi.org/10.1016/j.annals.2016.10.006

Kovács, A., & Király, A. (2021). Assessment of climate change exposure of tourism in Hungary using observations and regional climate model data. *Hungarian Geographical Bulletin, 70*(3), 215–231. https://doi.org/10.15201/hungeobull.70.3.2

Kozhokulov, S., Chen, X., Yang, D., Issanova, G., Samarkhanov, K., & Aliyeva, S. (2019). Assessment of tourism impact on the socio-economic spheres of the Issyk-Kul Region (Kyrgyzstan). *Sustainability, 11*(14), 3886. https://doi.org/10.3390/su11143886

Kumar, D. (2022). Economic and political uncertainties and sustainability disclosures in the tourism sector firms. *Tourism Economics, 29*(6), 1694–1699. https://doi.org/10.1177/13548166221113434

Kumar, S., Kumar, V., Bhatt, I. K., & Kumar, S. (2023). Mapping research trends on smart tourism: A bibliometric analysis. In J. Marques & R. P. Marques (Eds.), *Digital transformation of the hotel industry: Theories, practices, and global challenges* (pp. 87–109). Springer. https://doi.org/10.1007/978-3-031-31682-1_5

Kuzior, A., Lyulyov, O., Pimonenko, T., Kwilinski, A., & Krawczyk, D. (2021). Post-industrial tourism as a driver of sustainable development. *Sustainability, 13*(15), 8145. https://doi.org/10.3390/su13158145

Kwon, J., & Ahn, J. (2021). The effects of cruise attributes on impulse buying behavior: The mediating role of impulsive buying tendency. *International Journal of Quality and Service Sciences, 13*(3), 456–470. https://doi.org/10.1108/ijqss-09-2020-0141

Latapí Agudelo, M. A., Jóhannsdóttir, L., & Davídsdóttir, B. (2019). A literature review of the history and evolution of corporate social responsibility. *International Journal of Corporate Social Responsibility, 4*(1), 1–23. https://doi.org/10.1186/s40991-018-0039-y

Lecchini, D., Brooker, R. M., Waqalevu, V., Gairin, E., Minier, L., Berthe, C., Besineau, R., Blay, G., Maueau, T., & Sturny, V. (2021). Effects of COVID-19 pandemic restrictions on coral reef fishes at eco-tourism sites in Bora-Bora, French Polynesia. *Marine Environmental Research, 170*, 105451. https://doi.org/10.1016/j.marenvres.2021.105451

Li, G., & Liu, A. (2022). Editorial of the special issue on competitiveness in the visitor economy: New trends, issues, and perspectives. *28*(3), 661–664. https://doi.org/10.1177/13548166211065257

Li, H., Meng, F., & Pan, B. (2020). How does review disconfirmation influence customer online review behavior? A mixed-method investigation. *International Journal of Contemporary Hospitality Management, 32*(11), 3685–3703. https://doi.org/10.1108/ijchm-03-2020-0234

Li, X., Gong, J., Gao, B., & Yuan, P. (2021). Impacts of COVID-19 on tourists' destination preferences: Evidence from China. *Annals of Tourism Research, 90*, 103258. https://doi.org/10.1016/j.annals.2021.103258

Li, X., Liang, X., Yu, T., Ruan, S., & Fan, R. (2022). Research on the integration of cultural tourism industry driven by digital economy in the context of COVID-19—based on the data of 31 Chinese provinces. *Frontiers in Public Health, 10*, 780476. https://doi.org/10.3389/fpubh.2022.780476

Liang, T.-L., Chang, H.-F., Ko, M.-H., & Lin, C.-W. (2017). Transformational leadership and employee voices in the hospitality industry. *International Journal of Contemporary Hospitality Management, 29*(1), 374–392. https://doi.org/10.1108/ijchm-07-2015-0364

Liu-Lastres, B., Wen, H., & Huang, W.-J. (2023). A reflection on the Great Resignation in the hospitality and tourism industry. *International Journal of Contemporary Hospitality Management, 35*(1), 235–249. https://doi.org/10.1108/ijchm-05-2022-0551

Loehr, J., Becken, S., Nalau, J., & Mackey, B. (2022). Exploring the multiple benefits of ecosystem-based adaptation in tourism for climate risks and destination well-being. *Journal of Hospitality & Tourism Research, 46*(3), 518–543. https://doi.org/10.1177/1096348020944438

Luca, M., & Zervas, G. (2016). Fake it till you make it: Reputation, competition, and Yelp review fraud. *Management Science, 62*(12), 3412–3427. https://doi.org/10.1287/mnsc.2015.2304

Ma, G., Ma, J., Li, H., Wang, Y., Wang, Z., & Zhang, B. (2022). Customer behavior in purchasing energy-saving products: Big data analytics from online reviews of e-commerce. *Energy Policy, 165*(3), 112960. https://doi.org/10.1016/j.enpol.2022.112960

Macassa, G., da Cruz Francisco, J., & McGrath, C. (2017). Corporate social responsibility and population health. *Health Science Journal, 11*(5), 1–6. https://doi.org/10.21767/1791-809x.1000528

Madanaguli, A., Srivastava, S., Ferraris, A., & Dhir, A. (2022). Corporate social responsibility and sustainability in the tourism sector: A systematic literature review and future outlook. *Sustainable Development, 30*(3), 447–461. https://doi.org/10.1002/sd.2258

Mahadew, R., & Appadoo, K. A. (2019). Tourism and climate change in Mauritius: Assessing the adaptation and mitigation plans and prospects. *Tourism Review, 74*(2), 204–215. https://doi.org/10.1108/tr-12-2017-0203

Manuel, T., & Herron, T. L. (2020). An ethical perspective of business CSR and the COVID-19 pandemic. *Society and Business Review, 15*(3), 235–253. https://doi.org/10.1108/sbr-06-2020-0086

Marques, J., & Marques, R. P. (2023). Trends in the Hospitality Industry: A global perspective. In J. Marques & R. P. Marques (Eds.), *Digital transformation of the hotel industry: Theories, practices, and global challenges* (pp. 1–16). Springer. https://doi.org/10.1007/978-3-031-31682-1_1

Mayzlin, D., Dover, Y., & Chevalier, J. (2014). Promotional reviews: An empirical investigation of online review manipulation. *American Economic Review, 104*(8), 2421–2455. https://doi.org/10.1257/aer.104.8.2421

Medeiros, H. M. N., Guerreiro, Q. L. d. M., Vieira, T. A., Silva, S. M. S. d., Renda, A. I. d. S. A., & Oliveira-Junior, J. M. B. (2021). Alternative tourism and environmental impacts: Perception of residents of an extractive reserve in the Brazilian Amazonia. *Sustainability, 13*(4), 2076. https://doi.org/10.3390/su13042076

Melnychenko, S., Tkachenko, T., & Dupliak, T. (2021). Digitalisation as a tool of tourism recovery in European Union in post-COVID-19. *Financial and Credit Activity Problems of Theory and Practice, 6*(41), 427–436. https://doi.org/10.18371/fcaptp.v6i41.251471

Mihalic, T., Mohamadi, S., Abbasi, A., & Dávid, L. D. (2021). Mapping a sustainable and responsible tourism paradigm: A bibliometric and citation network analysis. *Sustainability, 13*(2), 853. https://doi.org/10.3390/su13020853

Modgil, S., Dwivedi, Y. K., Rana, N. P., Gupta, S., & Kamble, S. (2022). Has Covid-19 accelerated opportunities for digital entrepreneurship? An Indian perspective. *Technological Forecasting and Social Change, 175*, 121415. https://doi.org/10.1016/j.techfore.2021.121415

Mohamad, S. J. A. N. S., Nor, N. S. N. M., & Fikry, A. (2022). The effect of knowledge management capabilities towards organizational innovation in Malaysia's hotel: A conceptual analysis. *International Journal of Academic Research in Business and Social Sciences, 12*(12), 1040–1048. https://doi.org/10.6007/ijarbss/v12-i12/15514

Moher, D., Shamseer, L., Clarke, M., Ghersi, D., Liberati, A., Petticrew, M., Shekelle, P., & Stewart, L. A. (2015). Preferred reporting items for systematic review and meta-analysis protocols (PRISMA-P) 2015 statement. *Systematic Reviews, 4*(1), 1–9. https://doi.org/10.1186/2046-4053-4-1

Molina-Collado, A., Santos-Vijande, M. L., Gómez-Rico, M., & Madera, J. M. (2022). Sustainability in hospitality and tourism: A review of key research topics from 1994 to 2020. *International Journal of Contemporary Hospitality Management, 34*(8), 3029–3064. https://doi.org/10.1108/ijchm-10-2021-1305

Morrone, D., Raimo, N., Tarulli, A., & Vitolla, F. (2021). Digitalisation in the hospitality industry: Motivations, effects and role of COVID-19. *International Journal of Digital Culture and Electronic Tourism, 3*(3–4), 257–270. https://doi.org/10.1504/IJDCET.2021.116475

Nalau, J., Becken, S., Noakes, S., & Mackey, B. (2017). Mapping tourism stakeholders' weather and climate information-seeking behavior in Fiji. *Weather, Climate, and Society, 9*(3), 377–391. https://doi.org/10.1175/wcas-d-16-0078.1

Navío-Marco, J., Ruiz-Gómez, L. M., & Sevilla-Sevilla, C. (2018). Progress in information technology and tourism management: 30 years on and 20 years after the internet-Revisiting Buhalis & Law's landmark study about eTourism. *Tourism Management, 69*, 460–470. https://doi.org/10.1016/j.tourman.2018.06.002

Nguyen, C. P., Thanh, S. D., & Nguyen, B. (2022). Economic uncertainty and tourism consumption. *Tourism Economics, 28*(4), 920–941. https://doi.org/10.1177/1354816620981519

Nicholls, S., & Amelung, B. (2015). Implications of climate change for rural tourism in the Nordic region. *Scandinavian Journal of Hospitality and Tourism, 15*(1–2), 48–72. https://doi.org/10.1080/15022250.2015.1010325

Ntounis, N., Parker, C., Skinner, H., Steadman, C., & Warnaby, G. (2022). Tourism and Hospitality industry resilience during the Covid-19 pandemic: Evidence from England. *Current Issues in Tourism, 25*(1), 46–59. https://doi.org/10.1080/13683500.2021.1883556

Nusair, K., Butt, I., & Nikhashemi, S. (2019). A bibliometric analysis of social media in hospitality and tourism research. *International Journal of Contemporary Hospitality Management, 31*(7), 2691–2719. https://doi.org/10.1108/ijchm-06-2018-0489

Oduro, S., Alharthi Rami Hashem, E., & Alsharif, A. H. (2022). Organisational ambidexterity and social enterprise performance: A Ghanaian perspective. *South African Journal of Economic and Management Sciences, 25*(1), 1–13. https://doi.org/10.4102/sajems.v25i1.4635

Okumus, F. (2013). Facilitating knowledge management through information technology in hospitality organizations. *Journal of Hospitality and Tourism Technology, 4*(1), 64–80. https://doi.org/10.1108/17579881311302356

Padilla-Lozano, C. P., & Collazzo, P. (2021). Corporate social responsibility, green innovation and competitiveness–causality in manufacturing. *Competitiveness Review: An International Business Journal, 32*(7), 21–39. https://doi.org/10.1108/cr-12-2020-0160

Pahrudin, P., Liu, L.-W., & Li, S.-Y. (2022). What is the role of tourism management and marketing toward sustainable tourism? A bibliometric analysis approach. *Sustainability, 14*(7), 4226. https://doi.org/10.3390/su14074226

Pala, A. (2020). The relation between climate change and economic growth: The in-vestigation the regional differences with RCM model in EU-28 countries. *International Journal of Economic Sciences, 9*(1), 135–155. https://doi.org/10.20472/es.2020.9.1.008

Pathak, A., van Beynen, P. E., Akiwumi, F. A., & Lindeman, K. C. (2021). Climate adaptation within the tourism sector of a small Island developing state: A case study from the coastal accommodations subsector in the Bahamas. *Business Strategy & Development, 4*(3), 313–325. https://doi.org/10.1002/bsd2.160

Paul, J., & Bhukya, R. (2021). Forty-five years of International Journal of Consumer Studies: A bibliometric review and directions for future research. *International Journal of Consumer Studies, 45*(5), 937–963. https://doi.org/10.1111/ijcs.12727

Pavić, L., Blešić, I., Petrović, M. D., Radovanović, M. M., & Prišenk, J. (2019). Can the quality of hospitality services play a role in sustainable equestrian tourism in slovenia? Mediations, effects, and implications. *Sustainability, 11*(21), 6155–6167. https://doi.org/10.3390/su11216155

Pencarelli, T. (2020). The digital revolution in the travel and tourism industry. *Information Technology & Tourism, 22*(3), 455–476. https://doi.org/10.1007/s40558-019-00160-3

Pilelienė, L., Alsharif, A. H., & Alharbi, I. B. (2022). Scientometric analysis of scientific literature on neuromarketing tools in advertising. *Baltic. Journal of Economic Studies, 8*(5), 1–12. https://doi.org/10.30525/2256-0742/2022-8-5-1-12

Pilelienė, L., Grigaliūnaitė, V., & Bogoyavlenska, Y. (2024). A bibliometric review of innovations in sustainable tourism research: Current trends and future research agenda. *Sustainability, 16*(16), 7124. https://doi.org/10.3390/su16167124

Pilelienė, L., & Jucevičius, G. (2023). A decade of innovation ecosystem development: Bibliometric review of scopus database. *Sustainability, 15*(23), 16386. https://doi.org/10.3390/su152316386

Pimonenko, T., Lyulyov, O., & Us, Y. (2021). Cointegration between economic, ecological and tourism development. *Journal of Tourism and Services, 12*(23), 169–180. https://doi.org/10.29036/jots.v12i23.293

Popkova, E., DeLo, P., & Sergi, B. S. (2021). Corporate social responsibility amid social distancing during the COVID-19 crisis: BRICS vs. OECD countries. *Research in International Business and Finance, 55*, 101315. https://doi.org/10.1016/j.ribaf.2020.101315

Prasetio, A. P., Tai, T. D., Opulencia, M. J. C., Abbas, M., El-Ebiary, Y. A. B., Abbas, S. F., Bykanova, O., Samal, A., & Iswanto, A. H. (2022). Impact of the COVID-19 pandemic on religious tourism amongst Muslims in Iraq. *HTS Teologiese Studies/Theological Studies, 78*(4), 1–6. https://doi.org/10.4102/hts.v78i4.7565

Provenzano, D., & Baggio, R. (2020). Quantitative methods in tourism and hospitality: A perspective article. *Tourism Review, 75*(1), 24–28. https://doi.org/10.1108/tr-07-2019-0281

Qiao, G., & Gao, J. (2017). Chinese tourists' perceptions of climate change and mitigation behavior: An application of norm activation theory. *Sustainability, 9*(8), 1322–1335. https://doi.org/10.3390/su9081322

Radic, A., Law, R., Lück, M., Kang, H., Ariza-Montes, A., Arjona-Fuentes, J. M., & Han, H. (2020). Apocalypse now or overreaction to coronavirus: The global cruise tourism industry crisis. *Sustainability, 12*(17), 6968. https://doi.org/10.3390/su12176968

Razak, R. A., & Mansor, N. A. (2021). Instagram influencers in social media-induced tourism: Rethinking tourist trust towards tourism destination. In *Impact of new media in tourism* (pp. 135–144). IGI Global. https://doi.org/10.4018/978-1-7998-7095-1.ch009

Ren, X. (2020). Pandemic and lockdown: A territorial approach to COVID-19 in China, Italy and the United States. *Eurasian Geography and Economics, 61*(4–5), 423–434. https://doi.org/10.1080/15387216.2020.1762103

Ritika, K., Giri, I., & Khadka, U. R. (2021). Climate change and possible impacts on travel and tourism sector. *Journal of Tourism and Himalayan Adventures, 3*(1), 54–62. https://doi.org/10.3126/jtha.v3i1.39117

Rosati, F., Costa, R., Calabrese, A., & Pedersen, E. R. G. (2018). Employee attitudes towards corporate social responsibility: A study on gender, age and educational level differences. *Corporate Social Responsibility and Environmental Management, 25*(6), 1306–1319. https://doi.org/10.1002/csr.1640

Roson, R., & Sartori, M. (2014). Climate change, tourism and water resources in the Mediterranean: A general equilibrium analysis. *International Journal of Climate Change Strategies and Management, 6*(2), 212–228. https://doi.org/10.2139/ssrn.2078869

Rudyanto, R., Pramono, R., & Juliana, J. (2021). Perception of knowledge of the risk of the COVID-19 pandemic regarding touring intentions and tourism travel recommendations. *Journal of Environmental Management and Tourism, 12*(4), 929–947. https://doi.org/10.14505/jemt.v12.4(52).08

Saha, R., Shashi, C. R., Singh, R., & Dahiya, R. (2020). Effect of ethical leadership and corporate social responsibility on firm performance: A systematic review. *Corporate Social Responsibility and Environmental Management, 27*(2), 409–429. https://doi.org/10.1002/csr.1824

Scott, D. (2021). Sustainable tourism and the grand challenge of climate change. *Sustainability, 13*(4), 1966. https://doi.org/10.3390/su13041966

Scott, D., Gössling, S., & Hall, C. M. (2012). International tourism and climate change. *Wiley Interdisciplinary Reviews: Climate Change, 3*(3), 213–232. https://doi.org/10.1002/wcc.165

Scott, D. J., Lemieux, C. J., & Malone, L. (2011). Climate services to support sustainable tourism and adaptation to climate change. *Climate Research, 47*(1–2), 111–122. https://doi.org/10.3354/cr00952

Semenza, J. C., & Ebi, K. L. (2019). Climate change impact on migration, travel, travel destinations and the tourism industry. *Journal of Travel Medicine, 26*(5), taz026–taz039. https://doi.org/10.1093/jtm/taz026

Shariff, N. M. (2022). Impacts of climate change on tourism in Malaysia: Assessing the community perspectives. *Malaysian Journal of Social Sciences and Humanities, 7*(10), e001827–e001827. https://doi.org/10.47405/mjssh.v7i10.1827

Sharma, P., Sivakumaran, B., & Marshall, R. (2013). Exploring impulse buying in services: Toward an integrative framework. *Journal of the Academy of Marketing Science, 42*(2), 154–170. https://doi.org/10.1007/s11747-013-0346-5

Sharma, T., Chen, J., & Liu, W. Y. (2020). Eco-innovation in hospitality research (1998–2018): A systematic review. *International Journal of Contemporary Hospitality Management, 32*(2), 913–933. https://doi.org/10.1108/IJCHM-01-2019-0002

Shen, Z., Li, H., & Zhang, Y. (2021). Exploring the relationship between entrepreneurship and psychological characteristics, and corporate social responsibility under marketization. *Frontiers in Psychology, 12*, 693644. https://doi.org/10.3389/fpsyg.2021.693644

Shereni, N. C., & Rogerson, C. M. (2023). The end of covid-19 and implementation of the sustainable development goals by the hospitality sector in Sub-Saharan Africa. *GeoJournal of Tourism and Geosites, 49*(3), 1067–1074. https://doi.org/10.30892/gtg.49322-1106

Shin, H., & Perdue, R. R. (2022). Hospitality and tourism service innovation: A bibliometric review and future research agenda. *International Journal of Hospitality Management, 102*, 103176. https://doi.org/10.1016/j.ijhm.2022.103176

Siddiqi, U. I., Sun, J., & Akhtar, N. (2020). The role of conflicting online reviews in consumers' attitude ambivalence. *The Service Industries Journal, 40*(13–14), 1003–1030. https://doi.org/10.1080/02642069.2019.1684905

Spandre, P., François, H., Verfaillie, D., Lafaysse, M., Déqué, M., Eckert, N., George, E., & Morin, S. (2019). Climate controls on snow reliability in French Alps ski resorts. *Scientific Reports, 9*(1), 8043–8052. https://doi.org/10.1038/s41598-019-44068-8

Sroka, W., & Szántó, R. (2018). Corporate social responsibility and business ethics in controversial sectors: Analysis of research results. *Journal of Entrepreneurship, Management and Innovation, 14*(3), 111–126. https://doi.org/10.7341/20181435

Steiger, R. (2010). The impact of climate change on ski season length and snowmaking requirements in Tyrol, Austria. *Climate Research, 43*(3), 251–262. https://doi.org/10.3354/cr00941

Stoldt, R., Wellman, M., Ekdale, B., & Tully, M. (2019). Professionalizing and profiting: The rise of intermediaries in the social media influencer industry. *Social Media+ Society, 5*(1), 2056305119832587. https://doi.org/10.1177/2056305119832587

Suchánek, P., & Králová, M. (2019). Customer satisfaction, loyalty, knowledge and competitiveness in the food industry. *Economic research-Ekonomska istraživanja, 32*(1), 1237–1255. https://doi.org/10.1080/1331677x.2019.1627893

Susilo, A. (2020). Identifying factors that affect consumer satisfaction of Parklatz cafe in Ponorogo city, East Java, Indonesia: An application of exploratory factor analysis. *Falah: Jurnal Ekonomi Syariah, 5*(1), 1–14. https://doi.org/10.22219/jes.v5i1.11399

Szymkowiak, A., Gaczek, P., & Padma, P. (2021). Impulse buying in hospitality: The role of content posted by social media influencers. *Journal of Vacation Marketing, 27*(4), 385–399. https://doi.org/10.1177/13567667211003216

Tabash, M. I., Farooq, U., El Refae, G. A., Al-Faryan, M. A. S., & Athamena, B. (2023). Impact of religious tourism on the economic development, energy consumption and environmental degradation: Evidence from the Kingdom of Saudi Arabia. *Tourism Review, 78*(3), 1004–1018. https://doi.org/10.1108/tr-07-2022-0347

Tanoamchard, W., & Limjirakan, S. (2012). Impacts of Climate Change on Water Resources Affecting Tourism Industry in a Small Island: A Case Study of Koh Chang, Thailand. *International Journal of Environmental Science and Development, 3*(2), 109. https://doi.org/10.7763/ijesd.2012.v3.198

Thomas, S., White, G. R., & Samuel, A. (2018). To pray and to play: Post-postmodern pilgrimage at Lourdes. *Tourism Management, 68*, 412–422. https://doi.org/10.1016/j.tourman.2018.03.021

Titko, J., Svirina, A., Tambovceva, T., & Skvarciany, V. (2021). Differences in attitude to corporate social responsibility among generations. *Sustainability, 13*(19), 10944. https://doi.org/10.3390/su131910944

Townsend, L., Dowds, J., O'Brien, K., Sheill, G., Dyer, A. H., O'Kelly, B., Hynes, J. P., Mooney, A., Dunne, J., & Ni Cheallaigh, C. (2021). Persistent poor health after COVID-19 is not associated with respiratory complications or initial disease severity. *Annals of the American Thoracic Society, 18*(6), 997–1003. https://doi.org/10.1513/annalsats.202009-1175oc

Tranos, E., & Davoudi, S. (2014). The regional impact of climate change on winter tourism in Europe. *Tourism Planning & Development, 11*(2), 163–178. https://doi.org/10.1080/21568316.2013.864992

Varga, I. E., & Gabor, M. R. (2021). The influence of social networks in travel decisions. *Economics, 9*(2), 35–48. https://doi.org/10.2478/eoik-2021-0015

Vong, C., Rita, P., & António, N. (2021). Health-related crises in tourism destination management: A systematic review. *Sustainability, 13*(24), 13738. https://doi.org/10.3390/su132413738

Wang, C., Cui, W., Zhang, Y., & Shen, H. (2022). Exploring short video apps users' travel behavior intention: Empirical analysis based on SVA-TAM model. *Frontiers in psychology, 13*, 912177. https://doi.org/10.3389/fpsyg.2022.912177

Wong, I. A., Lu, M. V., Lin, S., & Lin, Z. (2023). The transformative virtual experience paradigm: The case of Airbnb's online experience. *International Journal of Contemporary Hospitality Management, 35*(4), 1398–1422. https://doi.org/10.1108/IJCHM-12-2021-1554

World Tourism Organization, W. (2020). *Iternational tourism continues to outpace the global economy*. https://www.e-unwto.org/doi/pdf/10.18111/9789284421152

Wu, J., Pierse, J., Orlandi, F., O'Sullivan, D., & Dev, S. (2023). Improving tourism analytics from climate data using knowledge graphs. *IEEE Journal of Selected Topics in Applied Earth Observations and Remote Sensing, 16*, 2402–2412. https://doi.org/10.1109/jstars.2023.3239831

Xie, X., Zhang, L., Sun, H., Chen, F., & Zhou, C. (2021). Spatiotemporal difference characteristics and influencing factors of tourism urbanization in China's major tourist cities. *International Journal of Environmental Research and Public Health, 18*(19), 10414. https://doi.org/10.3390/ijerph181910414

Xu, S., Stienmetz, J., & Ashton, M. (2020). How will service robots redefine leadership in hotel management? A Delphi approach. *International Journal of Contemporary Hospitality Management, 32*(6), 2217–2237. https://doi.org/10.1108/ijchm-05-2019-0505

Yallop, A., & Seraphin, H. (2020). Big data and analytics in tourism and hospitality: Opportunities and risks. *Journal of Tourism Futures, 6*(3), 257–262. https://doi.org/10.1108/jtf-10-2019-0108

Yeon, J., Song, H. J., Yu, H. C., Vaughan, Y., & Lee, S. (2021). Are socially responsible firms in the US tourism and hospitality industry better off during COVID-19? *Tourism Management, 85*, 104321. https://doi.org/10.1016/j.tourman.2021.104321

Yichuan, W., LeeAnn, K., & Terry Anthony, B. (2018). Big data analytics: Understanding its capabilities and potential benefits for healthcare organizations. *Technological Forecasting and Social Change, 126*(2), 3–13. https://doi.org/10.1016/j.techfore.2015.12.019

Yu, Y., Xu, S., Li, G., & Kong, H. (2020). A systematic review of research on abusive supervision in hospitality and tourism. *International Journal of Contemporary Hospitality Management, 32*(7), 2473–2496. https://doi.org/10.1108/IJCHM-12-2019-1004

Zhang, K. Z., Xu, H., Zhao, S., & Yu, Y. (2018). Online reviews and impulse buying behavior: The role of browsing and impulsiveness. *Internet Research, 28*(3), 522–543. https://doi.org/10.1108/intr-12-2016-0377

Zhang, K. Z., Zhao, S. J., Cheung, C. M., & Lee, M. K. (2014). Examining the influence of online reviews on consumers' decision-making: A heuristic–systematic model. *Decision Support Systems, 67*(3), 78–89. https://doi.org/10.1016/j.dss.2014.08.005

Zheng, C., Chen, Z., Zhang, Y., & Guo, Y. (2022). Does vivid imagination deter visitation? The role of mental imagery processing in virtual tourism on tourists' behavior. *Journal of Travel Research, 61*(7), 1528–1541. https://doi.org/10.1177/00472875211042671

Zhong, L., Sun, S., Law, R., Li, X., & Yang, L. (2022). Perception, reaction, and future development of the influence of COVID-19 on the hospitality and tourism industry in China. *International Journal of Environmental Research and Public Health, 19*(2), 991. https://doi.org/10.3390/ijerph19020991

Zhong, L., Yu, H., & Zeng, Y. (2019). Impact of climate change on Tibet tourism based on tourism climate index. *Journal of Geographical Sciences, 29*(12), 2085–2100. https://doi.org/10.1007/s11442-019-1706-y

Zopiatis, A., Pericleous, K., & Theofanous, Y. (2021). COVID-19 and hospitality and tourism research: An integrative review. *Journal of Hospitality and Tourism Management, 48*(3), 275–279. https://doi.org/10.1016/j.jhtm.2021.07.002

Sustainable Tourism and Its Role in Economic Development

Abhijit Chakraborty, Nisha Goyal, Abhijit Ranjan Das, and Priyanka Roy

Abstract Sustainable tourism has evolved as a significant attribute in the realm of economic development, environmental conservation, and social well-being. This chapter explores how sustainable tourism enhances positive environmental, social, and cultural impacts while balancing economic benefits and addressing challenges like regulations, infrastructure, and stakeholder engagement. The study adopts a theoretical approach, drawing on an extensive, systematic review of existing literature and secondary reports to gain insights into the subject. The findings reveal the economic paybacks associated with sustainable tourism, which includes income generation, employment opportunities, and foreign exchange earnings, and the inherent challenges, such as resource depletion, environmental degradation, and socio-cultural erosion, which can undermine long-term sustainability.

1 Introduction

Tourism is one of the major global sectors, fuelling the economic engines of various nations. In 2023, the tourism sector contributed 9.1 per cent to the global GDP, reflecting a substantial growth of 23.2 per cent compared to 2022.[1] As shown in Fig. 1, the number of jobs created by the travel and tourism industry worldwide has been on a constant rise since COVID-19 and this strengthens the contribution of this industry towards the economic development of a nation. Job creation has increased by more than 28 per cent from 2020 to 2024, with projections to reach 449 million by the year 2034. In the years ahead, tourism on an international scale is expected

[1] See: https://wttc.org/research/economic-impact

A. Chakraborty (✉) · N. Goyal · A. R. Das · P. Roy
School of Commerce and Management Studies, Dayananda Sagar University, Bengaluru, India
e-mail: abhijitchakraborty-scms@dsu.edu.in; nisha.goyal@dsu.edu.in; abhijit.das@dsu.edu.in; priyanka.roy@dsu.edu.in

J. Marques et al. (eds.), *Tourism and Climate Change*, Tourism, Hospitality & Event Management, https://doi.org/10.1007/978-3-031-99128-8_3

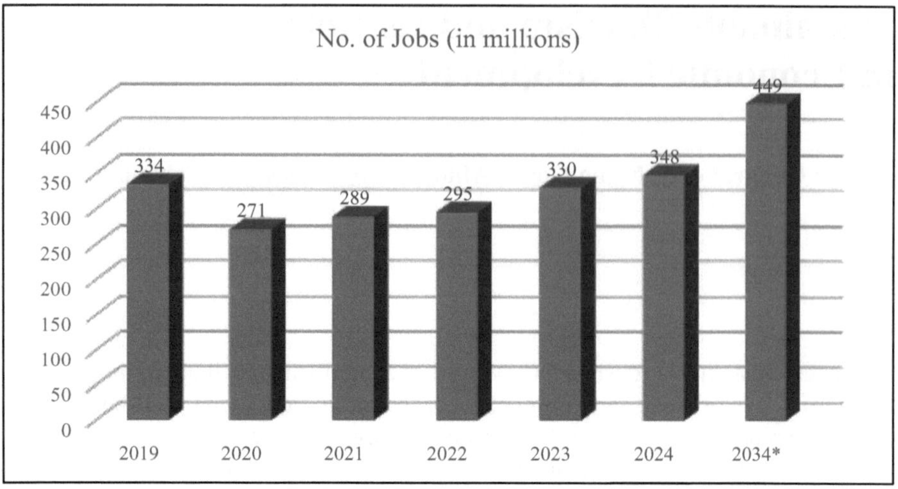

Fig. 1 Job creation by travel and tourism industry worldwide (in millions). (Source: Authors' compilation, data from World Travel & Tourism Council (WTTC))

to undergo a period of expansion.[2] It signifies that this sector plays a critical role in the global economy and has far-reaching impacts on societies and the environment.

As this sector experiences accelerated growth, its detrimental impact must not be overlooked. According to a report published by Nature Climate Change (NCC) in 2018, tourism is responsible for approximately 8 per cent of worldwide greenhouse gas (CO_2) emissions, which is considered a large percentage. In light of this, it is evident that shifting to a more sustainable approach to tourism is a pressing necessity. Hence, sustainable tourism gained prominence due to the growing awareness of the negative impacts of tourism on the environment, local communities, and cultural heritage. Further, sustainable tourism can help to achieve Sustainable Development Goals (SDGs) and inclusive growth. The World Tourism Organisation (WTO) defines sustainable tourism as:

> Tourism that takes full account of its current and future economic, social and environmental impacts, addressing the needs of visitors, the industry, the environment, and host communities.

The tourism sector is a sizeable contributor to India's economy. Before the COVID-19 pandemic, the sector contributed 5.18 per cent of India's GDP and had a share of 13.50 per cent in the total employment of the economy.[3] During 2020–2021, the sector's contribution to GDP declined to 1.50 per cent. However, the sector has already shown signs of recovery. According to the India Tourism Statistics Report, 2023 (Ministry of Tourism, India), the tourism sector contributed 1.77 per cent to India's GDP as of FY 2021–2022. During this period, the sector provided

[2] See: https://www.unwto.org/un-tourism-academy

[3] See: https://tourism.gov.in/sites/default/files/2023-12/SQ.110%20for%2011.12.2023.pdf

employment (direct and indirect) to over 70 million people, which amounts to a 12.66 per cent share in employment. The importance of the tourism sector further varies across the different states. In some states such as Himachal Pradesh, the share of tourism in Gross State Domestic Product (GSDP) is as high as 7.04 per cent ('Employment Scenario in the Tourism Sector' Report, 2024, Ministry of Tourism, India). Karnataka aims to increase tourism's contribution to GSDP from 14.8 per cent in 2019 to 20 per cent by 2025 (Report of 'Invest Karnataka').[4] Tourism also plays a key role as a foreign exchange earner for India. As of 2022, the Foreign Exchange Earnings (FEE) from tourism stood at USD 17.6 Bn, increasing 100 per cent year to year from 2021 and recovering over 57 per cent of the pre-pandemic levels. India ranked 17th in terms of International Tourist Arrivals (ITA) in 2021, up from 24th in 2019 (India Tourism Statistics Report 2023, Ministry of Tourism, India). The significance of the sector becomes even more evident after considering India's position in the World Travel & Tourism Council (WTTC) rankings. India ranks 6th and 2nd with respect to Travel & Tourism's total contribution to GDP and employment, respectively.[5] As per World Economic Forum's Travel & Tourism Development Index (TTDI), 2024, India has the largest travel & tourism sector in South Asia and ranks 39th globally. India is also the top lower-middle-income economy in the TTDI rankings.

Considering the strategic importance of the sector, the Government of India has set a vision to earn USD 56 billion Forex from tourism and create 140 million jobs in the sector by 2030.[6] Furthermore, the Ministry of Tourism aims to make India a global adventure tourism hub by 2047. Adventure tourism is envisioned to contribute USD 800 billion out of the total USD 3 trillion target set for the tourism economy in 2047.[7] The footprint of the travel industry is thus anticipated to grow multi-fold in the next few years in India. While an increase in tourism is good news for the economy, the same cannot be said about the environment. According to a report by WTTC, 2023,[8] at a global level the Travel & Tourism industry's greenhouse gas (GHG) emissions increased at an average annual rate of 2.5 per cent from 2010 through 2019. The sector accounted for 10.6 per cent and 5.8 per cent, respectively, of the global energy and water consumption. Investment in buildings and infrastructure created an increased demand for material extraction. The material requirement of the sector has increased by 64 per cent from 2010 to 2019, driven majorly by demand for non-metallic minerals (41 per cent). The United Nations Environment Programme (UNEP) posits[9] that in a 'business-as-usual scenario', the tourism sector's energy and water consumption will increase by 154 per cent and

---

[4] See: https://investkarnataka.co.in/wp-content/uploads/2020/07/Tourism.pdf

[5] See: https://www.investindia.gov.in/sector/tourism-hospitality

[6] See: https://pib.gov.in/PressReleaseIframePage.aspx?PRID=1892267

[7] See: https://pib.gov.in/PressReleaseIframePage.aspx?PRID=1988326

[8] See: https://globaltravelfootprint.wttc.org/reports/TheEnvironmentalImpactofGlobalTourismOct2023.pdf

[9] See: https://www.unep.org/explore-topics/resource-efficiency/what-we-do/responsible-industry/tourism

152 per cent, respectively, through 2050. During the same period, the industry's GHG and solid waste footprint will also increase by 131 per cent and 251 per cent, respectively. Nevertheless, tourism is the mainstay of many developing nations' economic development (Khan et al., 2020; Lal, 2024; Rasul & Manandhar, 2009) and the status quo may remain in the near future. Herein lays the criticality of sustainability in tourism, which therefore finds a mention in UN Sustainable Development Goals 5, 8, 12, and 14.[10]

The harsh impact of tourism on the environment has been witnessed, although sporadically, already in India. News about plastic pollution on India's beaches is abundant on print and social media. According to a study by Rotterdam-based Non-Profit organisation, The Ocean Clean-up, India ranks 2nd globally in marine plastic waste leaking into the ocean.[11] Annually, an estimated 0.6 million metric tons of plastic waste find their way into the Indian Ocean through various activities including tourism and recreation (Robin et al., 2023). Over-tourism has exposed the vulnerability and infrastructure inadequacy of hill stations in Northern India (Gupta et al., 2022). Tourism, if not managed sustainably, will therefore reduce the aesthetic appeal of Indian beaches and hill stations, eventually leading to the decline of tourism in those locations. Sustainable tourism, on the other hand, may help unlock the full economic potential of the tourism sector in India while ensuring minimal social and environmental impact. India has a long path to traverse in terms of sustainable tourism. As of 2020, India ranked 98 out of 99 countries surveyed to develop Euromonitor's Sustainable Travel Index Rankings.[12] The top 25 spots, barring two, were all taken up by European nations. New Zealand (23) and Canada (24) were the only non-European nations in the top 25. The EU recognises that the sustainability of its tourism industry will improve its competitiveness as the quality of tourist destinations is impacted by their natural and cultural environment.[13] The Euromonitor report concluded that post the pandemic, there has been a change in mindset from volume-driven tourism to sustainable tourism. The Government of India, taking cognizance of the situation, has already undertaken various initiatives to promote sustainable tourism such as waste management, energy conservation, etc. As a result, the concept is gaining traction as a tool for economic development which can also aid in preserving the country's rich cultural and natural heritage. By adopting sustainable practices and programmes in the tourism space, India can appeal to conscientious tourists from across the globe who seek authentic and enriching experiences (Jatav, 2023).

This chapter is an effort to understand how sustainable tourism maximises the positive environmental, social, and cultural impacts while balancing the economic benefits derived from the sector. In doing so, the authors have tried to analyse

[10] See: https://sustainabledevelopment.un.org/topics/sustainabletourism

[11] See: https://theoceancleanup.com/updates/mumbai-calling-the-ocean-cleanup-and-bharat-clean-rivers-foundation-join-forces-in-india/

[12] See: https://forumnatura.org/wp-content/uploads/2021/07/SustainableTravelIndex-v0.3.pdf

[13] See: https://single-market-economy.ec.europa.eu/sectors/tourism/eu-funding-and-businesses/funded-projects/sustainable_en

various challenges to implementing the concept including regulations, infrastructure, and stakeholder involvement, or the lack thereof. The role of technology and innovation is elucidated in this chapter, which also discusses the need for effective governance mechanisms. The research carried out culminates in the form of policy recommendations and strategies for fostering the idea of sustainability in the tourism sector of India.

2 Methodology

This chapter follows a theoretical approach through a systematic literature review. To gather, assess, and interpret the latest research on a specific topic, a systematic literature review is an appropriate approach (Chauhan et al., 2024). In this study, the 'Preferred Reporting Items for Systematic Reviews and Meta-Analyses (PRISMA)' guidelines were followed for data extraction (Moher et al., 2009). The Scopus database was used for data extraction due to its extensive global coverage of high-quality scientific publications (Chauhan et al., 2024; Ghosh, 2024).

The method begins with identifying relevant keywords for this study. The initial keyword search in the Scopus database included keywords such as '*Sustainable*', '*Sustainability*', '*Tourism*', '*Sustainable Tourism*', and '*Economic Development*'. Initially, only peer-reviewed articles published in the subject areas of Social Sciences, Business, Management, Accounting, Economics, Econometrics, and Finance were considered. A total of 609 articles were retrieved after applying all criteria. Articles categorised as literature reviews, conference proceedings, notes, letters, and erratum were excluded from the initial search, reducing the number of articles to 362. Subsequently, only articles published in English and those with a minimum of 5 citations were retained. After a manual review for duplicates, one article was eliminated from consideration, resulting in a final count of 192 articles. The flowchart for the method is shown in Fig. 2. Sections 3, 4, 5, 6, 7, and 8 reflect the findings from the comprehensive review of the studies conducted, Section 9 discusses the future outlook, and key implications are incorporated in Sect. 10.

3 Sustainable Development Goals and Tourism Industry

The travel and tourism industry has a significant role to play in ensuring the United Nations' Sustainable Development Goals (SDGs), both globally and within India. Investigating the key metrics across a few critical SDGs provides insights into the contributions of the tourism industry and indicates the scope for further improvement. Table 1 incorporates some of the important metrics of SDGs and exhibits the Indian scenario *vis-a-vis* the world. The SDGs include Gender equality (SDG 5); Clean water and sanitation (SDG 6); Affordable and clean energy (SDG 7); Decent work and economic growth (SDG 8); and Industry, innovation and infrastructure (SDG 9).

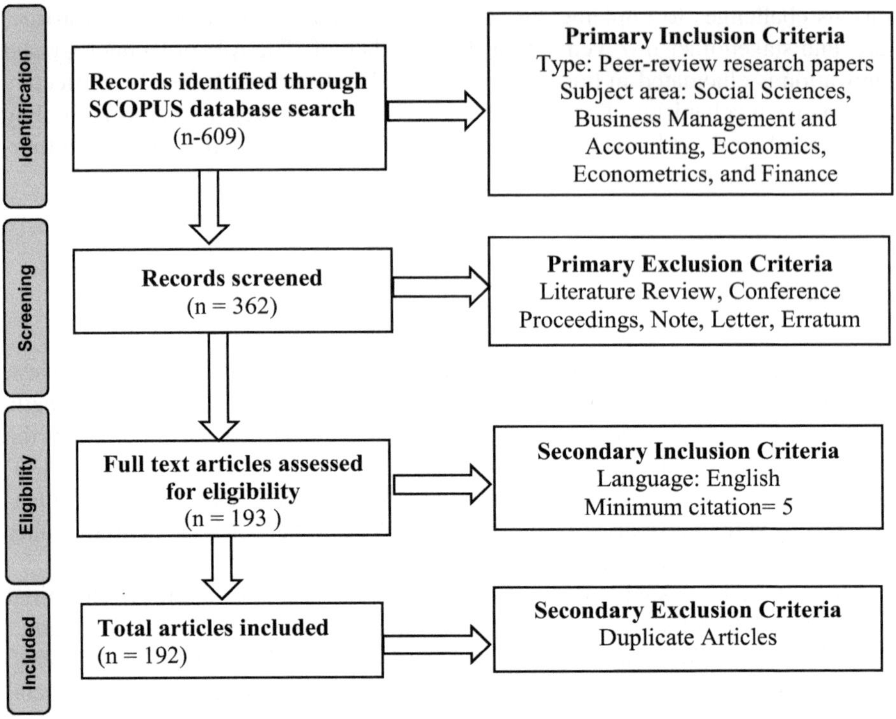

Fig. 2 PRISMA flowchart. (Source: Moher et al. (2009))

SDG 5: Gender Equality reveals that travel and tourism account for a substantial portion of female employment, accounting for 8 per cent of global female jobs and 5 per cent in India. This indicates potential for growth in women's employment in India's tourism industry.

SDG 6: Water usage efficiency is an important area where the tourism industry has made remarkable progress, reducing the water use intensity in the travel and tourism sector by 19 per cent globally and 48 per cent in India since 2000. While the industry accounts for 4 per cent of total water consumption for both the world and India, these efficiency gains exhibit a commitment to sustainable resource management.

SDG 7: The use of renewable energy by the tourism sector stands at 6 per cent globally, which accounts for 3 per cent of total global renewable energy consumption. In India, the industry's renewable energy share is comparatively lower at 3 per cent, but it contributes to a higher 4 per cent of the total renewable energy use. This indicates the scope for improving the sector's renewable energy mix as it is a crucial factor in achieving clean and affordable energy goals.

SDG 8: Economic growth, employment, and productivity are areas where travel and tourism have done remarkably well. The data in Table 1 highlight that tourism supports millions of jobs and drives GDP growth. The tourism industry provides 8

Table 1 SDG achievement through travel and tourism industry

SDG Goal	Particulars	Global	India
SDG 5: Achieve gender equality and empower all women & girls	*Women's equal rights to economic resources*		
	• Female employment supported by Travel & Tourism (thousands) • Share of total female employment	102,000 8%	5830 5%
SDG 6: Ensure availability and sustainable management of water and sanitation for all	*Water use efficiency*		
	• Travel & Tourism change in water use intensity (m^3 per thousand USD) since 2000 • Share of global/total water consumption	19% (negative) 4%	48% (negative) 4%
SDG 7: Ensure access to affordable, reliable, sustainable, and modern energy for all	*Renewable energy use*		
	• Renewable share of energy consumption by Travel & Tourism • Share of total global renewable energy consumption	6% 3%	3% 4%
	Improve energy efficiency		
	• Travel & Tourism energy consumption per unit GDP (MJ per $ of GDP) • National average energy consumption per unit GDP (MJ per $ of GDP)	7.2 6.1	16 13

(continued)

Table 1 (continued)

SDG Goal	Particulars	Global	India
SDG 8: Promote sustained, inclusive, and sustainable economic growth, full and productive employment and decent work for all	*Achieve higher economic productivity*		
	• Annual growth in real GDP per worker in Travel & Tourism • Global average annual growth in real GDP per worker	9% 3%	30% 4%
	Employment development and job creation		
	• Total employment supported by Travel & Tourism (thousands) • Share of total global employment	2,63,000 8%	37,200 8%
	Improve resource efficiency Improve resource efficiency		
	• Material footprint of Travel & Tourism, per unit GDP (tonnes/$m) • Global/Total economy-wide material footprint per unit GDP (tonnes per $m)	740 970	2400 2000
	Achieve decent work for all		
	• Share of Travel & Tourism jobs in high-wage sectors • Global average share of jobs in high-wage sectors • Total youth employment (15–24) supported by Travel & Tourism (thousands) • Growth since 2010	33% 30% 34,600 11%	66% 28% 4,690 29%
	Promote sustainable tourism		
	• Tourism direct GDP as a proportion of total GDP • Real growth in direct tourism GDP since 2010	2% 17%	2% 31%

SDG Goal	Particulars	Global	India
SDG 9: Build resilient infrastructure, promote inclusive and sustainable industrialisation and foster innovation	*Inclusive and sustainable industrialisation*		
	• Manufacturing GDP supported by Travel & Tourism ($bn) • Share of manufacturing GDP supported by Travel & Tourism	884 6%	28.3 6%
	Increase resource-use efficiency		
	• Travel & Tourism greenhouse gas emissions per unit GDP (Kg of CO_2e per $) • % change since 2010	0.44 13%	1.3 27%
	Technological development and industrial diversification		
	• Medium and high-tech industry GDP supported by Travel & Tourism ($bn) • % change since 2010	394 13%	12.3 27%

Source: Authors' compilation

per cent employment for both global and Indian territories, with annual productivity growth of 9 per cent globally and an outstanding 30 per cent in India. It also generates significant medium and high-tech industry GDP, supporting its contribution to fostering diversification and innovation.

SDG 9: Infrastructure development and environmental impact are additional dimensions where the travel and tourism sector can ensure significant contributions. Infrastructure and innovation reveals that the tourism industry supports manufacturing and technology development, with India marking visible growth in resource efficiency. The industry's material footprint and greenhouse gas emissions per unit of GDP support progress, but consistent efforts to improve resource efficiency and minimise environmental impacts will be required for installing resilient and sustainable systems.

The Sustainable Travel Report-2022 revealed some important data that provide insights into the attitudes and perceptions of travellers towards sustainable travel options, as shown in Fig. 3. The data reveal that 8 per cent of travellers found sustainable travel options as less luxurious than other travel options while 1 per cent of travellers feel that sustainable travel does not provide the expected level of luxury and comfort. This calls for the need to ensure quality enhancement in accommodation, food, and travel of the sustainable means. 10 per cent of the travellers believe that sustainable travel options are mostly available in remote areas, while 29 per cent opined that there were inadequate sustainable options in the destinations, they were looking for suggesting the availability of inadequate and wrong information among the travellers. 32 per cent of travellers indicated that non-sustainable travel options are more attractive and appeal more compared to sustainable travel options. This may be due to factors like availability, alternatives, cost, and convenience indicating space for improvement in these important areas.

This essentially calls for the need to educate travellers about the benefits of sustainable tourism. Appropriate responses to their concerns about luxury and comfort may help to shift the overall attitude of the tourists towards more sustainable travel options.

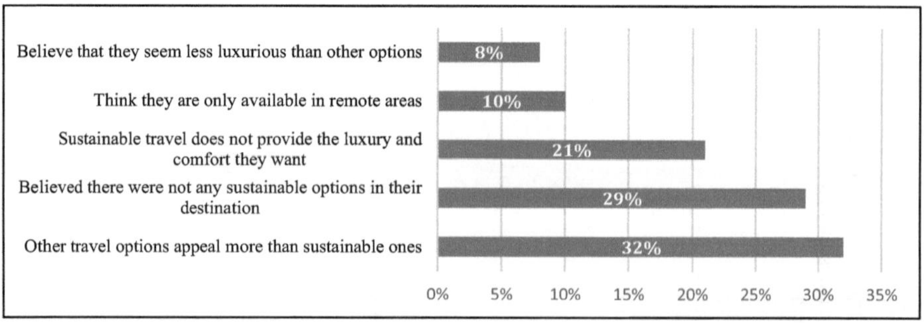

Fig. 3 Travellers' choice towards sustainable means for travel. (Source: Sustainable Travel Report-2022)

4 The Need for Sustainability in Tourism

The tourism industry has been a vital source of revenue in developing countries (World Tourism Organization, 2008). International tourism can become a catalytic agent in the process of economic growth. Sustainable environmental biodiversity is found to have a significantly positive relationship with the economic growth of a nation (Freytag & Vietze, 2013). To reap the benefits of tourism, the suppliers of tourism need to consider and take due care of the most important component—the environment. Nature is significantly influencing the demand for tourism (Valente, 2005). As a matter of fact, developing nations are often found to be relatively rich in biodiversity which can potentially be utilised as an enhancing factor for tourism promotion to ensure substantial contribution towards the economic growth and development of the economy in the long run. On the contrary, this race for development through the promotion of the tourism industry can negatively impact the biodiversity of the country (Berno & Bricker, 2001; Neto, 2003) and distort the ecological balance inviting mass destruction through natural calamities.

Tourism promotes other economic activities as tourists need to be supplied with food, accommodation, entertainment, souvenirs (mostly local handicrafts), transportation, guides, etc. All these goods and services required to serve the tourists can generate employment opportunities (Deloitte & Touch et al., 1999), and in that regard the tourism industry can be termed as more labour-intensive in nature. The potential that the tourism industry carries towards contributing to the economic well-being of the region demands action to ensure its continuous development which can be ensured only if there are efforts to make it sustainable.

Sustainability in the tourism sector is a need of the hour. With the escalating numbers of international tourists, the strain on natural resources, historical/cultural landmarks, and local communities has intensified. This swift growth which is much warranted for the economic benefits it brings also poses significant challenges that necessitate a shift towards more sustainable practices. The need for sustainable tourism is closely aligned with the following areas.

Environmental Preservation—One of the primary reasons for the need for sustainable tourism is the protection and preservation of the environment as irresponsible promotion of tourism activities may lead to:

(a) *Carbon emissions*: The transportation facility, particularly air travel, contributes significantly to global greenhouse gas emissions. These carbon emissions exacerbate climate change, leading to more extreme weather phenomena, in turn, these negative impacts deter tourists from choosing certain destinations for travelling. Excessive number of vehicles moving on hill roads puts disproportionate pressure on the unstructured roadways leading to landslides and road accidents. In the past few years, the state of Uttarakhand in India faced unprecedented flood situations in the past few years causing loss of life and property which, as per environmental experts, happened due to ecological imbalance caused by unorganised tourism. Sustainable tourism seeks to reduce this carbon footprint through more efficient transportation methods and carbon offset programmes.

(b) *Habitat destruction*: The development of tourism infrastructure often leads to the destruction of natural habitats and small set-ups are installed in nearby areas for commercial purposes. The homestay facilities and hotels are constructed all around causing extreme damage to natural habitats of wildlife and the ecology.

(c) *Resource depletion*: Popular tourist destinations often face challenges related to water scarcity, energy consumption, and waste management. There are areas with tourism prospects that are remotely located with limited means and infrastructure. A sudden surge in tourism adversely affects the resources available causing fast depletion that may lead to absolute scarcity.

(d) *Wildlife disturbance*: Unregulated tourism can disrupt ecosystems and animal behaviour. The need to develop infrastructure to ensure quality service to the tourists causes ecological imbalance disturbing wildlife. The construction of the golden quadrilateral that connects four extreme corner locations of India (Kashmir in the north to Kanyakumari in the south and Silchar in the east to Surat in the west) has passed through many forest areas including the Kaziranga National Park (The National Highway NH 715).

Traditional tourism often leads to environmental degradation through pollution, habitat destruction, and excessive resource consumption. Sustainable tourism practices, such as promoting 'environment-responsible' accommodations, reducing waste, and encouraging conscientious tourists, can help mitigate these impacts. By adopting sustainable practices, tourism can contribute to the conservation of natural resources and biodiversity, ensuring that destinations remain attractive and viable for future generations.

Economic Perspectives—The core purpose behind the promotion of tourism is to extract the economic benefits and support local communities that, in some cases, are deprived of the development process which often confines to the urban areas. Tourism, if let go in an unorganised manner, may not meet the desired goal of supporting the have-nots. The corporates and large business houses will share the most benefits leaving out scanty for the deprived community. On the other hand, it will also lead to the destruction of the finite resources ending up with a permanent stop to this industry in some places. The individual stakeholders often think about their own benefits ahead of the community and that personal benefit sometimes is ensured at the cost of community benefits. Construction of a multi-storied hotel complex on a hilltop can be considered as an example. Sonam Wangchuk, who is an engineer and education reformer, held a protest fast for 21 days in March 2024 to protect the Ladakh region in India. The mission was to highlight the detrimental effects of industrialisation on the ecology and glaciers of the Himalayan region (TOI, 2024).

Cultural Preservation—Tourism can have a profound impact on local cultures, often leading to the erosion of traditional customs and practices. The competition towards meeting the expectations of tourists and to attract them sometimes clashes with the local practices diluting the cultural heritage and tradition that a community belongs to. Chokhi Dhani village in Jaipur, India, is a classic effort to showcase the cultural heritage and tradition in the state of Rajasthan.

Sustainable tourism seeks to preserve and promote cultural heritage by encouraging tourists to engage with and respect local traditions. This can involve supporting cultural festivals, promoting traditional arts and crafts, and providing education about local customs. By doing so, sustainable tourism helps maintain cultural diversity and enriches the tourist experience. The Bhutanese government has implemented policies that limit the number of tourists and encourage high-value, low-impact tourism. This approach helps preserve the country's unique cultural heritage while providing economic benefits. Tourists are encouraged to participate in local cultural activities and learn about Bhutanese traditions.

Social Perspective—The tourism industry has a deep impact on the socio-cultural well-being of the inhabitants around the tourist destination. By adopting sustainable tourism, the fruits of the tourism business can be distributed in a balanced manner among the stakeholders which can ensure social equity. This includes creating opportunities for disadvantaged groups, ensuring fair wages and working conditions, and supporting community development initiatives. By addressing social inequalities, sustainable tourism can contribute to the overall well-being and empowerment of local communities. In Kenya, community-based tourism initiatives have been developed to ensure that local Maasai communities benefit from tourism activities. These initiatives provide employment opportunities, support education and healthcare projects, and empower the Maasai people to share their culture with visitors.

The Danger of Over-Tourism—If the tourist inflow exceeds the expectation and management capacity of the tourist destination, it results in the sufferings of over-tourism. Maintaining adequate facilities and control over the tourists' footfall is a strategic necessity for the long-term viability of the tourism industry. If the tourist destination fails to address the requirements of the tourists and the environmental needs adequately, it may lead to:

- Environmental degradation that reduces their attractiveness to tourists
- Social conflicts and resentment from local communities
- Breakdown of infrastructure facilities.

The sustained development of the tourism industry requires:

- Improved destination reputation and competitiveness
- Higher visitor satisfaction and increased repeat visits
- Greater resilience to economic and environmental shocks
- Diversification of tourism products and markets

Sustainable tourism strategies, such as implementing visitor caps, promoting off-season travel, and encouraging the exploration of lesser-known destinations, help manage tourist numbers and reduce the pressure on popular sites. Venice, Italy, has struggled with over-tourism for years. To address this issue, the city has introduced measures such as limiting the number of visitors to certain areas, promoting alternative attractions, and encouraging responsible tourism practices. These efforts aim to balance the needs of tourists with the preservation of the city's heritage and the well-being of its residents.

Meeting Changing Consumer Demands and Enhancing Visitor Experience— There is a growing trend among travellers, particularly younger generations, to seek out more responsible and authentic travel experiences. Tourists are now more concerned about the environment and appreciate initiatives towards environmental sustainability. Travellers are increasingly drawn towards enriching experiences that foster connections with nature and indigenous cultures. The tourism sector should address these changing dynamics and consider:

- Offering eco-friendly accommodations and activities
- Providing opportunities for volunteering and community engagement
- Ensuring transparency in sustainability practices

The alignment with these consumer preferences can ensure tourist destinations attract a growing segment of environmentally and socially conscious travellers. Sustainable tourism not only benefits the environment and local communities but also enhances the visitor experience. Sustainable tourism practices, such as promoting eco-friendly activities and warranting preservation of natural resources, help create memorable and fulfilling travel experiences.

5 Global Challenges and Need of the Hour

The rising mercury of the world requires every industry in every nation to work responsibly toward saving the earth. The tourism industry is not an exception and needs to respond to the call to save the globe. Sustainable tourism caters to achieving broader global objectives, such as the UN SDGs. Advancing the tourism sector with a focus on sustainability can contribute towards:

- Climate change mitigation and adaptation
- Biodiversity conservation
- Sustainable consumption and production patterns

By linking tourism development to these global priorities, the industry can become a powerful force for positive change. Governments and tourism organisations need to develop and enforce regulations that promote sustainable practices, such as environmental protection laws, zoning regulations, and sustainable tourism certifications. Public and private sector alliances alongside the involvement of indigenous communities are crucial for the development of comprehensive and effective sustainable tourism policies. The Maldives has implemented strict regulations to promote sustainable tourism, including limiting the number of resorts on each island, enforcing environmental protection laws, and promoting eco-friendly tourism practices. These measures help protect the country's fragile ecosystems while supporting the tourism industry.

The need for sustainability in the tourism sector arises from its potential for generating revenue and employment. It also helps in positioning India on the global map and market. The past few decades have shown the tourism sector burgeoning as the economic powerhouse. While this growth has brought numerous benefits, it has also led to significant challenges, particularly in terms of environmental degradation, cultural erosion, economic inequality, etc. By adopting sustainable practices, the tourism sector can minimise its negative impacts and maximise its benefits, contributing to environmental conservation, economic development, cultural preservation, and social equity. Sustainable tourism not only ensures the long-term viability of destinations but also enriches the travel experience for tourists. It is imperative for all the stakeholders including governments, tourism organisations, businesses, and travellers to work together to promote and implement sustainable tourism practices, ensuring that tourism remains a positive force for development and well-being taking simultaneous care for the preservation of the environment.

6 Tourism Industry in India

The tourism industry in India is rapidly growing in the current decade. Though India was ever a favourite tourist destination of the world, the potential remained untapped till recent years. The contribution of the tourism and hospitality industry to GDP as shown in Fig. 4 depicts the stupendous growth of this sector since 2017. In 2023 the contribution was over 231 billion USD and it is projected to reach approx. 523 billion USD by 2034. In Asia Pacific, India ranks second in the contribution of the tourism industry to GDP.

The tourism industry in India receives most of its revenue from domestic tourists. The number of domestic tourists in India rose from 220 million in 2000 to 1122 million in 2022, whereas the number of foreign tourists arriving for the same period rose from 2.65 million to 6.19 million. The Government of India has initiated new visa policies to allure foreign tourists and also investing hugely in developing tourist sites like the Statue of Unity, which is the world's tallest statue in the state of Gujarat, the Shri Ram Temple in Ayodhya, Rann of Kutch in Gujrat, Kashi Vishwanath Corridor, Varanasi, Uttar Pradesh, etc.

The tourism industry extensively supports the employment generation. Since tourism leans more towards services compared to products, more and more people can get involved in catering to those services required by the visitors. Direct employment in the tourism sector includes jobs in hotels, travel agencies, airlines, and other passenger transportation services (excluding commuter services), whereas indirect employment refers to jobs in supply chain industries supporting tourism, such as food and beverage suppliers, construction, and handicrafts. The government has been focusing on skill development in the tourism sector. The Skill India Mission has trained over five million people in tourism and hospitality skills in 2022. Figure 5 shows the direct and indirect employment generation by the Tourism industry in India from 2014 to 2019.

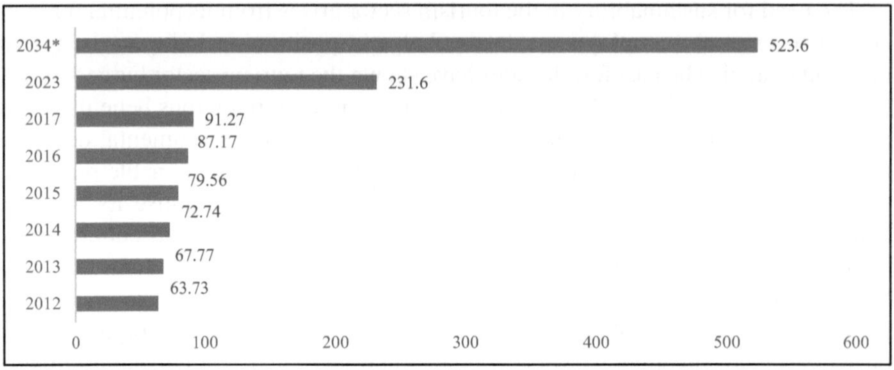

Fig. 4 Direct contribution of tourism and hospitality to GDP in India (in USD in Billion). (Source: Authors' compilation, data from World Travel & Tourism Council (WTTC))

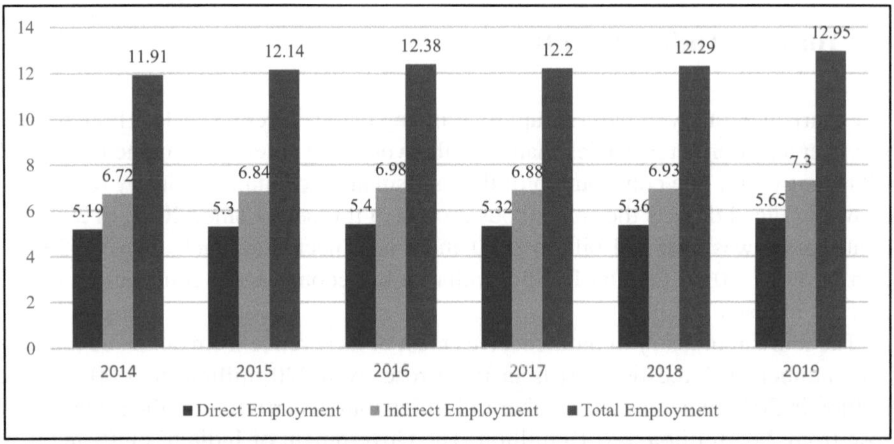

Fig. 5 Share of employment in the travel and tourism sector (in percentage). (Source: Authors' Compilation, Data from World Travel & Tourism Council (WTTC))

The generation of employment till 2019 can be seen as somewhat static ranging from 11 per cent to 13 per cent. This indicates the potential for growth of employment in this industry which can be ensured by promoting and developing this sector in an organised manner.

The tourist business in India plays a vital role in the country's economy, making substantial contributions to GDP, employment, foreign exchange profits, and community development. The sector comprises a wide array of attractions, including cultural heritage sites, natural landscapes, adventure activities, and spiritual tourism.

The Government of India implemented the following measures to foster the growth of the tourist sector:

- Paryatan Parv—This initiative comprises three key elements: 'Dekho Apna Desh' aims to stimulate domestic tourism by encouraging Indians to explore their own country, 'Tourism for All' involves organising tourism events in various locations across all states in the country, and 'Tourism and Governance' entails conducting interactive sessions and workshops with stakeholders on diverse topics.
- Swadesh Darshan Scheme—This scheme is designed to promote the comprehensive development of theme-based tourist circuits. Projects implemented across various circuits include Buddhist, Coastal, Desert, and Eco-tourism.
- PRASAD Scheme—This initiative is a step towards strengthening India's position as a centre for spiritual Tourism. Infrastructure improvements at key pilgrimage centres like Varanasi, Amritsar, and Ajmer were undertaken. It focuses on developing the pilgrimage sites to ensure Pilgrimage Rejuvenation and Spiritual Augmentation Drive (PRASAD).
- E-Visa Facility—This was initiated to promote foreign tourists' arrival by ensuring the simplification of visa processes and extension of e-visa to more countries and longer validity periods to boost international tourism. e-Visa facility is provided to the citizens of 163 countries under three categories—Tourist, Medical, and Business.
- The 'Heritage Trail' initiative aims to promote the World Heritage Sites in India.
- Digital Initiatives—India has been leveraging technology to enhance tourist experiences, including the development of audio guides and virtual tours for major attractions.
- Medical Tourism—By leveraging its advanced healthcare facilities and competitive prices, India is promoting itself as a destination for medical tourism.
- MICE Tourism—Efforts are being made to boost Meetings, Incentives, Conferences, and Exhibitions (MICE) tourism, with the development of world-class convention centres and ensuring the availability of modern infrastructure.

7 Steps Towards Sustainable Tourism in India

India is focusing on extensively promoting the tourism industry and has undertaken various initiatives to promote sustainable tourism, with a specific focus on environmental conservation, community involvement, and sustainable practices. Here are some of the significant steps:

i. Policy and Planning

- National Tourism Policy—National Strategy for Sustainable Tourism: In 2022, the Ministry of Tourism launched a comprehensive strategy focusing on responsible travel, environmental protection, and preserving cultural heritage. The Indian government has integrated sustainable tourism principles into its National Tourism Policy. This includes promoting responsible tourism and ensuring the sustainable use of resources.

ii. Community Involvement and Benefit Sharing

- Rural Tourism—Initiatives like the Rural Tourism Scheme encourage tourists to visit rural areas, providing economic benefits to local communities and promoting cultural exchange. Remote areas having tourism potential are developed in an eco-friendly manner to benefit the local communities.
- Homestays—Homestays are becoming popular among tourists as it provides an on-ground experience of the lifestyle and ambience of the tourist destination. Promoting homestays as an alternative accommodation option helps local communities benefit directly from tourism.

iii. Environmental Conservation

- Eco-tourism—Initiatives are taken to develop eco-friendly tourism infrastructure in national parks and wildlife sanctuaries to promote eco-tourism. Guidelines for sustainable tourism in protected areas have been issued by the Ministry of Environment, Forest and Climate Change. Steps are taken to promote eco-friendly accommodations and practices in popular tourist destinations. Eco-tourism circuits are developed particularly in biodiversity-rich areas.
- Water Conservation—Water-saving technologies are implemented in hotels and tourist facilities. Awareness campaigns for tourists about water conservation are being run to curb the wastage of water.
- Cleanliness Campaigns—Swachh Bharat Abhiyan (Clean India Campaign) includes efforts to keep tourist sites clean and waste-free, promoting environmental sustainability. Non-Govt Organisations (NGOs) and local bodies are encouraged to take up campaigns towards Swachh Bharat Abhiyan. Cleanliness in the Mawlynnong village, Meghalaya, is maintained by the local bodies only and it is known as Asia's Cleanest Village. It has attracted lots of tourists to that area.
- Waste Management—The government is implementing waste segregation and recycling programmes in tourist areas. Campaigns to reduce single-use plastics in tourism operations are frequently conducted. Initiatives such as Namami Gange are implemented to purify the river and achieve the two goals of reducing pollution and preserving and revitalising the National River Ganga. An allocation of Rs.22,500 crore was approved for this programme.

iv. Cultural Preservation

- Heritage Conservation—Government has taken steps to conserve and restore heritage monuments and cultural sites. Initiatives like the 'Adopt a Heritage Scheme' involve the private sector in the upkeep of historical monuments and promote responsible tourism practices at heritage sites.
- Intangible Cultural Heritage—Government is also keen on promoting traditional arts, crafts, and cultural festivals in order to preserve intangible cultural heritage while providing economic benefits to artisans and performers.

v. Sustainable Practices in the Tourism Industry

- Certification and Standards—Steps are taken to encourage hotels, resorts, and tour operators to obtain sustainability certifications such as Green Key and Earth Check. This promotes the adoption of environmentally friendly practices in the tourism industry.
- Biodiversity Protection—Strict surveillance and regulations on tourism activities in protected areas and wildlife reserves. The promotion of wildlife tourism needs to ensure that it supports conservation efforts.
- Climate Action—The climate-resilient tourism infrastructure is advocated and encouraged. To reduce the carbon footprint of the tourism sector specific steps are taken with regard to the construction of roadways, accommodation facilities, and other tourist-related services.
- Sustainable Transportation—Public transportation options for tourists are promoted to ensure a secure, cost-effective, convenient, and well-connected travel experience in tourist destinations. This also includes the use of electric vehicles.
- Responsible Tourism Initiatives—Awareness programmes are launched to initiate ethical tourism practices among travellers, including the observance of local customs, trash reduction, and support for local businesses.

vi. Economic Viability and Local Development

- Support for Local Enterprises—The local communities are encouraged and supported to take up enterprises related to tourism, such as handicrafts, local cuisine, and guided tours, to ensure economic benefits reach local communities.
- Community Involvement—Programmes are designed in a manner to promote community-based tourism models that directly benefit local populations. Tourists are also encouraged to engage in and support local and cultural activities.
- Infrastructure Development—Sustainable tourism infrastructure is developed including transport, accommodation, and tourist facilities, to support and promote tourism activities with consideration for environmental and social impacts.
- Capacity Building—With a long-term view to reap benefits and optimise the services towards tourists, training programmes on sustainable practices are undertaken regularly for tourism stakeholders. The sustainability principles in tourism and hospitality education are included in the curriculum.

vii. Research and Monitoring

- Data Collection and Analysis—Research activities are encouraged on tourism impacts, visitor satisfaction, environmental impact, heritage restoration, and community benefits to get insights that may help in framing policy decisions and improving tourism management. Tourist-friendly apps are developed to monitor tourist activities, manage tourist flows, and guide tourists on sustainable practices.
- Monitoring and Evaluation—Data collected through agencies are used to establish mechanisms for monitoring the environmental, social, and economic impacts of tourism so that the tourism experience can be improvised and corrective measures can be taken to address any negative impacts.

viii. Public–Private Partnerships

- Collaboration—Platforms are being developed for dialogue and cooperation among stakeholders and initiatives are taken to foster collaboration between government, private sector, and non-governmental organisations (NGOs) to promote sustainable tourism. Local communities in the tourist destinations are encouraged to work hand-in-hand with the government authorities to ensure balanced and sustainable growth of the tourism industry.
- Investment—To ensure easy finance to build up facilities in tourist destinations innovative financing mechanisms such as eco-tourism funds and public–private investment partnerships are developed and encouraged. Investments in the tourism sector are promoted through incentives and favourable policies.

India's efforts to promote sustainable tourism are multi-faceted, involving policy-making, community involvement, environmental conservation, cultural preservation, and the promotion of sustainable practices by the tourism industry. These initiatives aim to ensure that tourism development is both environmentally sustainable and socially inclusive, contributing to the well-being of local communities and the preservation of the country's rich natural and cultural heritage.

8 The Challenges in Promoting Sustainable Tourism

Developing countries rely on international tourism as a significant source of income. Simultaneously, the progress and administration of tourism would face numerous issues related to sustainability for policymakers and planners in developing nations (Pan et al., 2018). Over time, deterioration in environmental conditions may hurt economic development. These issues have gained more attention in development economics (Freytag & Vietze, 2013). India encounters multiple obstacles in the development of sustainable tourism, despite its abundant cultural history and varied natural settings. An important hurdle is insufficient infrastructure, especially in rural and environmentally vulnerable regions, which can result in environmental deterioration and excessive tourism in popular locations. Striking a balance between economic growth and conservation activities becomes challenging, as numerous local populations rely on tourism as their major source of income. The challenge of sustainable tourism is reducing the negative impacts while enhancing the beneficial results of tourism in the appropriate areas.

For instance, a financially lucrative and ecologically viable sector can offer enjoyable experiences for tourists and enhance the standard of living in nearby communities (Lim & McAleer, 2005). Moreover, India faces challenges in effectively managing trash and controlling pollution in popular tourist destinations, which are frequently worsened by a lack of understanding among both tourists and residents. The extensive dimensions and varied regions of the country further impede the execution of consistent sustainable tourism policy. The task of maintaining genuine cultural encounters while satisfying contemporary tourist demands presents an ongoing difficulty.

Additional challenges may encompass excessive energy consumption, inefficiencies in food utilisation, unfavourable business conditions, scarcity of qualified workforce, restricted availability of financial resources, and insufficient levels of investment. The UNEP has stated that the difficulties are increasing owing to many global crises, such as climate change, economic recession, fuel/energy crisis, food crisis, and water crisis (UNEP and UNWTO, 2012). In response to this, an increasing number of customers are showing a preference for more sustainable options. The potential problems in developing sustainable tourism during the transition to a green economy are as follows:

- **Inadequate Infrastructure**—The lack of sufficient infrastructure poses a major obstacle to the promotion of sustainable tourism in India's environmentally friendly economy. The absence of environmentally friendly modes of transportation, sustainable lodging establishments, and effective waste management systems impedes the development of responsible tourism. Moreover, the lack of adequate sewage treatment and solid waste management facilities in numerous tourist destinations results in environmental deterioration, as emphasised in the Ministry of Tourism's 2023 report on challenges to sustainable tourism. The situation is worsened by inadequate renewable energy infrastructure and poor water-saving measures in hotels and resorts. To fulfil its sustainable tourism goals and contribute significantly to its green economy initiatives, India needs to address these infrastructure deficiencies.
- **Limited Sustainable Option**—The scarcity of viable sustainable solutions poses a major obstacle to the development of sustainable tourism in India. This problem includes a deficiency of environmentally friendly lodging, transportation, and activities that adhere to the ideals of sustainable tourism. The scarcity of electric vehicle charging outlets in tourist locations intensifies the issue, deterring the adoption of environmentally friendly transportation. Tour packages that prioritise sustainability and activities that are environmentally beneficial are frequently restricted or inadequately promoted, resulting in a lack of knowledge among tourists. The lack of sustainable alternatives not only hampers the expansion of eco-friendly tourism but also poses challenges for environmentally aware travellers in making responsible decisions. It can obstruct India's advancement towards a more sustainable tourism industry.
- **Poor Strategic Planning**—Inadequate strategic planning is a major obstacle to the sustainable development of tourism in India. Inadequate collaboration and a lack of cohesive planning among different parties frequently lead to haphazard tourism development. It may disregard the importance of balancing economic progress with environmental conservation and societal well-being. Many tourism operations cannot properly undertake comprehensive environmental impact studies, effectively engage with local communities, and successfully execute sustainable resource management plans. Furthermore, there is frequently a lack of coherence between policies at the national level and their execution at the local level, resulting in disparate strategies being employed in various regions. It

is essential to rectify these shortcomings in planning in order to fully utilise the benefits of sustainable tourism and its impact on the environmentally friendly economy.

- **Community Involvement**—Active participation from the local community is a crucial yet difficult element of ensuring the long-term viability of tourism in India. Although local engagement is crucial for the sustainable sustainability of eco-friendly tourist initiatives, properly engaging communities continues to be challenging due to a range of problems. The factors contributing to these challenges are a lack of knowledge about sustainable practices, possible conflicts between traditional ways of making a living and the growth of tourism, and an unfair distribution of the advantages. Furthermore, the hierarchical social structure prevalent in numerous Indian communities might result in the exclusion of specific groups from participating in decision-making processes. Recent research has also emphasised the difficulty of finding a balance between the demands of the community and the goals of conservation, especially in places with high levels of biodiversity. To overcome these problems, it is necessary to employ new methods of empowering communities, enhancing their capabilities, and implementing mechanisms for sharing benefits that are in line with the local cultural contexts and ambitions.
- **Environmental Stress**—The environmental impact of tourism operations in India presents a major obstacle to achieving sustainable tourism, as the country's varied ecosystems are experiencing escalating strain. The exponential expansion of the tourism industry has resulted in the excessive utilisation of natural resources, deterioration of habitats, and contamination in numerous well-liked locations. For example, the Himalayan region has undergone rapid deforestation and soil erosion as a result of unregulated tourism development. Improper waste management and coral reef destruction pose significant dangers to coastal areas. Moreover, climate change intensifies these pressures, as increasing temperatures and shifting precipitation patterns impact biodiversity hotspots and modify tourist seasons. The problem is to achieve a balance between the growth of tourism and the conservation of the environment. This requires the use of new strategies for managing visitors, developing eco-friendly infrastructure, and implementing community-based conservation initiatives.

9 Future Outlook

The future of India's tourism industry looks promising with ongoing developments and initiatives. India has become a leading force in promoting sustainable tourism in the Asia Pacific region. According to KPMG (March 2024) report,[14] the sustain-

[14] See: https://assets.kpmg.com/content/dam/kpmg/in/pdf/2024/03/sustainability-in-tourism-reimagining-indias-sustainable-tourism-evolution.pdf

able tourism market in India was worth USD 26.01 million in 2022. It is expected to grow at a quick compound annual growth rate (CAGR) of 19.3 per cent from 2022 to 2032, reaching a value of USD 151.88 million by 2032.

However, to promote sustainable tourism in India, a comprehensive strategy is needed that effectively manages economic development while also prioritising environmental preservation and social accountability. It is important to create tools and strategies at the local level to direct the growth and progress of tourism in regional areas, to ensure the successful application of sustainable tourism practices. To progress, it is necessary to create extensive policies at both the national and state levels that encourage sustainable behaviour and ensure compliance with environmental rules. It is essential to invest in environmentally friendly infrastructure, such as sustainable transportation, renewable energy systems, and waste treatment facilities. Equally crucial is the task of increasing awareness and providing education to both tourists and local people regarding sustainable tourism practices. Promoting community engagement in the development and administration of tourism helps guarantee a fair allocation of advantages and the safeguarding of cultural heritage. By utilising technology to effectively manage resources, regulate visitor flow, and promote environmentally friendly places, sustainability efforts can be greatly enhanced.

Furthermore, the implementation of smart tourism involves the incorporation of technology to enhance the overall tourist experience. This can be achieved through the utilisation of mobile applications, digital payment systems, and virtual tours. The promotion of adventure tourism, wellness tourism, and rural tourism can attract a wide variety of travellers and encourage niche tourism experiences. Effective implementation of these policies requires collaborations among government agencies, commercial sector entities, NGOs, and local communities. In addition, the establishment of standardised sustainability criteria and the implementation of regular monitoring systems can facilitate the tracking of progress and enable the implementation of necessary adjustments. Incorporating sustainable methods into everyday operations, such as conserving energy, reducing waste, using environmentally friendly transportation, offering responsible tourism experiences, engaging with the community, and fostering a sustainable company culture, ensures that sustainability becomes an essential aspect of the tourism industry's values and actions. India can enhance the resilience and sustainability of its tourism sector by prioritising these specific areas. This will not only contribute positively to its green economy but also ensure the preservation of its abundant natural and cultural legacy for future generations.

In conclusion, the study underscores the critical importance of sustainable tourism as a catalyst for economic development, emphasising the need for holistic approaches that balance economic prosperity, environmental integrity, and social equity and uphold India's rich cultural heritage and diversity. By embracing sustainable tourism principles and practices, countries and destinations can unravel the optimum potential of tourism as a driver of positive change while preserving natural and cultural resources. This paradigm shift will pave the way towards ensuring economic development sustainably. The findings of the study may enlighten the path of various stakeholders across the tourism ecosystem to adopt a long-term perspective, prioritising the preservation of India's unique identity and natural glory, while simultaneously harnessing the numerous socio-economic benefits that sustainable tourism can ensure.

10 Key Implications

This chapter addresses crucial aspects impacting sustainable tourism such as sustainable industry practices, cultural preservation, environmental conservation, community involvement, government policies and planning. It also highlights the significant steps taken by India to promote sustainable tourism. The key implications of this research are:

- Policy Integration: The government policies initiating the integration of sustainable tourism principles into its National Tourism Policy envisage a strong commitment to responsible tourism and preservation of natural resources. This sets the foundation for coordinated efforts across various stakeholders.
- Community-Centric Approach: The promotion of homestays, rural tourism, and community-based tourism models warrants that local people inhibiting tourist spots directly benefit from tourism activities, nurturing cultural exchange and enhancing socio-economic empowerment.
- Ecological Sustainability: The growth of the tourism industry needs to protect and promote ecological sustainability. Government emphasis on waste management, water conservation and efficient use, promotion of eco-tourism, and biodiversity protection advocate India's sincere efforts and dedication to attain sustainable tourism and preserve its natural heritage.
- Industry Transformation: The tourism industry is shifting towards a more socially and environmentally conscious industry by adopting responsible tourism practices encouraged by climate-resilient infrastructure and sustainability certifications from different government agencies.
- Multi-Stakeholder Collaboration: The attainment of sustainable tourism requires coordinated efforts from different stakeholders including government, local populations, tourists, and the tourism industry. This coordinated effort is strengthened by public–private partnerships encouraged by the government and development of the community as a whole.

References

Berno, T., & Bricker, K. (2001). Sustainable tourism development: the long road from theory to practice. *International Journal of Economic Development, 3*(3), 1–18. https://spaef.org/article/1061/Sustainable-Tourism-Development-The-Long-Road-from-Theory-to-Practice

Chauhan, A. S., Sikarwar, T. S., & Raj, M. (2024). Trends and patterns in behavioral accounting research: 53 years of bibliometric analysis. *Accounting Research Journal, 37*(6), 595–614. https://doi.org/10.1108/ARJ-09-2023-0246

Deloitte & Touch, International Institute for Environment and Development (IIED) & Overseas Development Institute (ODI). (1999). Sustainable tourism and poverty elimination study. A report to the Department for International Development, London: Deloitte &Touch, IIED, ODI. https://odi.org/en/publications/sustainable-tourism-and-poverty-elimination-study/

Freytag, A., & Vietze, C. (2013). Can nature promote development? The role of sustainable tourism for economic growth. *Journal of Environmental Economics and Policy, 2*(1), 16–44. https://doi.org/10.1080/21606544.2012.754611

Ghosh, M. (2024). Financial inclusion studies bibliometric analysis: Projecting a sustainable future. *Sustainable Futures, 7*, 1–10. https://doi.org/10.1016/j.sftr.2024.100160

Gupta, S., Gupta, N., & Chauhan, P. (2022). Controlling the uncontrolled: Over tourism in hill stations of Northern India. *Journal of Positive School Psychology, 6*(5), 6549–6568. https://journalppw.com/index.php/jpsp/article/view/8181

Jatav, S. (2023). Current trends in sustainable tourism in the Indian Context. In H. Sezerel & B. Christiansen (Eds.), *Handbook of research on sustainable tourism and hotel operations in global hypercompetition* (pp. 391–412). IGI Global. https://doi.org/10.4018/978-1-6684-4645-4.ch018

Khan, A., Bibi, S., Lorenzo, A., Lyu, J., & Babar, Z. U. (2020). Tourism and development in developing economies: A policy implication perspective. *Sustainability, 12*(4). https://doi.org/10.3390/su12041618

Lal, B. S. (2024). Sustainable tourism and economic growth in India—An analysis. *Journal of World Economy, 3*(1), 34–42. https://www.pioneerpublisher.com/jwe/article/view/702

Lim, C., & McAleer, M. (2005). Ecologically sustainable tourism management. *Environmental Modelling & Software, 20*(11), 1431–1438. https://doi.org/10.1016/j.envsoft.2004.09.023

Moher, D., Liberati, A., Tetzlaff, J., Altman, D. G., & PRISMA Group*, T. (2009). Preferred reporting items for systematic reviews and meta-analyses: the PRISMA statement. *Annals of internal medicine, 151*(4), 264–269. https://doi.org/10.7326/0003-4819-151-4-200908180-00135

Neto, F. (2003). A new approach to sustainable tourism development: Moving beyond environmental protection. *Natural Resources Forum, 27*(3), 212–222. https://doi.org/10.1111/1477-8947.00056

Pan, S. Y., Gao, M., Kim, H., Shah, K. J., Pei, S. L., & Chiang, P. C. (2018). Advances and challenges in sustainable tourism toward a green economy. *Science of the Total Environment, 635*, 452–469. https://doi.org/10.1016/j.scitotenv.2018.04.134

Rasul, G., & Manandhar, P. (2009). Prospects and problems in promoting tourism in South Asia: A regional perspective. *South Asia Economic Journal, 10*(1), 187–207. https://doi.org/10.1177/139156140901000108

Robin, R. S., Karthik, R., Nithin, A., & Purvaja, R. (2023). Removal of marine litter and its impact along the coast of India. *Records of the Zoological Survey of India: A Journal of Indian Zoology, 123*(1S), 67–86. https://doi.org/10.26515/rzsi/v123/i1S/2023/172453

UNEP and UNWTO. (2012). *Tourism in the Green Economy: Background Report*. United Nations Environment Programme (UNEP) and World Tourism Organization (UNWTO), Spain and Kenya, p. 167. https://doi.org/10.18111/9789284414529

Valente, S. (2005). Growth, conventional production and tourism specialisation: technological catching-up versus terms-of-trade effects. *Fondazione Eni Enrico Mattei Working Paper Series*, NOTA DI LAVORO 140.2005. https://ideas.repec.org/p/ags/feemnr/12139.html

World Tourism Organization (2008). UNWTO Tourism Highlights, 2008 Edition, UNWTO, Madrid. https://doi.org/10.18111/9789284413560

The European Union's Potential in Mitigating Climate Change Caused by Tourism: Classification of the 27 EU Member States Tourists' Sustainable Behaviours

Micaela Pinho and Fátima Leal

Abstract Climate change is now a global phenomenon with severe social and economic implications, including for tourism. Tourism is currently one of the most dynamic economic sectors in the world and one of the main ones responsible for greenhouse gas emissions. At the same time as it contributes to global warming, the tourism sector is also one of the primary victims of climate change. Strengthening climate change mitigation measures is paramount for the sectors' resilience. These changes require a collective and concerted effort by all stakeholders, including tourists, to transition to a tourism sector that minimises its environmental footprint. The question arises whether tourists are willing to change their habits towards an environmentally sustainable tourism. This study proposes a system for classifying citizens from the 27 European Union (EU) member states as future sustainable tourists. The data used were retrieved from the European survey—Flash Eurobarometer 499, involving 25711 European citizens. The proposed method relies on a decision tree-based algorithm—Random Forest, to achieve intelligent classifications based on Machine Learning. The implementation of this classification system was preceded by the exploration of respondents' willingness to change, in the future, their travel and tourist habits towards environmental sustainability by gender, age, and region of residence. The main findings suggest a notable trend in EU countries towards greater tourism sustainability among women belonging to Generation Y (aged

M. Pinho (✉)
Research on Economics Management and Information Technologies, REMIT, Portucalense University, Porto, Portugal

IJP – Portucalense Legal Research, IJP, Portucalense University, Porto, Portugal

Research Unit in Governance, Competitiveness and Public Policies, GOVCOPP, Aveiro University, Aveiro, Portugal
e-mail: mpinho@upt.pt

F. Leal
Research on Economics Management and Information Technologies, REMIT, Portucalense University, Porto, Portugal
e-mail: fatimal@upt.pt

J. Marques et al. (eds.), *Tourism and Climate Change*, Tourism, Hospitality & Event Management, https://doi.org/10.1007/978-3-031-99128-8_4

between 30 and 44) and from the Eastern European region. With this information, the implemented algorithm can classify the environmental sustainability of EU citizens as future tourists with an accuracy of 63%. This study enriches the theoretical understanding of the intention of tourists' pro-environmental behaviour. It highlights the need to adopt specific measures to increase awareness of human action in tourism for male citizens, for older generations, and for south, west, and north European regions.

1 Introduction

Man-made environmental degradation is a much-discussed topic, but there are hardly any solutions. Although the problem has been well recognised, a coordinated and effective solution is lacking. Sustainability has gained prominence in many discussions in all social and political spheres, but serious policies that reduce or reverse our environmental impact have yet to be adopted. In this context, the tourism sector plays a key role due to its interdependent relationship with the environment. Tourism is currently one of the most dynamic economic sectors in the world, prompting more and more countries and regions to invest in it. Although the tourism industry is a driver of economic growth, accounting for 10% of the global Gross Domestic Product (Liu et al., 2023), its environmental impact is similarly high. Tourism is responsible for around 8% of global carbon emissions, divided into transport, accommodation, construction, food, and shopping (Liu et al., 2023). Visitors from industrialised countries emit the largest share of the carbon footprint. The World Tourism Organization's report on transport-related CO2 emissions from the tourism sector (UNWTO, 2019) reports that, in 2016, 5% of human-induced CO2 emissions were attributable to tourism and that by 2030, this proportion will increase to 5.3%. Equally important is to emphasise the issue related to the use of plastics in the tourism sector. According to recent data (One Planet Network, 2024), 8 million tons of plastic end up in the oceans, increasing by 40% during holiday seasons. These statistics represent a considerable challenge for governments and companies regarding the actions needed to improve their relationship with the environment.

Tourism is not only a significant contributor to global warming but is also highly susceptible to environmental degradation. Climate change, characterised, for example, by rising temperatures and an increase in the frequency and severity of extreme weather events, affects both the demand for and the supply of tourism. While on the supply side, climate change threatens tourism infrastructure, natural attractions, leisure opportunities, and the accessibility of destinations, on the demand side, climate change alters visitor comfort levels, health risks, and the seasonality of destinations (Tanrisever et al., 2024).

The impact of climate change on tourism is of particular concern in the European Union since it is a significant generator of growth, employment, and social development. The latest available data show that in 2019 Europe was the most visited region in the world, accounting for 51% of all international arrivals and 41% of tourism receipts (UNWTO, 2020). Europe relies heavily on tourism as an economic driver and a powerful tool in fighting economic decline and unemployment, as well as an instrument that strengthens Europe's image in the world, projecting European values and promoting the attractions of the European model. However, the European tourism industry will face considerable stress due to climate change. In the 2100 timeframe, climate changes are expected to trigger regional changes in travel patterns as we know them today. With global warming, a north–south pattern in tourism demand changes is predicted, with northern regions benefiting from climate change and southern regions facing significant reductions in tourism demand; this pattern becomes more pronounced for higher warming scenarios (Matei et al., 2023). In addition, the seasonal distribution of tourist demand will also change, with relative reductions in summer and increases in the intermediate and winter seasons (Matei et al., 2023).

Therefore, the need to strengthen climate change mitigation measures in tourism is paramount. The tourism sectors' resilience hinges on the measures taken to combat climate change. Key tourism stakeholders recognise that the future resilience of tourism depends on the sectors' ability to take action to reduce carbon emissions by 50% by 2030 (UNWTO, 2024). Support for this global commitment is reflected in the Glasgow Declaration on Climate Action in Tourism, officially launched at the United Nations (UN) Climate Change Conference COP26. The signatories to this declaration, recognising that the cost of inaction on climate will be greater than the cost of any other crisis over time, committed to act immediately and take action to accelerate climate change action to halve carbon emissions over the next decade and achieve net zero emissions by 2050 (UNWTO, 2024). Recently, in March 2024, the UN issued policy guidance to help government agencies involved in tourism make this low-carbon transition.

In order to mitigate tourism's contribution to climate change, sustainable consumption and production must be promoted, energy consumption minimised, renewable energy used, efficiency improved, and more sustainable modes of transport used, among other things. These changes require a collective and concerted effort by all stakeholders, including tourists, to transition to a tourism sector that continues to deliver socio-economic benefits and minimises its environmental footprint. In this regard, sustainable behaviour on the part of tourists is crucial to accelerate the decarbonisation process. Tourists' behavioural intentions are one of the important factors influencing the selection of tourist destinations; then, changing tourists' behaviour and attitudes could lead to the change of tourism into a sustainable sector (Aman et al., 2021).

In this regard, it remains unclear, on the one hand, whether tourists are actually interested in engaging in environmentally friendly behaviour when on holiday and, on the other hand, which individual demographic variables affect this predisposition towards environmentally friendly tourism habits. About the first concern, a recent 2023 Sustainable Travel Report from Booking (World Economic Forum, 2023) found that from 33 000 travellers around the world, 76% want to travel sustainably over the next year, while other studies indicate that tourists are not so committed to the environment (Bilynets & Cvelbar, 2022; Wilkinson & Coles, 2024). As recognised elsewhere (European Travel Commission, 2022), the desire expressed by individuals to adopt more sustainable travel behaviour does not always materialise, making these changes difficult to predict. Regarding the second concern, there is increasing literature exploring the predictors of tourist environmentally friendly behaviours (Butnaru et al., 2022; MacInnes et al., 2022), but the results are not consensual. However, little is known about cross-national differences in both issues since most studies have been conducted in a single or small number of countries, jeopardising comparisons between studies due to different study designs and methodologies (Poortinga et al. 2019).

The present paper intends to address these empirical deficits. Focusing on the European region and based on secondary data collected in the Eurobarometer survey 499, this study explores the willingness of citizens of the 27 EU countries to change their travel and tourism habits in order to be more environmentally sustainable and whether this predisposition is influenced by their gender, age, and residence region. From this analysis we propose a system that classifies individuals as having more or less potential to become promoters of more environmentally friendly tourism, depending on their demographic characteristics and place of residence. Thus, our analysis begins with a statistical analysis of citizens' sustainable tourism behavioural trends stratified by gender, age (divided into four generation: Z, Y, X, and BB) and European region (North, Centre, East, and West) of origin. Then, Machine Learning was used to implement a decision tree-based algorithm—Random Forest—that classifies people from the 27 EU Member States according to their level of sustainability. In addition, the proposed method includes explanations which help to understand why an individual has been classified as sustainable.

This study contributes to the sustainable tourism literature by developing a classification system of individuals based on their gender, age, and region of living. With this classification method, decision makers know in real time whether or not a male or female citizen belonging to a certain age group and residing in one of the regions of Europe has the potential to become an environmentally sustainable tourist.

2 Literature Background

2.1 Troubling Tourism Sustainability and Tourists' Sustainable Environmental Behaviours

The last few years have been fruitful in discussing sustainable development in general and tourism sustainability in particular. Human concern for the impact of economic activity on the environment began to be formalised in 1972 with the publication of the Meadows Report. This document postulates that the pace of economic growth is not, in the long term, compatible with the availability of natural resources. Pollution and environmental deterioration, together with economic growth and population growth, can, if the necessary measures are not taken, trigger long-term imbalances with adverse effects on the quality of life of human beings (Meadows et al., 2018). Later in 1987, the Brundtland Report advanced the most widely used definition of sustainable development today by highlighting the need to balance economic growth, social development, and environmental care so that current needs are met without compromising the needs of future generations (WCED, 1987). Five years later, in 1992, an event held in Brazil called the Earth Summit highlighted the importance of political powers being involved in environmental preservation and implementing alternatives for the development of society. Thus, sustainability, analysed from the perspective of the environment, aims to balance human needs and the ecosystems' capacity to fulfil these needs without diminishing biological diversity (Brown et al., 1987).

The combination of sustainability and tourism was first used in the seminal held in 1990 in Bali, Indonesia (Wall, 1990). Subsequent works have been undertaken with an emphasis on the substantial increase in publications on sustainable tourism from 2011 onwards [for literature reviews, see, for example, Zolfani et al., 2015; Huang et al., 2019; García-Garcia et al., 2023; Bassi & Martín, 2024]. The sustainable tourism topic gained increasing preponderance with the commitment made in 2015 by United Nations countries to pursue seventeen sustainable development goals by 2030 set out in the United Nations Agenda 2030. Although the tourism sector impacts all SDGs, its most direct contribution is to the eighth, twelfth, and fourteenth objectives.

Tourists have a leading role in environmental sustainability, primarily through adopting environmentally responsible behaviours (Loureiro et al., 2022). Pro-environmental behaviours, in general, are understood as an individual who consciously seeks to minimise the negative impact of their actions on the environment by carrying out positive actions that are favourable to the environment (e.g., recycling, minimising the consumption of resources and energy) while avoiding actions that are harmful to the environment (e.g., using disposal bags in the supermarket) (Parzonko et al., 2021). Sustainable environmental tourist behaviour is therefore understood to be that which does not negatively impact the natural environment, either globally or at the destination level (Juvan & Dolnicar, 2016), nor does it harm the community's well-being (European Travel Commission, 2022).

Understanding sustainable tourism behaviours has become a focus of interest among researchers (Li et al., 2023). The complexity and uniqueness of the issues that affect different tourist destinations, as well as the variety of tourist typologies, make the concept of sustainable tourist behaviour elusive. Indeed, many labels have been attributed to these tourist behaviours (Juvan & Dolnicar, 2016), from "ethical", "pro-environmental", "environmentally responsible", "environmentally significant" to "green", "environmentally friendly", or "low carbon", (Kim & Filimonau, 2017; Kuo & Dai, 2012; Lee & Jan, 2019).

In the sustainable tourist behaviour research, more than 50 theories have been used, with the theory of planned behaviour, the norm-activation theory, and value-belief-norm theory, and Schwartz's Value Theory being the most applied theories (Li et al., 2023 for a review). However, it must be taken into account that changing tourists' habits is challenging. It isn't easy because the attitude of people on vacation is to relax and want to benefit from everything, they rightfully understand without being willing to give up amenities. On the other side, tourists rely on economic self-interest when making choices, and they may feel reluctant to change them unless the adoption of sustainable solutions brings them additional benefits (Bolderdijk et al., 2013). It is also difficult because, despite the benefits that sustainable behaviours bring immediately, they will mostly be felt in the future, that is, by future generations. In their choices, tourists make assessments based on self-interest, which they resist changing, unless they perceive additional benefits from introducing sustainable solutions (Bolderdijk et al., 2013). Therefore, achieving sustainable tourism depends on tourists' feelings and their moral obligation about environmental conservation (Wu et al., 2021).

2.2 Nexus Between Sustainable Environmental Behaviours and Tourists' Age and Gender

Age and gender are components of personal factors. It is long overdue for sustainable tourism research to move beyond the fundamental questions of whether age and gender matter because there is no human phenomenon without age and gender dimensions. Moreover, integrating gender and age perspectives into the discussion of tourism sustainability is particularly important as the tourism industry is a major employer of women and a significant employer of young people. Actually, according to recent data (Eurostat, 2024), in 2023, the tourism industries in Europe Union employed more female workers than male workers (with the highest proportions in travel agencies and tour operators and in accommodation), and 11 % of people employed in the tourism industries were young workers (15–24).

Engaging in sustainable behaviour involves an awareness of climate change and the fact that it has human origins. Many studies on climate change perceptions show a consistent pattern across different demographic groups, with men and older people being more sceptical about the reality and anthropogenic nature of climate change (Milfont et al., 2015; Poortinga et al., 2011, 2019; Echavarren, 2016), and therefore less concerned about the impacts of climate change (Shi et al., 2016).

In general, gender studies show that women report stronger environmental behaviours and attitudes than men (Zelezny et al., 2000 for a review). It was found that compared to males, females were more socially responsible and exhibited higher levels of socialisation, demonstrating more empathy and a stronger ethic of care and responsibility for environmental problems. Psychological literature considers "responsibility" as one of the pillar principles of sustainable tourism (Passini & Morselli, 2006). In this regard, there is evidence that women are more socially responsible and less oriented towards social dominance than men (Donohoe & Needham, 2006). Pilot research concluded that men have a more selfish value orientation than women (Cavagnaro & Staffieri, 2015) and, with this value orientation considered antagonistic to sustainable attitudes and behaviours (Lindenberg & Steg,

2007) it may explain men's weaker pro-environmental attitudes. The relevance of women to sustainability was recognised in 1999, in Agenda 21, when the need for them to be involved in defining the environmental sustainability agenda was made explicit by having specific environmental concerns and perspectives based on their social and biological roles and by are traditionally disadvantaged and excluded from decision-making (Buckingham-Hatfield, 1999).

The recognition of gender differences is of utmost importance in the tourism sector since all aspects related to tourism activities and development incorporate gender relations (Kinnaird & Hall, 1994). According to Swain (1995), gender is a primary factor in tourism social sciences. Thus, for social scientists involved in tourism research, gender is fundamental in human resource studies, economic development projects, marketing strategies, site and infrastructure planning, and policy development (Swain, 1995). Even less pronounced, gender differences in travel and tourism remain substantial (Collins & Tisdell, 2002). Although few studies (Dietz et al., 2002; Stevens, 2010) analysed gender impact on sustainable tourism, the existing ones show a greater attitude of women in managing hospitality and in promoting good practices (Meera, 2014). Therefore, the rise of women's role in the tourism industry, as well as the greater recognition of men's and women's interdependence, conflicts, and value differences in tourism development, is essential.

Age is also an essential factor in sustainable tourism. The relationship between age and sustainable tourism is multifaceted, reflecting generational differences in values, behaviours, and preferences. Understanding these distinctions is crucial for developing and promoting sustainable tourism practices that can appeal to a diverse range of travellers (Yeoman & McMahon-Beattie, 2019). Tourism demand and future travel patterns are believed to be driven by age and take on a unique character for each generation. Individuals of different ages have different attitudes and perceptions about tourism and, thus, different demands for tourism services. For example, preferences for travel destinations, entertainment, cooking, and specific activities change over time and with age (Pemani & Massie, 2017). Evidence shows that while younger tourists were more motivated to contribute to sustainable hospitality than their adult counterparts this pattern of concern is not maintained when choosing travelling options where they value above all economic and psychological advantages (Fermani et al., 2016). Thus, understanding tourists' behaviours and trends of different generations is paramount in a globalised world (Haddouche & Salomone, 2018).

The idea of generational cohorts originated in Sociology and has been used in tourism to understand tourists past, present, and future behaviours. A generation consists of individuals born within a specific period who share everyday experiences and values that shape lifestyles and attitudes (Strauss & Howe, 2009), differentiating them from the preceding and succeeding generations. Research on generational cohorts is crucial for achieving tourism sustainability since it allows the identification of tourists' profiles and habits, as well as destination selection processes. Recent studies have focused on generational differences in pro-environmental behaviour (PEB) in general (Song et al., 2020) and in environmentally friendly tourism in particular (Prayag et al., 2022; Pinho & Gomes, 2023; Seyfi et al., 2023; Sharma et al., 2023) with very different results. For example, Song et al. (2020) found that younger generations, like Gen Z, are more environmentally conscious and engaged than their older counterparts. Prayag et al. (2022) find that Generation Z tourists differ from Generation Y, X, and Baby Boomers in their environmental attitudes and travel behaviour. However, Generation Z tourists focus more on resource conservation, community-building activities, and buying local food and environmentally friendly products when travelling they use public transport, bicycles, and electric vehicles less than other generations (Prayag et al., 2022). Sharma et al. (2023) found that food waste avoidance behaviours tend to be high among Generation Z tourists compared to older generations, but Qiu et al. (2022) show that Generation Z has significantly less environmentally responsible behavioural intentions, especially in terms of environmental protection issues, social contact, damage reporting, and waste recycling. Pinho and Gomes (2023) using a sample of 305 young Portuguese (Generation Z) conclude that respondents have neither pro-environmental tourists' behaviours nor care about keeping destinations sustainable.

Since recent research on gender and generational cohorts' effects on tourism eco-friendly support comes up with diverse findings, it would be useful to proceed with future studies. In this regard, the present study examines how the demographic factors of gender and age are associated with the attitude of tourists towards future environmentally friendly behaviours across the 27 EU Member States.

3 Methodology

3.1 Dataset

The Eurobarometer survey on Europeans' Attitudes towards tourism examines how COVID-19 has influenced EU citizens' travel behaviours over the long term. Additionally, the survey investigates the significance of sustainability in enhancing the tourism experience. This Flash Eurobarometer interviewed a representative sample of individuals aged 15 and above from all 27 EU member states. The survey was conducted from October 18 to October 28, 2021, by telephone (both landline and mobile). A total of 27 valid answers were obtained. The present study relies on one question from Flash Eurobarometer 499 that pertains exclusively to sustainable tourism—the fifth question of the Eurobarometer survey. This question assesses whether respondents were prepared to change their travel and tourism habits to become more sustainable. Details can be found in Table 1.

Table 1 Description of the dataset

Variable Description	Categories	Scale
"Are you prepared to change your travel and tourism habits to be more sustainable? If so, in which of the following ways?" 1. Pay more to protect the natural environment	0- No 1- Yes	Nominal
2. Pay more to the benefit of the local community		
3. Take holidays outside of the high tourist season		
4. Travel to less visited destinations		
5. Contribute to carbon-offsetting activities (e.g. plant trees)		
6. Consume locally sourced products on holiday		
7. Choose transport options based on ecological impact		
8. Reduce waste while on holiday		
9. Reduce water usage on holiday		

3.2 Proposed Method

Given the critical importance of climate change for the stability and growth of the tourism sector, it is imperative to develop thorough analysis and innovative studies in this domain. The proposed method comprehends a comprehensive quantitative analysis utilising data from a substantial sample of 25,711 participants across the 27 EU Member States, derived from the Flash Eurobarometer 499—"Attitudes of Europeans towards Tourism" (GESIS, 2022).

We propose a method that explores and classifies sustainable behavioural trends among EU tourists. This methodology involved a two-step approach. Initially, we performed a descriptive analysis based on statistical techniques to examine the willingness of the citizens of the 27 EU countries to change their travel and tourism habits to assume a more sustainable behaviour based on both demographic factors, namely gender, and age, and regional divisions within Europe. Subsequently, we employed Machine Learning (ML) techniques, specifically a decision tree-based Random Forest algorithm, to classify these EU citizens according to their potential level of tourists' sustainability. Furthermore, the proposed ML approach provides explanations for understanding the output. This dual-step methodology enabled understanding sustainable tourism behaviours across the European Union.

Below is a brief description of each phase of the method employed.

1. *Descriptive analysis*. This step of the method examines the sustainability behaviours of participants by calculating the percentages of sustainable practices across different demographics. An individual is marked as sustainable if the average of the answers given in the survey is higher than 0.5. Based on this assumption, the analysis starts by analysing the data by gender, revealing distinct trends in sustainability between males and females. Next, we segment the participants by age into generational cohorts. Following Koksal (2019) and Shams et al. (2020), we divided the participants' age into four generations to uncover generational differences in tourism sustainable behaviours: Generation Z comprising individuals born between 1995 and 2010; Generation Y, comprising participants born between 1980 and 1994, Generation X includes respondents born between 1965 and 1979, and Generation Baby-Boomer comprising respondents born before 1965. Finally, we categorise the participants by European regions divided according to the four points of the compass: Northern, Southern, Eastern, and Western Europe. This regional division allows the identification of regional variations in citizens' future sustainability practices when acting as tourists. Northern Europe includes Denmark, Estonia, Finland, Ireland, Latvia, Lithuania, and Sweden. Southern Europe comprises Cyprus, Greece, Italy, Malta, Portugal, and Spain. Eastern Europe encompasses Bulgaria, Croatia, Czech Republic, Hungary, Poland, Romania, Slovakia, and Slovenia. Western Europe consists of Austria, Belgium, France, Germany, Luxembourg, and Netherlands.

This comprehensive descriptive analysis provided a detailed overview of how tourists' sustainable behaviours vary across gender, age groups, and regions within

Europe, offering valuable insights into the demographic and residential factors influencing tourism sustainability.

2. *Classification*. The second step of our approach relies on the Random Forest algorithm, a robust ML technique, to classify individuals from the 27 EU Member States according to their level of tourist sustainability. The Random Forest algorithm constructs multiple decision trees during training and outputs the mode of the classes for classification tasks. It is an ensemble learning model combining multiple Decision Trees (Parmar et al., 2019) to solve complex problems. By incorporating demographic factors (gender and age) and region of residence data the algorithm effectively classified participants into different levels of tourism sustainability. This classification process highlighted key predictors of sustainable behaviour and allowed us to identify distinct patterns and trends across the European Union member states, providing actionable insights for promoting sustainable tourism practices.
3. *Explainability*. In the methods' explanation stage of the method, we enhance the transparency and interpretability of our classification results. Explanations are relevant to describing and understanding the outputs generated by ML-based applications. We employed the Local Interpretable Model-agnostic Explanations (LIME) technique (Ribeiro et al., 2016). LIME is a powerful tool designed to explain individual predictions made by complex ML models. By approximating the model locally with an interpretable model, LIME provides insights into the specific features contributing to each prediction. In our analysis, LIME helped elucidate how age, gender, and regional attributes influenced the classification of participants' sustainability levels. This explainability phase was relevant for validating the models' outputs and ensuring that stakeholders could understand the outcomes. Therefore, the explanations provide greater confidence in the proposed method and enable to analyse and promote tourism sustainable behaviours among citizens of the European Union member states
4. *Evaluation*. The last step of the methodology evaluates the model to determine the feasibility and effectiveness of the proposed solutions. Our method involves offline processing to analyse tourist behaviour. The data were divided into two sets: a training set, used to develop the initial model, and a testing set, used to evaluate the quality of the classifications. The performance of these classifications is measured using standard evaluation metrics: classification accuracy and F-measure. Classification Accuracy measures the overall performance of ML model considering the number of true positives and true negatives. In turn, the F-measure computed in both macro and micro-averaging scenarios assesses the model's effectiveness by combining Precision and Recall. Precision measures the percentage of correct classifications, while Recall assesses the model's ability to correctly identify positive cases in the dataset. The combination of macro and micro averages provides a comprehensive evaluation of the model's performance across all target classes, either by giving equal weight to each class or considering them individually.

3.3 Experiments and Results

To assess the proposed method, we have conducted offline experiments utilising data from the European survey Flash Eurobarometer 499—"Attitudes of Europeans towards Tourism". Our system is equipped with the following hardware specifications:

- Operating System: Windows 64-bit
- Processor: Intel(R) Core(TM) i7-8565U CPU @ 1.99 GHz
- RAM: 16 GB
- Disk: 500 GB SSD

The experiments comprise: (i) descriptive analysis to understand the data; (ii) classification; and (iii) explainability. The classification models rely on scikit-learn from Python.

3.3.1 Descriptive Analysis

The analysis focuses on categorising respondents by gender, generation (age), and regions within Europe according to their willingness to change behaviours towards sustainability in travel and tourism. Individuals are classified as sustainable if their average responses to the questions exceed 0.5 on a scale where 0 denotes low sustainability awareness, and 1 signifies high sustainability awareness.

Table 2 presents the results of the gender analysis. Among respondents, females showed the highest willingness to follow tourism sustainable practices. Indeed, 37% of females revealed interest in tourism sustainability compared to 33% of male respondents. Individuals identified as being from the "Other" gender comprised a small sample size, and therefore, we remove them from the analysis so as not to bias the results.

Table 3 presents the results of the Generation analysis. According to the results, respondents belonging to the Millennial (Generation Y) demonstrated the highest (37%) willingness to change tourists' behaviours towards sustainability. Millennial were followed by respondents from Generation X and Generation Z, each with 35%. The Baby Boomers (BB) proved to be the most reluctant to change their habits as tourists. These findings suggest that younger generations, particularly Millennial, are more engaged with sustainability in tourism, reflecting their greater awareness and prioritisation of environmental issues compared to older generations.

Table 2 Percentage of sustainable individuals based on gender

	Gender	Total	Marked Sustainable	Percentage
0	Female	13,017	4773	37
1	Male	12,375	4051	33
2	Other	128	48	38

Table 3 Percentage of sustainable individuals based on age generation

Generation	Total	Marked as sustainable	Percentage
BB	8133	2653	33
X	7583	2683	35
Y	6796	2488	37
Z	3008	1048	35

Table 4 Percentage of sustainable individuals based on region of Europe

	Region	Total	Marked as sustainable	Percentage
0	North	5987	1987	33
1	South	5140	1489	29
2	East	8151	3526	43
3	West	6241	1870	30

Table 4 summarises the results for the European region. The results denote significant regional variation in the willingness to adopt sustainability attitudes in tourism. The eastern region of Europe has the highest percentage (43%) of sustainable tourists. In contrast, respondents from the Southern European region were just less likely to become sustainable tourists, with only 29% being willing to change travel and tourism habits. The difference in attitudes between the southern citizens and their counterparts from the eastern was substantial—24 percentage points. Southern European countries are followed very closely by participants from Western countries (29% against 30%) and more distantly by respondents from Northern Europe (33%).

In sum, Fig. 1 highlights difference between European countries in terms of their ability to become sustainable tourists.

3.3.2 Classification

The ML classification method estimates if an individual has the potential to become a sustainable tourist considering his/her gender, generation, and region of origin in Europe. Class #0 represents respondents who do not show a desire to be sustainable tourists, while class #1 stands for those who reveal a desire to change tourists' behaviours to help achieve sustainable tourism. Table 5 depicts the results of the classification.

The classifier performance results for the Random Forest model, with an accuracy of 0.63, indicate moderate predictive power. The macro *F*-measure of 0.47 highlights the challenge of achieving balanced performance across both classes (#0 and #1). The F-measure for class #0 is 0.76, suggesting that the model performs well in identifying the majority class. However, the significantly lower F-measure of 0.18 for class #1 reveals that the model struggles with accurately predicting the minority class. This disparity underscores the need for strategies to address class imbalance, such as resampling techniques or adjusting class weights to ensure fairer treatment of both classes.

Fig. 1 Colour European map concerning sustainable habits

Table 5 Performance of the classification

Classifier	Accuracy	F-measure		
		Macro	#0	#1
Random Forest	0.63	0.47	0.76	0.18

3.3.3 Explanations

The LIME technique identified the features that had the most impact on the classification. Figure 2 displays the LIME explanations for a scenario of unsustainability. This simulation involves a 39-year-old male from southern Europe. According to the classification system the individual was considered a potential unsustainable

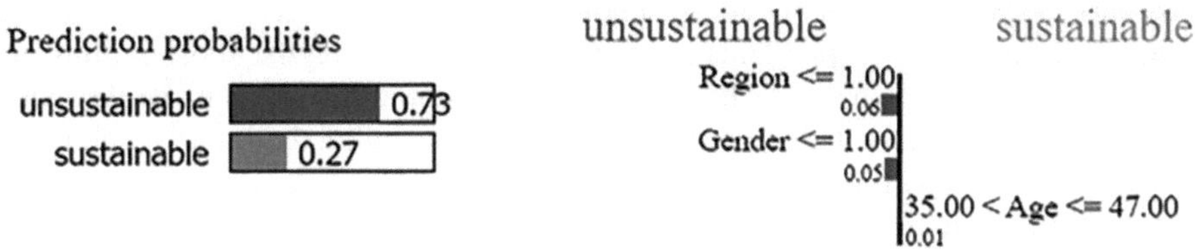

Fig. 2 LIME explanations

tourist with a probability of 73 %. From the LIME explanations, we can conclude that gender and region of inhabitancy were the factors that contributed most to the unsustainable classification. This classification reveals good predictive power as it is aligned with previous results. This simulacrum uses an individual from southern Europe (the region with the lowest level of sustainability among its citizens) and a man (men are also less sustainable than women) with age belonging to the millennial generation (the one considered to have the most potential for contributing to sustainable tourism). In short, even though age attenuated the individuals' negative classification it was not sufficient in the face of the other two variables that classified him negatively.

4 Discussion

The present paper examined the association of EU citizens' willingness to change tourist and travel habits towards environmentally sustainable tourism with individual demographic variables of gender, age (generation), and region of residence and how these three factors can be combined into an environmentally sustainable classification scheme. Expanding on earlier research regarding the individual determinants of sustainable tourism behaviours (Poortinga et al., 2019; Pinho & Gomes, 2023), we demonstrate that both demographic factors, translated in gender and age and European residence region, are important predictors of tourists' willingness to change habits to become more environmentally sustainable across 27 European Union member states.

The Random Forest algorithm classified with 67% accuracy the respondents from the 27 EU countries as more or less willing to change habits to become environmentally friendly tourists according to their gender, the generation they belong, and the region of living. It is worth saying that a comprehensive comparison of the present findings with other studies is not possible since the only three studies (Bassi & Martín, 2024; Gomes & Lopes, 2023; Sharma et al., 2023) that covered the same territory and used the same database used different methodologies or analysed different associations. Thus, we only made fragmentary comparisons.

We found gender differences in the willingness to change to environmentally conscious tourist behaviours. Women are more committed to these behavioural changes in the future than men. Identical results were found elsewhere (Skanavis &

Sakellari, 2008; Fermani et al., 2016; Schönherr & Pikkemaat, 2024). Moreover, we found that European respondents from Generation Y were the most willing to change their future tourists' behaviours towards ecological tourism, while older respondents (Generation BB) were more resistant to changing their tourist habits. In between, with an equal willingness to change, were participants from Generation Z and X. While our results align with some empirical evidence (Cavagnaro et al., 2018; Sharma et al., 2023), they contradict others (Pinho & Gomes, 2023; Ribeiro et al., 2023).

These age and gender effects may be explained by the motivation to maintain prevailing social structures. As older people are more integrated into existing social orders, they lose more from the changes needed to meet environmental challenges. On the other hand, climate change is on the agenda and, therefore, part of formal education may explain young people's greater perception and fear of climate change, making them more conscious of the need to change tourists' habits.

Finally, we found great heterogeneity regarding the willingness to change towards more environmentally friendly behaviours in tourist practices among citizens from different regions of the European Union. Thus, converging environmentally sustainable tourist habits among the 27 EU member states still appear to be far from realised. Indeed, significant differences were found among the European Union regions. Eastern European residents demonstrate the greatest aptitude for changing habits as tourists. Next come residents of the north of the European Union, although with a substantial difference of 10 percentage points compared to the eastern region. The findings for northern Europe are somewhat surprising if we put them in a cultural context. Culture has a significant impact on sustainability behaviour. In Northern Europe environmental values and respect for nature are already built-in from the childhood, thus we would expect that tourists do not have to change their behaviour when travelling as they have a sustainable value base. The citizens least likely to change their tourist habits are those from Southern Europe, immediately preceded by those from Eastern Europe (with a difference of just one percentage point). These findings partially contradict empirical evidence that suggests that many countries of Eastern Europe, as well as Mediterranean countries, are not very much involved in sustainable practices (Bassi & Dias, 2019). If we corroborate this result for the Mediterranean countries, the same is not valid for Eastern countries. The findings for southern European residents are worrying for several reasons. Firstly, they are popular tourist destinations that suffer greatly from the ecological footprint left by tourists. Residents should therefore be more aware of the need to adopt more sustainable behaviours when they travel. Secondly, because they are the region expected to suffer most from the negative impacts of climate change.

Our findings demonstrate that both demographic factors and region of residence are needed to understand individuals' availability to change tourism habits towards environmental sustainability.

One key strength of our study is the application of the robust Machine Learning technique known as the Random Foster algorithm, which proved to have a relevant accuracy in the classification of participants from the 27 EU member states according to their potential tourists' contribution to environmental sustainability by considering their gender, age, and residence region. Notwithstanding, we are aware of the limitations of our study and want to point them out. First, we used secondary data from the Eurobarometer survey; however, this data is not collected periodically. It would be important to collect data so that it is possible to analyse the evolution of citizens' behaviours regarding tourism sustainability. Second, the data was collected in 2021 at the height of the pandemic, which will certainly have affected the perception of the respondents. This is yet another reason why it would be interesting to ask European citizens about this issue again. Third, we notice that the database presents a bias problem in the question we used (question 5 of the original questionnaire) as most respondents did not show a desire to change their behaviour in favour of environmental sustainability. Therefore, reducing these limitations will allow future research to identify the actions necessary to promote greater convergence of environmentally friendly habits in tourism among EU member states. Fourth, we are aware that using just one question to assess potential predispositions to change tourist behaviour may be insufficient. Future work should use more questions to assess tourists' sensitivity to the need to change their behaviour when on holiday, such as their daily sustainable behaviours and their belief in climate change. Finally, future work should deepen understanding of the differences in sustainable behaviour profiles between regions of residence in Europe, since there are various dimensions related to the region that can explain different behaviours, such as cultural aspects.

5 Conclusions and Practical Implications

From our study, it is possible to conclude that individual availability to change tourist behaviour in favour of environmental sustainability is explained by individual demographic variables such as gender and age as well as the European region they inhabit. Thus, gender, generation cohort, and region are important drivers for developing tourism environmental sustainability. Based on these factors, the environmental sustainability classification system adequately predicts whether a given European Union citizen has the potential to become an environmentally sustainable tourist.

The conclusions of this study are of the utmost importance to tourism professionals and policymakers. First, taking into account the harmful consequences of climate change in the tourism sector, ecological behaviours in travel and tourism become crucial (Li & Wu, 2020), forcing companies and political decision-makers to develop new strategies and campaigns to instigate environmental awareness (Sharma & Gupta, 2020) and thus boost sustainable traveller behaviour. Second, this study identifies the factors that influence individuals' pro-environmental behaviour when placed in the role of tourists. Knowing these factors is extremely important for political decision-makers and tourism professionals to seek to promote pro-environmental behaviours among different citizens during their trips. The conclusions suggest that campaigns and strategies should be created aimed at men, citizens of older and younger generations (Generation Z), and citizens of member states belonging to the south, west, and north of Europe, which increase their sense of value and responsibility for your travel behaviours and habits. The offer of future tourist solutions must be diverse because the perception of tourism is very different among future tourists of varying ages, genders, and residences.

In future work, we intend to add new data to the study to balance the dataset and explore textual data, such as open-answers and textual reviews in tourism platforms.

References

Aman, S., Hassan, N. M., Khattak, M. N., Moustafa, M. A., Fakhri, M., & Ahmad, Z. (2021). Impact of tourist's environmental awareness on pro-environmental behavior with the mediating effect of tourist's environmental concern and moderating effect of tourist's environmental attachment. *Sustainability, 13*(23), 12998. https://doi.org/10.3390/su132312998

Bassi, F., & Dias, J. G. (2019). Circular economy implementation practices in SMEs. *Resources Conservation and Recycling, 146*, 523–533. https://doi.org/10.1016/j.resconrec.2019.03.019

Bassi, F., & Martín, J. A. M. (2024) Drivers of sustainable tourism in Europe: How to design efficient business strategies. *Quality and Quantity.* https://doi.org/10.1007/s11135-023-01819-0

Bilynets, I., & Cvelbar, L. (2022). Tourist pro-environmental behaviour: The role of environmental image of destination and daily behaviour. *Annals of Tourism Research Empirical Insights, 3*(2), 100070. https://doi.org/10.1016/j.annale.2022.100070

Bolderdijk, J., Steg, L., Geller, E., Lehman, P. K., & Postmes, T. (2013). Comparing the effectiveness of monetary versus moral motives in environmental campaigning. *Nature Climate Change, 3*, 413–416. https://doi.org/10.1038/nclimate1767

Brown, B. J., Hanson, M. E., Liverman, D. M., & Merideth, R. W. (1987). *Global sustainability: Toward definition. Environmental Management, 11*, 713–719. https://doi.org/10.1007/BF01867238

Buckingham-Hatfield, S. (1999). Gendering Agenda 21: Women's involvement in setting the environmental agenda. *Journal of Environmental Policy & Planning, 1*(2), 121–132. https://doi.org/10.1080/714038529

Butnaru, G. I., Niţă, V., Melinte, C., Anichiti, A., & Brînză, G. (2022). The Nexus between sustainable behaviour of tourists from generation Z and the factors that influence the protection of environmental quality. *Sustainability, 14*(19), 12103. https://doi.org/10.3390/su141912103

Cavagnaro, E., & Staffieri, S. (2015). A study of students' travellers values and needs in order to establish futures patterns and insights. *Journal of Tourism Futures, 1*(2), 94–107. https://doi.org/10.1108/JTF-12-2014-0013

Cavagnaro, E., Staffieri, S., & Postma, A. (2018). Understanding millennials' tourism experience: Values and meaning to travel as a key for identifying target clusters for youth (sustainable) tourism. *Journal of Tourism Futures, 4*(1), 31–42. https://doi.org/10.1108/JTF-12-2017-0058

Collins, D., & Tisdell, C. (2002). Gender and differences in travel life cycles. *Journal of Travel Research, 41*(2), 133–143. https://doi.org/10.1177/004728702237413

Dietz, T., Kalof, L., & Stern, P. C. (2002). Gender, values, and environmentalism. *Social Science Quarterly, 83*(1), 353–364. https://doi.org/10.1111/1540-6237.00088

Donohoe, H. M., & Needham, R. D. (2006). Ecotourism: The evolving contemporary definition. *Journal of Ecotourism, 5*(3), 192–210. https://doi.org/10.2167/joe152.0

Echavarren, J. M. (2016). From objective environmental problems to subjective environmental concern: A multilevel analysis using 30 indicators of environmental quality. *Society & Natural Resources, 30*(2), 145–159. https://doi.org/10.1080/08941920.2016.1185555

European Travel Commission. (2022). Sustainable travel in an era of disruption: Impact of Covid-19 on sustainable tourism attitudes. Brussels. https://etc-corporate.org/uploads/2022/01/

Sustainable-Travel-in-an-Era-of-Disruption-Impact-of-COVID-on-Sustainable-Tourism-Attitudes.pdf

Eurostat. (2024). Tourism industries – employment. https://ec.europa.eu/eurostat/statistics-explained/index.php?title=Tourism_industries_-_employment

Fermani, A., Crespi, I., & Stara, F. (2016). Sustainable hospitality and tourism at different ages: Women's and men's attitudes in Italy. *Research in Hospitality Management, 6*(1), 83–92. https://doi.org/10.2989/RHM.2016.6.1.11.1299

García-García, A. R., Jacobo-Hernández, C. A., Ochoa-Jiménez, S., & Valdez-del, R. S. (2023). Sustainable development and tourism: A review of the literature in WoS from 2001 to 2020. *Sustainability, 15*(24), 16805. https://doi.org/10.3390/su152416805

GESIS. (2022). Flash Eurobarometer 499 (Attitudes of Europeans towards Tourism, Autumn 2021). Available online: https://search.gesis.org/research_data/ZA7807 (accessed on 28 June 2024)

Gomes, S., & Lopes, J. M. (2023). Insights for pro-sustainable tourist behavior: The role of sustainable destination information and pro-sustainable tourist habits. *Sustainability, 15*(11), 8856. https://doi.org/10.3390/su15118856

Haddouche, H., & Salomone, C. (2018). Generation Z and the tourist experience: Tourist stories and use of social networks. *Journal of Tourism Futures, 4*(1), 69–79. https://doi.org/10.1108/JTF-12-2017-0059

Huang, R.-Y., Chang, W.-J., Chung, Y.-C., Yin, Y.-S., & Yen, J. C. (2019). A literature review of sustainable tourism (1990–2016). *International Journal of Tourism & Hospitality Reviews, 6*, 20–49.

Juvan, E., & Dolnicar, S. (2016). Measuring environmentally sustainable tourist behaviour. *Annals of Tourism Research, 59*, 30–44. https://doi.org/10.1016/j.annals.2016.03.006

Kim, S., & Filimonau, V. (2017). On linguistic relativity and pro environmental attitudes in tourism. *Tourism Management, 63*, 158–169. https://doi.org/10.1016/j.tourman.2017.06.012

Kinnaird, V., & Hall, D. (1994). Conclusion: The way forward. In V. Kinnaird & D. Hall (Eds.), *Tourism: A gender analysis* (pp. 210–216). Wiley.

Koksal, M. (2019). Differences among baby boomers, generation X, millennials, and generation Z wine consumers in Lebanon: Some perspective. *International Journal of Wine Business Research, 31*(3), 456–472. https://doi.org/10.1108/IJWBR-09-2018-0047

Kuo, N. W., & Dai, Y. Y. (2012). Applying the theory of planned behavior to predict low-carbon tourism behavior: A modified model from Taiwan. *International Journal of Technology and Human Interaction, 8*(4), 45–62. https://doi.org/10.4018/jthi.2012100103

Lee, T. H., & Jan, F.-H. (2019). The low-carbon tourism experience: A multidimensional scale development. *Journal of Hospitality & Tourism Research, 43*(6), 890–918. https://doi.org/10.1177/1096348019849675

Li, J., Coca-Stefaniak, A., Nguyen, H., & Morrison A. (2023). Sustainable tourist behavior: A systematic literature review and research agenda. *Sustainable Development*. https://doi.org/10.1002/sd.2859

Li, Q., & Wu, M. (2020). Tourists' pro-environmental behaviour in travel destinations: Benchmarking the power of social interaction and individual attitude. *Journal of Sustainable Tourism, 28*(9), 1371–1389. https://doi.org/10.1080/09669582.2020.1737091

Lindenberg, S., & Steg, L. (2007). Normative, gain and hedonic goal frames guiding environmental behavior. *Journal of Social Issues, 63*(1), 117–137. https://doi.org/10.1111/j.1540-4560.2007.00499.x

Liu, D., Ji, J., & Wu, M. (2023). Tourism carbon emissions: A systematic review of research based on bibliometric methods. *Journal of Quality Assurance in Hospitality & Tourism*, 1–21. https://doi.org/10.1080/1528008X.2023.2266861

Loureiro, S. M. C., Guerreiro, J., & Han, H. (2022). Past, present, and future of pro-environmental behavior in tourism and hospitality: A text-mining approach. *Journal of Sustainable Tourism, 30*(1), 258–278. https://doi.org/10.1080/09669582.2021.1875477

MacInnes, S., Grün, B., & Dolnicar, S. (2022). Habit drives sustainable tourist behaviour. *Annals of Tourism Research, 92*, 103329. https://doi.org/10.1016/j.annals.2021.103329

Matei, N. A., García-León, D., Dosio, A., Batista e Silva, F., Ribeiro Barranco, R., & Císcar Martínez, J. C. (2023). *Regional impact of climate change on European tourism demand.* Publications Office of the European Union, Luxembourg. https://doi.org/10.2760/899611.JRC131508.

Meadows, D. H., Meadows, D. L., Randers, J. & Behrens, W. W. (2018). *Limits to growth: He limits to growth: A report for the Club of Rome's Project on the Predicament of Mankind.* Universe Books. ISBN 9780444641304.

Meera, S. (2014). Challenges of women entrepreneurs: With specific reference to tourism industry. *International Journal (Toronto, Ont.), 1*(1) www.tmimtjournal.org

Milfont, T. L., Milojev, P., Greaves, L. M., & Sibley, C. G. (2015). Socio-structural and psychological foundations of climate change beliefs. *New Zealand Journal of Psychology., 44*(1), 17–30.

One Planet Network. (2024). Global plastic initiatives. https://www.oneplanetnetwork.org/programmes/sustainable-tourism/global-tourism-plastics-initiative/tourisms-plastic-pollution-problem

Parmar, A., Katariya, R., & Patel, V. (2019). A review on random forest: An Ensemble classifier. In J. Hemanth, X. Fernando, P. Lafata, & Z. Baig (Eds.), *International Conference on Intelligent Data Communication Technologies and Internet of Things (ICICI) 2018. ICICI 2018* (Lecture Notes on Data Engineering and Communications Technologies) (Vol. 26). Springer. https://doi.org/10.1007/978-3-030-03146-6_86

Parzonko, A. J., Balinska, A., & Sieczko, A. (2021). Pro-environmental behaviors of generation Z in the Context of the concept of homo socio-oeconomicus. *Energies, 14*, 1597. https://doi.org/10.3390/en14061597

Passini, S., & Morselli, D. (2006). La responsabilità tra autoritarismo e valori. *Psicologia e Sociedade, 3*, 501–516. https://hdl.handle.net/11585/41562

Pemani, P. O., & Massie, J. D. (2017). The effect of personal factors on consumer purchase decision (Case study: Everbest Shoes). *Jurnal EMBA: Jurnal Riset Ekonomi, Manajemen, Bisnis Dan Akuntansi, 5*(1). https://doi.org/10.35794/emba.5.1.2017.15451

Pinho, M., & Gomes, S. (2023), Generation Z as a critical question mark for sustainable tourism – an exploratory study in Portugal, *Journal of TourismFutures*. https://doi.org/10.1108/JTF-07-2022-0171

Poortinga, W., Spence, A., Whitmarsh, L., Capstick, S., & Pidgeon, N. F. (2011). Uncertain climate: An investigation into public scepticism about anthropogenic climate change. *Global Environmental Change, 21*(3, SI), 1015–1024. https://doi.org/10.1016/j.gloenvcha.2011.03.001

Poortinga, W., Whitmarsh, L., Steg, L., Böhm, G., & Fisher, S. (2019). Climate change perceptions and their individual-level determinants: A cross-European analysis. *Global Environmental Change, 5*, 25–35. https://doi.org/10.1016/j.gloenvcha.2019.01.007

Prayag, G., Aquino, R. S., Hall, C. M., Chen, N. & Fieger, P. (2022). Is Gen Z really that different? Environmental attitudes, travel behaviours and sustainability practices of international tourists to Canterbury, New Zealand. *Journal of Sustainable Tourism.* https://doi.org/10.1080/09669582.2022.2131795

Qiu, H., Wang, X., Morrison, A.M., Kelly, C. & Wei, W. (2022). From ownership to responsibility: Extending the theory of planned behavior to predict tourist environmentally responsible behavioural intentions. *Journal of Sustainable Tourism*, Online-First, 1–24. https://doi.org/10.1080/09669582.2022.2116643

Ribeiro, M., Seyfi, S., Elhoushy, S., Woosnam, K., & Patwardhan V. (2023). Determinants of generation Z pro-environmental travel behaviour: The moderating role of green consumption values, *Journal of Sustainable Tourism.* https://doi.org/10.1080/09669582.2023.2230389

Ribeiro, M. T., Singh, S. & Guestrin, C. (2016, August). Why should i trust you?" Explaining the predictions of any classifier. In *Proceedings of the 22nd ACM SIGKDD international conference on knowledge discovery and data mining* (pp. 1135–1144).https://doi.org/10.1145/2939672.2939778

Schönherr, S., & Pikkemaat, B. (2024). Young peoples' environmentally sustainable tourism attitude and responsible behavioral intention. *Tourism Review, 79*(4), 939–952. https://doi.org/10.1108/TR-01-2023-0022

Seyfi, S., Sharifi-Tehrani, M., Hall, C. M., & Vo-Thanh, T. (2023). Exploring the drivers of Gen Z tourists' buycott behaviour: A lifestyle politics perspective. *Journal of Sustainable Tourism*. https://doi.org/10.1080/09669582.2023.2166517

Shams, G., Rehman, M., Samad, S., & Oikarinen, E. (2020). Exploring customer's mobile banking experiences and expectations among generations X, Y and Z. *Journal of Financial Services Marketing, 25*(1–2), 1–13. https://doi.org/10.1057/s41264-020-00071-z

Sharma, N., Goel, P., Nunkoo, R., Sharma, A., & Rana, N. P. (2023). Food waste avoidance behavior: How different are generation Z travelers? *Journal of Sustainable Tourism*. https://doi.org/10.1080/09669582.2023.2187741

Sharma, R., & Gupta, A. (2020). Pro-environmental behaviour among tourists visiting national parks: Application of value-belief-norm theory in an emerging economy context. *Asia Pacific Journal of Tourism Research, 25*(8), 829–840. https://doi.org/10.1080/10941665.2020.1774784

Shi, J., Visschers, V. H. M., Siegrist, M., & Arvai, J. (2016). Knowledge as a driver of public perceptions about climate change reassessed. *Nature Climate Change, 6*, 759. https://doi.org/10.1038/nclimate2997

Skanavis, C., & Sakellari, M. (2008). Gender and sustainable tourism: Women's participation in the environmental decision-making process. *European Journal of Tourism Research, 1*(2), 78–93. https://doi.org/10.54055/ejtr.v1i2.13

Song, Y., Qin, Z., & Qin, Z. (2020). Green marketing to gen Z consumers in China: Examining the Mediating factors of an eco-label–informed purchase. *Sage Open, 10*(4). https://doi.org/10.1177/2158244020963573

Stevens, C. (2010). Are women the key to sustainable development? *Sustainable Development Insights, 3*, 1–8. https://hdl.handle.net/2144/22685

Strauss, W., & Howe, N. (2009). *The fourth turning: What the cycles of history tell us about America's next rendezvous with destiny*. Broadway Books.

Swain, M. B. (1995). Gender in tourism. *Annals of Tourism Research, 22*(2), 247–266.

Tanrisever, C., Pamukçu, H., & Baydeniz, E. (2024). Climate change in tourism: Understanding the impacts and opportunities for sustainability. In C. Tanrisever, H. Pamukçu, & A. Sharma (Eds.), *Future tourism trends volume 1 (Building the future of tourism)* (pp. 33–45). Emerald Publishing Limited. https://doi.org/10.1108/978-1-83753-244-520241003

UNWTO. (2019). World Tourism Organization and International Transport Forum (2019), *Transport-related CO_2 Emissions of the Tourism Sector – Modelling Results*, UNWTO, Madrid. https://doi.org/10.18111/9789284416660.

UNWTO. (2020). *International tourism highlights 2020 edition*. https://www.e-unwto.org/doi/epdf/10.18111/9789284422456.

UNWTO. (2024). Transforming tourism for climate actions. https://www.unwto.org/sustainable-development/climate-action

Wall, G. (1990). Human ecology, tourism and sustainable development. *Annals of Tourism Research, 17*, 633–635. https://doi.org/10.1016/0160-7383(90)90040-X

WCED. World Commission on Environment and Development. (1987). *Our common future*. World Commission on Environment and Development: New York, NY, USA.

Wilkinson, T., & Coles, T. (2024). Do tourists want sustainability transitions? Visitor attitudes to destination trajectories during COVID-19. *Tourism Geographies, 26*(2), 274–291. https://doi.org/10.1080/14616688.2023.2249423

World Economic Forum. (2023). Travel and tourism. https://www.weforum.org/agenda/2023/04/survey-of-travelers-finds-76-want-more-sustainable-options/

Wu, J., Font, X., & Liu, J. (2021). Tourists' pro-environmental behaviors: Moral obligation or disengagement? *Journal of Travel Research, 60*(4), 735–748. https://doi.org/10.1177/0047287520910787

Yeoman, I., & McMahon-Beattie, U. (Eds.). (2019). *The future past of tourism: Historical perspectives and future evolutions*. Channel View Publications.
Zelezny, L. C., Chua, P.-P., & Aldrich, C. (2000). Elaborating on gender differences in environmentalism. *Journal of Social Issues, 56*(3), 443–457. https://doi.org/10.1111/0022-4537.00177
Zolfani, S., Sedaghat, M., Maknoon, R., & Zavadskas, E. K. (2015). Sustainable tourism: A comprehensive literature review on frameworks and applications. *Economic Research-Ekonomska Istraživanja, 28*(1), 1–30. https://doi.org/10.1080/1331677X.2014.995895

Educating for the Circular Economy: The Key Competencies of Future Tourism Professionals

Sanna-Mari Renfors

Abstract Circular economy is increasingly regarded as a pathway to sustainable tourism development and a tool to help the industry to tackle climate change. This means that also higher tourism education should renew its curricula and provide students with a set of relevant competencies. Therefore, this article introduces education for the circular economy (ECE) for the first time in higher tourism education and aims at identifying the key circular economy competencies in tourism. In this qualitative study, the data were mainly collected by semi-structured interviews, which were analysed by content analysis to develop a competence profile. The findings indicate that the competence profile for tourism jobs of tomorrow consists of circular management, systems thinking, resource management, circular collaboration and supply chain management, circular innovation, and circular communication as the key circular economy competencies. Based on this profile, tourism teachers can design industry-oriented courses and make informed decisions about their contents and learning outcomes. Hopefully, the article will also inspire a dialogue on education for the circular economy in tourism and support higher education institutions and their teachers in contributing to its circular transition.

1 Introduction

The tourism industry is a significant source of carbon emissions and an important contributor to climate change (Lee & Brahmasrene, 2013). This means that the industry should urgently be part of the solution and take action to reduce these emissions (Mishra et al., 2022). To accomplish this, circular economy is regarded as a possible tool. Its practices support reducing carbon emissions and achieving carbon neutrality by improving resource-efficiency (Einarsson & Sorin, 2020; Vargas-Sánchez, 2021; Yang et al., 2023). Indeed, circular economy is seen as a means to

S.-M. Renfors (✉)
Lapland University of Applied Sciences, Rovaniemi, Finland
e-mail: sanna-mari.renfors@lapinamk.fi

J. Marques et al. (eds.), *Tourism and Climate Change*, Tourism, Hospitality & Event Management, https://doi.org/10.1007/978-3-031-99128-8_5

apply sustainable development in practice and an integrative framework that can be used to boost tourism stakeholders' progress in their sustainability journey (Manniche et al., 2021; Nocca et al., 2023; Padilla-Rivera et al., 2020). It accelerates the industry's sustainability-transition, helps to attain Sustainable Development Goals (SDGs), and use resources more efficiently (Davies & Egas, 2022; Einarsson & Sorin, 2020).

Circular economy has recently begun to attract more interest not only from policy makers, but from higher education institutions. It is deemed necessary that higher education institutions support tackling climate change with their curricula and increase their students' competencies because everyone applies circular practices at work and will be a circular economy expert in the future (de la Torre et al., 2021; Tiippana-Usvasalo et al., 2023). This requires focusing on the needed competencies as the transition towards a circular economy is subject to their availability (Sumter et al., 2021).

So far, circular economy in tourism education has remained an unexplored topic even though jobs in tourism are greening. A closer look at the literature shows that even education for sustainability in tourism and integration of sustainability into teaching have not been systematically investigated (Boyle et al., 2015; Farsari, 2022; Moscardo, 2015). Because different aspects of the circular economy are emphasised by different disciplines and circular transformations are industry-specific (Kirchherr & Piscicelli, 2019; Renfors & Wendt, 2024), understanding the contextual circular economy competencies in tourism is required. This makes it meaningful to explore the relevant circular economy competencies in the tourism industry further.

The article introduces education for the circular economy (ECE) for the first time in the context of higher tourism education. The aim of the article is to identify the key circular economy competencies and develop a competence profile for tourism jobs of tomorrow. Hopefully, these offer insight and support tourism teachers in their curriculum and course design and development. As the dialogue on education for the circular economy is just starting in higher tourism education, the article inspires teachers to contribute to this dialogue and address the current issues in their teaching.

The article starts with an introduction to circular economy in tourism and education for the circular economy. The methodology section is followed by the findings of the study. Lastly, the discussion and conclusions section is presented with implications drawn for tourism teachers.

2 Circular Economy in Tourism

During the last decade, many researchers have attempted to define the concept of circular economy, but it still lacks consensus. Many researchers propose that it is an umbrella concept and represents a paradigm shift in how human society interacts with nature (e.g., Homrich et al., 2018; Moraga et al., 2019; Prieto-Sandoval et al.,

2018). Most researchers also agree that the word more closely associated with circular economy is resources (Homrich et al., 2018; Preston, 2012; Prieto-Sandoval et al., 2018; Urbinati et al., 2017; Velenturf & Purnell, 2021). Thus, circular economy is an economic model to save and make better use of resources, use them responsibly and cyclically (Geissdoerfer et al., 2017; Moraga et al., 2019; Murray et al., 2017; Velenturf & Purnell, 2021). Indeed, it is about fundamental changes in resource use: they are reused and kept in a loop of production and usage (Preston, 2012; Urbinati et al., 2017).

Although there are various definitions of circular economy, defining it in the tourism context is still developing as it is an emerging field of study (Kaszás et al., 2022; Nocca et al., 2023). According to Girard and Nocca (2017) and Nocca et al. (2023), it is a model to operationalise sustainable principles in tourism and to manage resources sustainably. Circular economy provides various possibilities to innovate processes, products, and services as well as business models, but as highlighted by Kaszás et al. (2022), its implementation in tourism is still focused on a limited number of solutions.

Circular transition requires a systemic approach (Gusmerotti et al., 2024). It concerns all societal actors and demands collaboration and changes at multiple levels. As said, its implementation is often categorised into three systemic levels: micro (consumers, single company activities, products), meso (supply chains, company networks, and clusters), and macro level (city, region, nation, government, society) (Ghisellini et al., 2016; Kirchherr et al., 2017; Merli et al., 2018). The micro level relates to the particular needs of individual companies and the meso level to the specific needs of clustered companies to increase their circularity (Merli et al., 2018). The macro level addresses circularity at the societal level and in the entire economy by public decision-makers (Kirchherr et al., 2017; Merli et al., 2018). In addition, circular economy is about understanding the connection between its ecological, economic, and social objectives and between the needs of the existing and future generations (Korhonen et al., 2018; Nocca et al., 2023).

Most tourism research on the circular economy is implemented at the micro level, i.e., in tourism companies. They often deal with large hotel chains and their circular strategies and practices (e.g., Khodaiji & Christopoulou, 2020; Rodríguez-Antón & Alonso-Almeida, 2019; Sorin & Sivarajah, 2021). In tourism companies, environmental sustainability, especially the practices related to the use of energy, water, and waste, has been explored (e.g., Bux & Amicarelli, 2023; Camilleri, 2021; Ioannidis et al., 2021). As stated by Kaszás et al. (2022), the key points in circular tourism have been energy efficiency, waste management, and food waste. Researchers have also found appealing topics like environmental certifications and their adoption in tourism SMEs (e.g., Khan et al., 2021) and circular economy and digital technologies (e.g., del Vecchio et al., 2022; Saura et al., 2024). Since tourists play a major role in the circular transition, tourist researchers have also been examining tourists' circular consumption behaviour, their awareness, and attitudes about circular economy (Bosone & Nocca, 2022; Julião et al., 2020; Nocca et al., 2023; Patti, 2017; Rodríguez et al., 2020; Sørensen & Bærenholdt, 2020). Indeed, tourist' awareness and changes in their behaviour are one of the key prerequisites of circular transition (Gusmerotti et al., 2024; Nocca et al., 2023).

Although tourism is strongly linked with several industries, the meso level has not been widely studied in tourism. Some studies concern rural tourism and agriculture (e.g., Immacolata, 2018; Joshi et al., 2020) but more examples are needed to understand how different clustered companies can jointly increase their impact on circularity. At the macro level, studies relate to tourism destinations moving towards a circular economy (Gusmerotti et al., 2024; He & Mai, 2021; Pongsakornrungsilp & Pongsakornrungsilp, 2021; Rudan et al., 2021; Schumann, 2020). These mostly focus on the changes needed in the administrative and regulatory frameworks, assessing circularity in tourism destinations, establishing local partnerships, networks, and a collaborative culture to implement circular economy at the destination level.

3 Education for the Circular Economy

There is not yet a lot of literature on education for the circular economy (ECE) as it is an emerging field of study (de la Torre et al., 2021; Giannoccaro et al., 2021). The concept of education for the circular economy was introduced for the first time by Kirchherr and Piscicelli (2019), who illustrated how to design a course in circular economy for bachelor students. Education for the circular economy defines the teaching approaches to encourage students and involve them in the transition towards circular economy (Kirchherr & Piscicelli, 2019). It promotes systemic and circular thinking, cultivates environmental and social responsibility as well as integrates different bodies of knowledge, experiences, and viewpoints (Marouli, 2016). The focus is on solving real problems related to implementing circular economy. Indeed, circular economy is an interdisciplinary framework, which identifies the interrelatedness of problems and the interconnectedness of solutions (Hall & Velez-Colby, 2018). This means that education for the circular economy should promote understanding how individual actions, social problems, and global concerns are linked (Marouli, 2016).

Circular economy curricula and courses define the learning outcome i.e., the competencies the students should have when the course ends (Giannoccaro et al., 2021). In this case, competence is defined as the capacity to perform the key occupational tasks satisfactorily in a profession. So far, some researchers (Burger et al., 2019; Giannoccaro et al., 2021; Janssens et al., 2021; Sumter et al., 2021) have identified and defined a set of competencies for circular economy in different contexts. They have concluded that circular economy demands more competencies in resource management and systems thinking compared to the rest of the economy (Janssens et al., 2021). In addition, circular business models, circular economy communication and collaboration, as well as production and product design, green marketing and market knowledge, circular impact assessment, circular user engagement, and circular materials and manufacturing are defined as other key circular economy competencies by previous studies. However, these studies have mainly been conducted in the context of productive industries, which shows in the identified competencies. For this reason, it is important to define and increase understanding of these competencies also in the tourism industry context.

4 Methodology

The data of this qualitative study consisted of mixed data sets. To collect primary data, 19 in-depth, semi-structured one-to-one interviews (Table 1) were conducted during spring and autumn 2023 in Finland. This country was chosen since it has a reputation as a leader in environmental policy and sustainable development (OECD, 2021). It has been ranked number one in an international comparison assessing how countries advance in implementing the 2030 Agenda for Sustainable Development among 193 UN member states (Sachs et al., 2023). Since sustainability is also a priority in the country's national tourism strategy, its national tourism board, Visit Finland, has developed the Sustainable Travel Finland (STF) Programme as a toolkit to tourism enterprises to support them in adopting sustainable practices.

Purposive sampling was used to select the interviewees. These interviewees worked in the restaurant, accommodation, programme service, visitor attraction, and destination management sectors. In addition, representatives of business

Table 1 Interview participants

Person	Job title	Sector	Gender
1	Entrepreneur	Restaurant and programme service	F
2	Entrepreneur	Restaurant	M
3	Manager	Visitor attraction	F
4	Entrepreneur	Restaurant, accommodation, and programme service	M
5	Entrepreneur	Café and restaurant	F
6	Manager	Destination management organisation	M
7	Manager	Accommodation	F
8	Entrepreneur	Cafe	F
9	Manager	Restaurant and accommodation	M
10	Entrepreneur	Café	F
11	Entrepreneur	Restaurant	F
12	Lecturer	Higher education institution	F
13	Project Manager	Business development organisation	F
14	Project Manager	Business development organisation	F
15	Lecturer	Higher education institution	F
16	Entrepreneur	Restaurant	F
17	Entrepreneur	Restaurant	F
18	Manager	Restaurant	M
19	Project Manager	Business development organisation	F

development organisations and higher education institutions related to tourism were interviewed to get a more holistic perspective. For example, the following open-ended questions were asked from the interviewees representing the tourism industry: How have you integrated circular practices in your daily operations? What actions are taken in your organisation to increase resource-efficiency? What would you consider circular economy competencies? Because circular economy as a concept is not yet very familiar in the tourism industry, questions and the used terms were explained and adapted according to the respondent's understanding of circular economy. From the interviewees of business development organisations and higher education institutions, the following questions were asked: What are the key characteristics of the circular economy in the tourism industry? How should the tourism industry apply the principles and practices of circular economy in their daily operations? What are the key circular economy competencies in the tourism industry? The interviews were also shaped by emerging points raised by the interviewees. The interviews took place online for practical reasons and lasted between 45 minutes and 1 hour. The interviewees are referred to by the letter I and their number in the findings section.

In addition, secondary sources were regarded as a necessary option for data collection to achieve the aim of this study, explore the key research question, and augment the primary data. This is because circular economy is an emerging and not yet a very well-known topic among tourism stakeholders. Using secondary data sources ensured a wider perspective on the novel topic and a larger geographical scope of the study. The secondary data included online reports, handbooks, and policy papers (n = 11) produced by the projects funded by the European Union and implemented in various European countries. The criteria for selecting the secondary data included: 1) project's relevance to accelerating and managing circular transition in the tourism industry and 2) the quality of the data source's content in relation to the research question and presenting best circular practices, recommendations, and frameworks for tourism stakeholders. The relevant secondary data were identified by using EU project databases (keep.eu and cordis.europa.eu), Google, and projects' own websites.

The data were analysed by qualitative content analysis to develop a competence profile of the key circular economy competencies in the tourism industry. Content analysis is a research technique for making replicable and valid inferences from texts to the contexts of their use (Krippendorf, 2013). The selected texts were transformed systematically into competence categories by using qualitative content analysis. The analysis was inductive, and it used the process of abstraction to reduce and group data (cf. Kyngäs, 2020). In this process, the secondary data were analytically filtered, i.e., the data were filtered according to their relevance to the research question (Gibson & Brown, 2011). The analysis was also assorted (cf. Heaton, 2008), in other words the analysis of secondary data was combined with primary data analysis.

First, these texts were studied several times in their entirety to acquire a sense of the whole and to identify the essential features of the content for further analysis. Then, the texts were divided into meaning units. Each identified meaning unit was labelled with a preliminary code while reading the text. The codes were generated from the phrases and sentences related to necessary competencies (i.e., the codes were derived from the data). In the following stage, categories (i.e., groups of content) that share a commonality (Krippendorf, 2013) were generated by grouping similar codes under higher-order headings that described the necessary circular economy competencies. At the same time, the preliminary codes were revised and refined, if necessary. Subcategories related to each other through their content were sorted into broader categories. The aim of grouping the data was to reduce the number of categories by incorporating those which were similar or dissimilar into broader higher-order categories. As a result, mutually exclusive categories of the key circular economy competencies were generated.

5 Findings

First, it is certainly important to have an overall understanding of the characteristics of circular economy in tourism: the concept of circular economy, its purpose, and the role of tourism in circular transition. This involves competencies in understanding and assessing the main characteristics of circular transformation in different tourism sectors and market contexts. The so-called asset-heavy businesses (cf. Einarsson & Sorin, 2020), e.g., hotels and restaurants, own physical assets and impact resource circularity. However, the asset-light businesses (cf. Einarsson & Sorin, 2020), e.g., tour operators, deliver non-tangible services. It is important to note that they can play a vital role in shaping circular business models, processes, and supplier selection even if they do not own physical assets. In addition, understanding the requirements of circular transformation in tourism, including its opportunities, barriers, and challenges, is needed.

A major competence that emerged from the results is *circular management*. This includes competencies in managing circular performance by establishing a circular strategy, clarifying the objectives, and setting up the key performance indicators to measure the progress. It is impossible to save resources without knowing how resources are used and waste generated, which makes it crucial to constantly monitor and evaluate the circular transition and the circular development impacts as well as introduce changes based on the collected data. Thus, circular management also involves impact management because the company should manage the impacts of its decisions and activities across the system. It should be noted that competencies in measuring environmental, economic, and social impacts are all relevant. In addition, an attitude check of the managers is highly recommended as they play a significant role: Do they perceive waste as avoidable and share circular values? The interviewees commented different issues in circular management as follows:

"*Nowadays, one of the most important thing seems to be management: to have a management plan, manage and monitor it holistically. I bet these kinds of occupational profiles will emerge in tourism companies.*" I19

"*First, it is important to identify the starting level of the business. The company moves from one level to next, as this is a continuous process of development that requires constant effort. The company must choose their direction and circular goals at the beginning; it is not necessary to aim for everything at once.*" I14

"*Monitoring is the key issue for us, for example monitoring the amount of waste we generate. We would also like to learn to calculate our CO2 emissions as we are not yet skillful enough to do so.*" I5

Systems thinking is deemed as another essential competence in the tourism industry. It refers to understanding a company's position and role as a part of the system. It is necessary to understand the requirements of a system-level change, the context of a larger whole with all the levels and its connected parts. This means that companies should apply systems thinking, take a holistic approach to decision-making and activities within the wider system. In practice, individual companies operate in a specific system formed of stakeholders and complex value chains. They are also affected by external factors such as legislation, policies, infrastructure, tax systems, changes in technology, and customer behaviour since transitioning to circular economy takes place in a societal context. Infrastructure provided by public authorities allows circular practices, and policy making has an essential role in establishing the business environment. In addition, systems thinking competence is not only about implementing circular economy at different levels but also focuses on the economy's environmental, economic, and social dimensions as a whole and their interdependence. The interviewees illustrated how they have applied systems thinking in their company:

"*We have a compost where we put our food waste. The compost is taken back to the producers, we purchase our ingredients from, to fertilise their field.*" I2

"*In our café, we prevent food waste by distributing the surplus products to the local parish, Red Cross, and youth employment services, which use them or further distribute them to those in need. We also organise art exhibitions with the refugees living in the area where they can present their art to the community and feel belongingness.*" I10

The findings emphasise *resource management* as one of the key circular economy competencies in the tourism industry. Resource management refers to the competencies in analysing, optimising, and rethinking resource use in the whole system, value chain, and company levels. In particular, competencies in applying circular economy principles of reducing, reusing, and recycling resources are deemed critical. Thus, waste prevention and reduction is a major competence. There is no concept of waste, and the use of non-renewable resources should be avoided. For example, it is a necessity to minimise the use of plastic and prevent food waste, keep products in use, and re-use items, e.g., furniture, cutlery, equipment, and building materials. Secondary raw materials/by-products can be converted into new, locally produced materials or products of better quality. In addition, the findings highlight the importance of ensuring energy-efficient processes and products, efficient energy use, and the use of energy from renewable sources. For example, the interviewees presented the following examples of managing their resources:

"*We cut down the trees in poor condition from the yard, after which we took them to the sawmill and the sawyer made us our own fences from these existing trees.*" I4

"*Our company is in an old building, 90 years old. The only thing that is new is the oven. Everything is kind of recycled. From flea markets, all service utensils and coffee cups. The service counter is made of boards from an old barn.*" I10

"*When we hire new employees, first we teach them how to utilise surplus food. If there are some leftovers from the lunch buffet, how to use them later. You must think beforehand what to do from the leftovers next day.*" I16

"*Renewable energy is the biggest investment we have. We have geothermal energy and we have solar panels on the roof.*" I17

Another major competence, closely connected with resource management, is *circular collaboration and supply chain management.* Since tourism is an ecosystem of highly diverse and complex supply chains, this also highlights the central role of circular collaboration and related competencies in the tourism industry. Circular economy requires a network approach, and circular transition cannot be achieved without collaboration and shared objectives as collaboration enables and ensures the efficient use of resources. Thus, it is necessary to understand the connection between tourism and other industries and how to connect with local actors. Collaboration includes establishing multi-sectoral networks, building partnerships with local suppliers, communities, and authorities to achieve a better impact together. Circular solutions are developed in collaboration with local entrepreneurs and members of the supply chain, e.g., in green procurement and circulating materials by building personal relationships. It is about rethinking and redesigning products and services in supply chain collaboration considering the entire product and service life cycle. It is also about serving customers better by coordinating the needed resources at the destination level. The interviewees showcased many great examples of their collaboration activities to ensure better use of resources:

"*We, as the central museum, created a list of our surplus materials for local museums around the region. Whenever there are some materials, we don't need, we give them to others to use. If they are such that they would be suitable as a decorative element, they are sold in a museum shop.*" I3

"*Craftsmen bring added value to tourism companies by converting waste into, for example, decorative items and support textile recycling. Hotel curtains or sheets can be made into decorative items, work clothes, business gifts.*" I11

"*Collaboration with the distributors is essential. They transport ingredients in boxes that circulate and don't cause any packaging waste.*" I16

As said, circular economy demands innovation built on circular principles to function, which makes *circular innovation* a key competence. Circular economy promotes innovations that increase resource-efficiency and design out waste using new ideas, methods, and improved solutions. This requires creativity. Mainly, innovation is related to business model and product innovation to create business value and to advance circular transition. In tourism, innovative business models can be related to the use of sharing platforms, new consumption models, and resource-efficiency. Sharing platforms increase product and service utilisation, e.g., bike and car sharing or selling surplus food as technological innovations are needed to ensure circular patterns of production and consumption. Many tourism companies take also advantage of the product as a service business model. In addition, buildings and spaces are used for multiple purposes to increase their occupancy rate. Therefore, a relevant competence is to understand how companies can design their business models in coherence with circular economy principles and in collaboration to contribute to their profitability and understanding their environmental impacts. The interviewees explained their outlook as follows:

> "*With creativity and imagination, you can maximise the use of resources. For example, we create delicious dishes from by-products of other companies in our restaurant.*" I2
>
> "*We own quite a few textiles, we rent them all.*" I7
>
> "*One possibility is the shared use of buildings as a part of the circular economy. How, for example, could the hotel facilities be used most optimally? For which purposes?*" I14
>
> "*Different online platforms, such as ResQ, are important to share resources. By selling our leftovers with this app to customers, our business is more profitable. In addition, as our newest innovation, we have started to sell our coffee grounds to a company that produces soil out of them.*" I6

The last competence is *circular communication*. Competencies are required to report the progress towards circularity in a clear, timely, and honest manner. Companies should communicate their process and concrete actions as well as the benefits of these actions. It is also vital to increase staff engagement and awareness at all levels of the company to facilitate the circular transition as staff should embrace circular values in their daily work. In addition, customers are co-responsible for this transition, which makes them a key target group for communication activities. A successful company increases the customer's awareness, e.g., by designing awareness-raising campaigns and telling their stories to customers. They educate customers and give them new perspectives on resource use. They can also ask for their feedback to improve their circularity. The interviewees commented their communication activities as follows:

> "*Tourism is an operating environment familiar to the customer, in which she/he is a participant. Companies create a movement towards the circular economy by raising customers' awareness.*" I8
>
> "*Over the next three years we are focusing on energy and waste management and communication. We are just designing material for customers to make them aware of our environmental actions. We also train our new employees how to perform sustainably in their daily work. We even ask this question already in the job interview.*" I7

The competence profile of the key circular economy competencies in the tourism industry is summarised in Table 2.

6 Discussion and Conclusions

The article clarifies the key circular economy competencies for future professionals working in the tourism industry. The findings indicate that these key competencies include circular management, systems thinking, resource management, circular collaboration and supply chain management, circular innovation, and circular communication. It is imperative to promote the efficient use of resources by applying circular economy principles, to understand the context of a larger whole and all its

Table 2 The competence profile of the key circular economy competencies in the tourism industry

Competence	Learning outcomes—At the end of the course, the student is able to
Characteristics of circular economy in tourism	• understand the concept of circular economy, its purpose, and the role of the tourism industry in circular transition. • assess the main characteristics of circular transformation in different tourism sectors and market contexts. • understand the requirements of circular transformation in tourism, including its opportunities, barriers, and challenges.
Circular management	• integrate circular values and culture in everyday company operations. • establish a circular strategy, clarify the objectives, and set up the key performance indicators to measure the progress. • monitor and evaluate the circular transition and the circular development impacts, make changes based on the collected data. • manage the impacts of company's decisions and activities across the system.
Systems thinking	• understand the context of a larger whole and all its connected parts. • analyse the company's position and role as a part of the system. • understand the role of local authorities in circular economy. • understand the demand side perspective and new market requirements in circular economy. • understand the linkage between resource use, economic and societal value.
Resource management	• understand the concept of resource-efficiency and analyse, optimise, and rethink resource use in the whole system, value chain, and company levels. • apply circular economy principles of reducing, reusing, and recycling resources in the tourism industry. • prevent and reduce waste management by identifying the waste types and volumes, setting targets, monitoring and improving the practices. • focus on preventing plastic and food waste, reducing energy and water use, and avoiding the use of non-renewable resources.

(continued)

Table 2 (continued)

Competence	Learning outcomes—At the end of the course, the student is able to
Circular collaboration and supply chain management	• understand the connection between tourism and other industries in circular supply chains. • establish multi-sectoral networks and build partnerships with local suppliers, communities, and authorities to achieve a better impact together. • develop circular solutions in collaboration with the members of the supply chain.
Circular innovation	• promote innovations that increase resource-efficiency and design out waste using new ideas, methods, and improved solutions. • rethink and redesign creative products and services and business models in collaboration considering the entire product and service life cycle. • understand the role of technological innovations (incl. sharing platforms) to ensure circular patterns of production and consumption.
Circular communication	• report the progress towards circularity in a clear, timely, and honest manner. • communicate the company's process and concrete actions towards circularity as well as the benefits of these actions. • increase staff engagement and awareness at all levels of the company to facilitate the circular transition. • increase the customer's awareness of circular economy, educate them, and ask for their feedback.

connected parts as well as to improve company's circular performance and manage the impacts of its decisions across the system. In addition, establishing networks and partnerships and developing circular solutions in collaboration with the members of the supply chain to achieve a better impact together is required. Creativity and innovation are needed to rethink and redesign products, services, and business models. Furthermore, it is vital to create an organisational culture embracing circularity and communicate the company's process and concrete actions towards circularity internally and externally. In sum, circular economy is a mindset, an attitude of how to operate, which should be integrated across all company's operations.

Indeed, the findings increase understanding on how circular economy and increasing relevant competencies can act as a possible tool to address the climate change issue in the tourism industry. The circular economy approach can reduce carbon emissions across the industry and its value chains by improving resource-efficiency. As the tourism industry is very closely linked to the food system, built environment, logistics and their resource flows, it can act as a driver and enabler in circular economy. The industry can prevent and eliminate waste, circulate and recycle materials, and shift consumption patterns to help reduce emissions. The members of the value chains can jointly increase their impact on circularity: the greater the impact, the more they cooperate.

In general, the findings shed more light on the key circular economy competencies. When comparing the findings to those of previous studies (Giannoccaro et al., 2021; Janssens et al., 2021; Sumter et al., 2021), it can be stated that resource management, systems thinking, circular collaboration, and communication are the common circular economy competencies to different industries. So far, researchers (e.g., Burger et al., 2019; Giannoccaro et al., 2021; Janssens et al., 2021; Sumter et al., 2021) examining the circular economy competencies have focused their attention on productive industries and highlighted competencies related to physical product and production design and development. This is a significant difference between the competence requirements defined previously in the education for circular economy literature and the necessary competencies in the tourism industry.

Furthermore, the previous studies usually approach circular economy from the supply-side perspective. Only Sumter et al. (2021) paid attention to circular user engagement as a key competence to share and care for products and engage users in the product design process to speed up their adoption. However, in the tourism industry, tourists play a key role in promoting circular transformation. Their role entails much more than participating in the planning process or sharing products and services. Therefore, competencies in applying circular economy principles at the demand side are of utmost importance to incorporate in curricula and course planning in higher tourism education.

The findings also stress that it is important to take a systemic approach and understand all its connected parts to truly include the multiple levels of the circular economy in curricula and course contents in higher tourism education. As the previous studies related to circular economy in tourism show, the focus is too often only on the micro level, i.e., case studies related to circular economy implementation in companies and its environmental dimensions. However, the context of a larger whole with the meso and macro levels of implementation to promote systems thinking competence should be integrated into the curricula and course contents. It is equally necessary to incorporate the social dimension of circular economy in the curricula and course contents, which is often lacking. Sharing resources in the community, educating staff and customers are just a few examples.

As Kirchherr and Piscicelli (2019) have pointed out, a dialogue on education for the circular economy (ECE) supports higher education institutions and their teachers in contributing to circular transition and increases the quality of teaching circular economy in their classrooms. Hopefully, this article inspires tourism teachers to contribute to teaching for a sustainable future through circular economy and integrate circular economy into their curricula and courses. Based on the competence profile, they can make informed decisions about the contents of the courses and their learning outcomes. In sum, they can design industry-oriented courses that concentrate on up-to-date and relevant competencies required when working in tourism companies.

Moreover, the information can be used when upskilling the current tourism workforce. This so-called green training boosts employee understanding of resource-efficiency and enables them to adopt and engage in circular economy behaviours at work. Therefore, tourism companies can use the results in staff training to ensure that their employees have the relevant circular economy competencies. They can compare their existing job profiles with the new circular economy competence profile, identify the competence gaps, and integrate the missing competencies in their existing job profiles. Tourism companies could also evaluate the necessity of a separate Circular Economy Manager job profile in their company by comparing the contents of the competence profile to their future needs.

Limitations are related to its context, i.e., one country and its current state of circular economy and sustainable tourism development. However, the mixed data sets enlarged the geographical scope of the study. To further advance education for the circular economy in tourism, studies are needed to connect the relevant teaching and learning approaches with the identified key competencies. For example, to move this agenda forward, further studies could explore what teaching and learning approaches can be applied to improve the key circular economy competencies in higher tourism education. Certainly, to be able to teach the future workforce, teachers need more training in circular economy as well as teaching and learning approaches that consider its specific characteristics.

References

Bosone, M., & Nocca, F. (2022). Human circular tourism as the tourism of tomorrow: The role of travellers in achieving a more sustainable and circular tourism. *Sustainability, 14*(19), 12218. https://doi.org/10.3390/su141912218

Boyle, A., Wilson, E., & Dimmock, K. (2015). Transformative education and sustainable tourism: The influence of a lecturer's worldview. *Journal of Teaching in Travel & Tourism, 15*(3), 252–263. https://doi.org/10.1080/15313220.2015.1059303

Burger, M., Stavropoulos, S., Ramkumar, S., Dufourmont, J., & van Oort, F. (2019). The heterogeneous skill-base of circular economy employment. *Research Policy, 48*(1), 248–261. https://doi.org/10.1016/j.respol.2018.08.015

Bux, C., & Amicarelli, V. (2023). Circular economy and sustainable strategies in the hospitality industry: Current trends and empirical implications. *Tourism and Hospitality Research, 23*(4), 624–636. https://doi.org/10.1177/14673584221119581

Camilleri, M. A. (2021). Sustainable production and consumption of food. mise-en-place circular economy policies and waste management practices in tourism cities. *Sustainability, 13*(17), 9986. https://doi.org/10.3390/su13179986

Davies, S. J., & Egas, J. L. (2022). Do social corporate responsibility initiatives help to promote circular economic activity and quality of work life for employees? *Worldwide Hospitality and Tourism Themes, 14*(3), 221–235. https://doi.org/10.1108/WHATT-02-2022-0018

de la Torre, R., Onggo, B. S., Corlu, C. G., Nogal, M., & Juan, A. A. (2021). The role of simulation and serious games in teaching concepts on circular economy and sustainable energy. *Energies, 14*(4), 1138. https://doi.org/10.3390/en14041138

del Vecchio, P., Malandugno, C., Passiante, G., & Sakka, G. (2022). Circular economy business model for smart tourism: The case of Ecobnb. *EuroMed Journal of Business, 17*(1), 88–104. https://doi.org/10.1108/EMJB-09-2020-0098

Einarsson, S., & Sorin, F. (2020). *Circular Economy in travel and tourism: A conceptual framework for a sustainable, resilient and future proof industry transition*. CE360 Alliance.

Farsari, I. (2022). Pedagogy for sustainable tourism: Reflections on the curriculum space of a master programme in Sweden. *Journal of Teaching in Travel & Tourism, 22*(1), 6–35. https://doi.org/10.1080/15313220.2021.1978127

Geissdoerfer, M., Savaget, P., Bocken, N. M., & Hultink, E. J. (2017). The circular economy–a new sustainability paradigm? *Journal of Cleaner Production, 143*, 757–768. https://doi.org/10.1016/j.jclepro.2016.12.048

Ghisellini, P., Cialani, C., & Ulgiati, S. (2016). A review on circular economy: The expected transition to a balanced interplay of environmental and economic systems. *Journal of Cleaner Production, 114*, 11–32. https://doi.org/10.1016/j.jclepro.2015.09.007

Giannoccaro, I., Ceccarelli, G., & Fraccascia, L. (2021). Features of the higher education for the circular economy: The case of Italy. *Sustainability, 13*(20), 11338. https://doi.org/10.3390/su132011338

Gibson, W., & Brown, A. (2011). Using documents in research. In W. Gibson & A. Brown (Eds.), *Working with qualitative data* (pp. 65–83). SAGE.

Girard, L. F., & Nocca, F. (2017). From linear to circular tourism. *Aestimum, 70*, 51–74. https://doi.org/10.13128/Aestimum-21081

Gusmerotti, N. M., Carlesi, S., Iannuzzi, T., & Testa, F. (2024). The role of tourism in boosting circular transition: A measurement system based on a participatory approach. *Journal of Sustainable Tourism, 32*(5), 961–985. https://doi.org/10.1080/09669582.2023.2190056

Hall, N., & Velez-Colby, F. (2018). AMFI's reality school: A circular economy agenda for fashion education. *Art, Design & Communication in Higher Education, 17*(1), 11–24. https://doi.org/10.1386/adch.17.1.11_1

He, J., & Mai, T. H. T. (2021). The circular economy: A study on the use of Airbnb for sustainable coastal development in the Vietnam Mekong Delta. *Sustainability, 13*(13), 7493. https://doi.org/10.3390/su13137493

Heaton, J. (2008). Secondary analysis of qualitative data. In P. Alasuutari, J. Brannen, & L. Bickman (Eds.), *The SAGE handbook of social research methods* (pp. 506–519). SAGE.

Homrich, A. S., Galvão, G., Abadia, L. G., & Carvalho, M. M. (2018). The circular economy umbrella: Trends and gaps on integrating pathways. *Journal of Cleaner Production, 175*, 525–543. https://doi.org/10.1016/j.jclepro.2017.11.064

Immacolata, V. (2018). Agriculture, rural tourism and circular paradigm. *Calitatea, 19*(S1), 556–562.

Ioannidis, A., Chalvatzis, K. J., Leonidou, L. C., & Feng, Z. (2021). Applying the reduce, reuse, and re-cycle principle in the hospitality sector: Its antecedents and performance implications. *Business Strategy and the Environment, 30*(7), 3394–3410. https://doi.org/10.1002/bse.2809

Janssens, L., Kuppens, T., & Van Schoubroeck, S. (2021). Competences of the professional of the future in the circular economy: Evidence from the case of Limburg. *Belgium. Journal of Cleaner Production, 281*, 125365. https://doi.org/10.1016/j.jclepro.2020.125365

Joshi, S., Sharma, M., & Singh, R. K. (2020). Performance evaluation of agro-tourism clusters using AHP–TOPSIS. *Journal of Operations and Strategic Planning, 3*(1), 7–30. https://doi.org/10.1177/2516600X20928646

Julião, J., Gaspar, M., & Alemão, C. (2020). Consumers' perceptions of circular economy in the hotel industry: Evidence from Portugal. *International Journal of Integrated Supply Management, 13*(2–3), 192–209. https://doi.org/10.1504/IJISM.2020.107849

Kaszás, N., Keller, K., & Birkner, Z. (2022). Understanding circularity in tourism. *Society and Economy, 44*(1), 65–82. https://doi.org/10.1556/204.2021.00025

Khan, O., Marrucci, L., Daddi, T., & Bellini, N. (2021). Adoption of Circular Economy and Environmental Certifications: Perceptions of Tourism SMEs. *Journal of Management and Sustainability, 11*(1), 218–218. https://doi.org/10.5539/jms.v11n1p218

Khodaiji, J. D., & Christopoulou, D. (2020). Sustainable development and the circular economy in Greece: Case examples from Costa Navarino and Grecotel. *Worldwide Hospitality and Tourism Themes, 12*(5), 609–621. https://doi.org/10.1108/WHATT-06-2020-0048

Kirchherr, J., & Piscicelli, L. (2019). Towards an education for the circular economy (ECE): Five teaching principles and a case study. *Resources, Conservation and Recycling, 150*, 104406. https://doi.org/10.1016/j.resconrec.2019.104406
Kirchherr, J., Reike, D., & Hekkert, M. (2017). Conceptualizing the circular economy: An analysis of 114 definitions. *Resources, Conservation and Recycling, 127*, 221–232. https://doi.org/10.1016/j.rescon-rec.2017.09.005
Korhonen, J., Honkasalo, A., & Seppälä, J. (2018). Circular economy: The concept and its limitations. *Ecological economics, 143*, 37–46. https://doi.org/10.1016/j.ecolecon.2017.06.041
Krippendorf, K. (2013). *Content analysis. An introduction to its methodology*, SAGE.
Kyngäs, H. (2020). Inductive content analysis. In H. Kyngäs, K. Mikkonen, & M. Kääriäinen (Eds.), *The application of content analysis in nursing science research* (pp. 13–21). Springer. https://doi.org/10.1007/978-3-030-30199-6_2
Lee, J. W., & Brahmasrene, T. (2013). Investigating the influence of tourism on economic growth and carbon emissions: Evidence from panel analysis of the European Union. *Tourism Management, 38*, 69–76. https://doi.org/10.1016/j.tourman.2013.02.016
Manniche, J., Larsen, K. T., & Broegaard, R. B. (2021). The circular economy in tourism: Transition perspectives for business and research. *Scandinavian Journal of Hospitality and Tourism, 21*(3), 247–264. https://doi.org/10.1080/15022250.2021.1921020
Marouli, C. (2016). Moving towards a circular economy: The need to educate—why and how. In *Proceedings of the 4th International Conference on Sustainable Solid Waste Management*, Limassol, Cyprus, pp. 23–25.
Merli, R., Preziosi, M., & Acampora, A. (2018). How do scholars approach the circular economy? A systematic literature review. *Journal of Cleaner Production, 178*, 703–722. https://doi.org/10.1016/j.jclepro.2017.12.112
Mishra, H. G., Pandita, S., Bhat, A. A., Mishra, R. K., & Sharma, S. (2022). Tourism and carbon emissions: A bibliometric review of the last three decades: 1990–2021. *Tourism Review, 77*(2), 636–658. https://doi.org/10.1108/TR-07-2021-0310
Moraga, G., Huysveld, S., Mathieux, F., Blengini, G. A., Alaerts, L., Van Acker, K., et al. (2019). Circular economy indicators: What do they measure? *Resources, Conservation and Recycling, 146*, 452–461. https://doi.org/10.1016/j.resconrec.2019.03.045
Moscardo, G. (2015). The Importance of Education for Sustainability in Tourism. In G. Moscardo & P. Benckendorff (Eds.), *Education for sustainability in tourism: A handbook of processes, resources, and strategies* (pp. 1–21). Springer.
Murray, A., Skene, K., & Haynes, K. (2017). The circular economy: An interdisciplinary exploration of the concept and application in a global context. *Journal of Business Ethics, 140*, 369–380. https://doi.org/10.1007/s10551-015-2693-2
Nocca, F., Bosone, M., De Toro, P., & Fusco Girard, L. (2023). Towards the human circular tourism: Recommendations, actions, and multidimensional indicators for the tourist category. *Sustainability, 15*(3), 1845. https://doi.org/10.3390/su15031845
OECD. (2021). *OECD environmental performance reviews: Finland 2021*. https://www.oecd.org/environment/oecd-environmental-performance-reviews-finland-2021-d73547b7-en.htm
Padilla-Rivera, A., Russo-Garrido, S., & Merveille, N. (2020). Addressing the social aspects of a circular economy: A systematic literature review. *Sustainability, 12*(19), 7912. https://doi.org/10.3390/su12197912
Patti, S. (2017). Circular economy and sharing consumption: Attitudes towards low-carbon tourism. *Economics and Policy of Energy and the Environment, 1–2*(1–2), 219–234. https://doi.org/10.3280/EFE2017-001011
Pongsakornrungsilp, P., & Pongsakornrungsilp, S. (2021). Mindful tourism: Nothing left behind–creating a circular economy society for the tourism industry of Krabi, Thailand. *Journal of Tourism Futures, 9*(3), 366–380. https://doi.org/10.1108/JTF-01-2021-0001
Preston, F. (2012). *A global redesign? Shaping the circular economy. Energy, environment and resource governance.* Briefing Paper. London: Chatham House

Prieto-Sandoval, V., Jaca, C., & Ormazabal, M. (2018). Towards a consensus on the circular economy. *Journal of Cleaner Production, 179*, 605–615. https://doi.org/10.1016/j.jclepro.2017.12.224

Renfors, S. M., & Wendt, T. (2024). Restaurants without Bins: How Does a circular restaurant operate? *Sustainability, 16*(6), 2312. https://doi.org/10.3390/su16062312

Rodríguez, C., Jacob, M., & Florido, C. (2020). Socioeconomic profile of tourists with a greater circular attitude and behaviour in hotels of a sun and beach destination. *International Journal of Environmental Research and Public Health, 17*(24), 9392. https://doi.org/10.3390/ijerph17249392

Rodríguez-Antón, J. M., & Alonso-Almeida, M. D. M. (2019). The circular economy strategy in hospitality: A multicase approach. *Sustainability, 11*(20), 5665. https://doi.org/10.3390/su11205665

Rudan, E., Nižić, M. K., & Grdić, Z. Š. (2021). Effect of circular economy on the sustainability of cultural tourism (Croatia). *Ekonomia i Środowisko-Economics and Environment, 76*(1), 120–138. https://doi.org/10.34659/2021/1/7

Sachs, J. D., Lafortune, G., Fuller, G., & Drumm, E. (2023). *Implementing the SDG stimulus. Sustainable development report 2023*. Dublin University Press.

Saura, J. R., Ribeiro-Soriano, D., & Palacios-Marqués, D. (2024). Adopting digital reservation systems to enable circular economy in entrepreneurship. *Management Decision, 62*(8), 2388–2408. https://doi.org/10.1108/MD-02-2022-0190

Schumann, F. R. (2020). Circular economy principles and small island tourism Guam's initiatives to transform from linear tourism to circular tourism. *Journal of Global Tourism Research, 5*(1), 13–20. https://doi.org/10.37020/jgtr.5.1_13

Sørensen, F., & Bærenholdt, J. O. (2020). Tourist practices in the circular economy. *Annals of Tourism Research, 85*, 103027. https://doi.org/10.1016/j.annals.2020.103027

Sorin, F., & Sivarajah, U. (2021). Exploring Circular economy in the hospitality industry: Empirical evidence from Scandinavian hotel operators. *Scandinavian Journal of Hospitality and Tourism, 21*(3), 265–285. https://doi.org/10.1080/15022250.2021.1921021

Sumter, D., de Koning, J., Bakker, C., & Balkenende, R. (2021). Key competencies for design in a circular economy: Exploring gaps in design knowledge and skills for a circular economy. *Sustainability, 13*(2), 776. https://doi.org/10.3390/su13020776

Tiippana-Usvasalo, M., Pajunen, N., & Maria, H. (2023). The role of education in promoting circular economy. *International Journal of Sustainable Engineering, 16*(1), 92–103. https://doi.org/10.1080/19397038.2023.2210592

Urbinati, A., Chiaroni, D., & Chiesa, V. (2017). Towards a new taxonomy of circular economy business models. *Journal of Cleaner Production, 168*, 487–498. https://doi.org/10.1016/j.jclepro.2017.09.047

Vargas-Sánchez, A. (2021). The new face of the tourism industry under a circular economy. *Journal of Tourism Futures, 7*(2), 203–208. https://doi.org/10.1108/JTF-08-2019-0077

Velenturf, A. P., & Purnell, P. (2021). Principles for a sustainable circular economy. *Sustainable Production and Consumption, 27*, 1437–1457. https://doi.org/10.1016/j.spc.2021.02.018

Yang, M., Chen, L., Wang, J., Msigwa, G., Osman, A. I., Fawzy, S., et al. (2023). Circular economy strategies for combating climate change and other environmental issues. *Environmental Chemistry Letters, 21*(1), 55–80. https://doi.org/10.1007/s10311-022-01499-6

Instruments, Models and Strategies for Sustainability in Tourism

Enhancing Green Capability in the Hospitality Sector Through Nudging: A Conceptual Framework

Marco Martins, Ricardo Jorge da Costa Guerra, Lara Santos, Luísa Lopes, and Ana Rita Conde

Abstract Tourism activities bring on considerable environmental costs, consequently, tourism is not only a climate change victim but also responsible for a significant percentage of the world's carbon emissions. Due to people's growing concern with environmental issues, the hospitality sector has been trying to reduce its environmental impact and at the same time managing to improve their green capability. One of the initiatives involved implementing a choice-architecture tool known as nudge. Nudging allows for effortlessly steering people in a desired direction; therefore, how can we promote 'greener' consumer behaviour in the hospitality industry through nudging and thereby reduce the industry's impact on climate change. In order to answer to this question a conceptual methodology was chosen. Such method enables to raise questions and hypotheses that will enlarge the discussion's scope on green capabilities and on nudge in the hospitality sector. Moreover, a novel theoretical framework for green capabilities is presented. This framework will help to implement nudge in a way that this can work as an accelerator for green capacity building in the hospitality sector. However, we must also refer to the fact

M. Martins (✉)
CiTUR, Coimbra, Portugal
e-mail: marco.mpm@gmail.com

R. J. da Costa Guerra
Polytechnic University of Guarda, TECHN&ART, CEGOT and CiTUR, Seia, Portugal
e-mail: ricardoguerra@ipg.pt

L. Santos
Polytechnic University of Cávado and Ave, Barcelos, Portugal
e-mail: laramgsantos@gmail.com

L. Lopes
Bragança Polytechnic University, UNIAG and CiTUR, Mirandela, Portugal
e-mail: luisa@ipb.pt

A. R. Conde
University Lusófona and HEI-Lab, Porto, Portugal
e-mail: rita.conde@ulusofona.pt

J. Marques et al. (eds.), *Tourism and Climate Change*, Tourism, Hospitality & Event Management, https://doi.org/10.1007/978-3-031-99128-8_6

that our study being conceptual is like all conceptual studies heavily dependent on the researcher's own perspectives and experiences.

1 Introduction

Tourism activities always end up causing extensive environmental damage to destinations (Juvan & Dolnicar, 2017) and lessening the damage caused by them is a complex endeavour (Martins et al., 2024). Industries worldwide are trying to move towards the goal of zero emissions by 2050 in line with the Paris Agreement, albeit unevenly (UNEP, 2023), the tourism industry is moving against that tendency, and between 2010 and 2050, it is expected to triple its emissions (Gössling & Scott, 2018).

The current trend has led the hospitality sector to find new strategies based on behavioural theories that may help the industry to lessen its impact on climate change, among which they discovered the applicability of choice architecture and of nudging (Martins et al., 2024). However, as pointed out by Schmidt (2024, pp. 415–416) 'choice architecture' consists of finding solutions that often are "created by identifying key moments of decision-making or behaviour and altering the environment in which options are presented with the intent of encouraging more desirable choices when people's preferences are known but their follow-through is lagging". Although nudging has proven to contribute to societal challenges and achieved quite a success (OECD, 2017), passing from the micro level (tactical interventions) to the macro level (strategical interventions) has been a challenge, primarily due to the intervention's heterogeneous nature and to a wide tendency to seek only optimisation of individual contexts (Kuehnhanss, 2019; Bryan et al., 2021). Organisations normally use choice architecture as a quick 'bandage' and not as a long run strategic framework. Thus, this research is concerned with the how environmental management improves the hospitality industry´s competitiveness and also seeks to demonstrate how a sustainable competitive advantage can be achieved by developing their green capability. Hence, our aim is to contribute to the existing corpus of research on green capabilities, green behaviour, and nudge. More specifically, this chapter explores their interconnections in the hospitality business sector.

In an effort to shed light on the remarks and questionings made before, this chapter was organised according to the following scheme. After the introduction, the chosen methodology and the main reasons for choosing it in this context are presented. Afterwards, the meaning of choice architecture and of nudging was addressed as well as green capability in the hospitality sector. Due to its importance, we make an incursion on the significance of green values when wanting to change guest's behaviours. Moving on, we discuss the importance of green marketing strategies when aiming to reduce the sector's impact on climate change. Next, we analyse the nudge effectiveness in the hospitality sector recurring to examples. Leaning further, we can aim to develop green capability in the hospitality sector through the implementation of nudging, where our proposed framework (see Fig. 1) is presented and discussed. After exploring the aforementioned thematic, we discussed the results

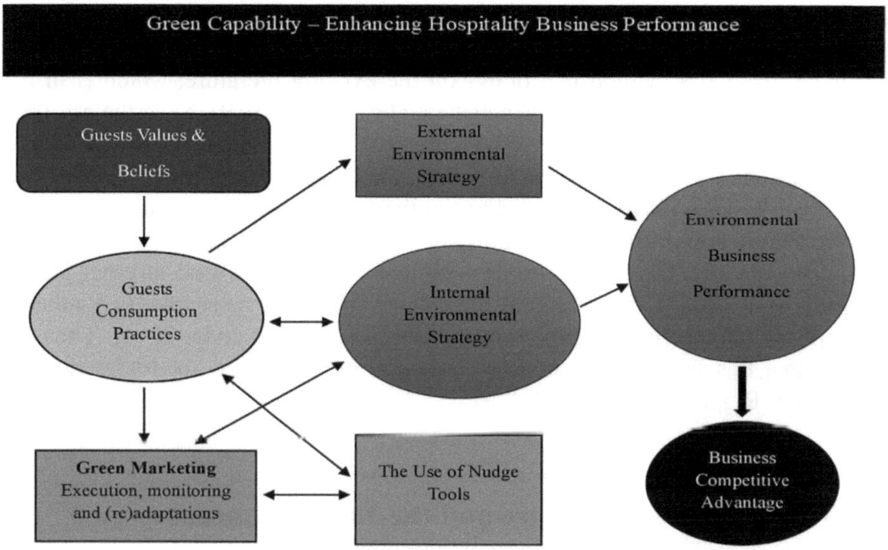

Fig. 1 Framework based on developing green capability in the hospitality sector. (Source: Developed by the authors)

and ended up with the conclusion where, besides the study itself, we discuss possible future avenues to research in this field.

2 Methodology

As conceptual study that it is, the objective of this chapter is to develop a discussion about green capability in the hospitality sector through nudge. Thus, this study follows the idea that conceptual analysis sometimes consists in disintegrating concepts into their constituent parts in order to improve overall understanding (Martins et al., 2023). Thereby, this conceptual research, in particular, falls under the constructivism-interpretivism paradigm and provides opportunities for critical engagement (Alharahsheh & Pius, 2020). One recognises that "reality is not a singular, fixed entity but an array of diverse constructs that are continually shaped and reshaped by individual experiences and social interactions. This understanding of reality underscores the paradigm's commitment to exploring the intricacies of individual perspectives, subjective meanings and personal experiences" (Lim, 2023, p. 13).

Therefore, we intended to raise questions and hypotheses on nudge as a tool to help improve or even change behaviours in the hospitality sector aiming for improving green capability. According to Mollick (2014), one should move away from the formal testing of hypotheses and attempt to analyse the key issues around the core concepts that we are concerned with in this chapter, i.e. green capability, green marketing and nudge and their links to theory. Therefore, the aim of this study is to take a broader look at nudge and on its transformational potential, thereby expanding the discussion around this topic. Specifically, in this study, we used a semi-systematic secondary research method that draws on the existing literature, which is in line with the literature review approach developed by Snyder (2019). Nevertheless, there are some disadvantages when we conduct an exploratory study, because as it is an interpretation of a new reality, we only have the opportunity to move forward with some hypotheses without the chance to test them. Essentially, it is a reflexive work, using different approaches and meanings to define conceptual analytical dimensions and, consequently, to create a framework that can later be used in other studies (Martins et al., 2023). A conceptual framework allows the researcher to synthesise concepts or perspectives drawn from multiple sources (Imenda, 2014). Thus, the idea is that this framework can serve as a guide and support for future research (Ravitch & Riggan, 2016).

3 The Choice Architecture and the Nudge Concept

Martins et al. (2024, p. 255) stated that "over the years, behavioural researchers and practitioners have struggled to deliver a framework that recognises human behaviour for what it is, rather than based on simplistic economic models". The choice architecture was brought to the surface when scholars began to take notice on some cognitive particularities present on human decision making based upon existing judgement and decision-making research (Münscher et al., 2016). Nevertheless, choice architecture only gains real impact among academics and practitioners after the release of the 2008 work by Thaler and Sunstein 'Nudge'. In it the authors suggest that scholars should study the predictability of deviations from rational behaviour and how they can be used to 'nudge' individuals into socially desirable directions. The concept of nudge is then in line with the behavioural sciences, which draw on the vision from the cognitive and social psychology, enhancing the rational nature of humankind and highlighting the substance of behavioural preconceptions and the decision circumstances, i.e., the environment in which individuals make choices (Lehner et al., 2016). A "new branch to the 'prescriptive' part of the field of judgment and decision making" was consequently added (Baron, 2010, p. 224). However, we must also be aware that from a behavioural economics view, every situation may always represent a possible choice architecture, even if it is not clearly designed for a particular intent (Kahneman, 2011). Despite this, and according to Thaler et al. (2013), choice architecture refers to the idea that alterations in the decision environment can affect individual decision-making and behaviour while preserving their freedom of choice. Thus, it is a concept that consists in defining the physical, social, and psychological context in which decisions are made (Thaler et al., 2013).

When Thaler and Sunstein coined the concept of 'nudge' in 2008, they argued that nudge in itself explores choice architecture and could be defined as "the tactic of skilfully modifying a behaviour. It is then any aspect of the choice architecture that alters people's behaviour in a predictable way without forbidding any options or significantly changing their economic incentives" (Thaler & Sunstein, 2008, p. 6). Hausman and Welch (2010) added a 'constructive' condition to the concept of nudge, elucidating that nudges are necessary because of errors in individual decision-making and that the choice architect's role is to explore these same errors. Hansen (2016, p. 158) stated that nudge can be explained as "any attempt at influencing people's judgement, choice or behaviour in a predictable way (1) this is made possible because of cognitive boundaries, routines and habits in individual and social decision-making posing barriers for people to perform rationally in their own declared self-interests and which (2) working by making use of those boundaries, biases, routines, and habits as an integral part of such attempts". Thus, nudging "help individuals to make what, on reflection, they themselves would recognise as the better choice" (Infante et al., 2016, p. 4).

To consolidate existing definitions and his own (in epigraph), Hansen (2016) puts on some conditions under which a nudge must be a nudge. According to this, a nudge would be any aspect of the choice architecture that alters people's behaviour in a foreseeable manner without:

1. Forbidding any option;
2. Significantly changing incentives, whether regarded in terms of time, trouble, social sanctions, economic, and so forth;
3. Nudges are called for because of cognitive boundaries, biases, routines, and habits in individual and social decision-making;
4. Work by making use of those binderies, biases, routines, and habits as integral parts of the choice architecture.

Resuming, individuals often make decisions that are not in their best interest because they procrastinate or lack self-control, because they are influenced by the context in which decisions are made, or because they are overwhelmed by the information (Reisch & Gwozdz, 2013).

Decision-making integrates cognitive processes that result in choosing a course of action or behaviour among several alternatives. The traditional perspective postulated the human being as rational, who analyses each option's benefits/advantages and costs/disadvantages. However, decisions are influenced by cognitive shortcuts and biases, emotional impulses, and social pressures (Ramos, 2019). Biases and cognitive shortcuts can lead to unfavourable or suboptimal decision-making without the person knowing it. Some examples are anchoring, where decisions are inadequately influenced by initial information, or the confirmation or phenomenon of self-confirmatory expectations, where one unconsciously seeks information or acts to confirm pre-existing beliefs. These cognitive biases shape perception and impact decision-making and subsequent behaviours (Ramos, 2019; Korteling et al., 2023). Thus, cognitive preconceptions and biases impact not only the individual level but also various areas of human life, such as business, marketing, governance, and public policies. The notion of nudge recognises and seeks to address (mitigate or capitalise) these biases, contributing to promoting better decision-making results and contributing to a more equitable and sustainable society (Korteling et al., 2023)

Hence, nudging works by supplying contextual attributes (i.e. a change in decision architecture occurs) that take predominance due to the human tendency to act automatically (e.g. a heuristic) but diverge from a biased heurist in that they guide people to make (presumably) good decisions (Tagliabue et al., 2019), moreover, green nudges have the potential to contribute to the well-being of both current and future generations (Hage, 2019).

4 Green Capability in the Hospitality Sector and Climate Change

The hospitality's businesses detrimental operations and unsustainable practices are resulting in a rise in environmental disturbances. Focusing on the business green capability may be an encouraging step forward that can relieve environmental issues and attain a greener ideology in businesses (Khan et al., 2022). Business green capability refers to the business's ability to bring into effective action resources and to achieve specific objectives (Nkrumah et al., 2021). So, when addressing green capabilities in the hospitality business, we are speaking about a notion that was defined by Lee and Klassen (2008), i.e., all the assets, technologies, and expertise that an organisation can explore to administrate the multiple environmental needs of customers and of other stakeholders. Hence, the umbrella concept of green capability considers the integration, construction, and refiguration of both internal and external resources aiming towards environmental protection. And although green capabilities have proven to help achieve a competitive advantage, much less attention has been paid to the extent to which green capability may be determinant in achieving greater environmental performance (Qiu et al., 2020). The existent literature, however, recognises that green innovation significantly improves environmental performance (Kraus et al., 2020), and that researchers worldwide are now paying much more attention to environmental sustainability orientation that allows to improve organisational performance (Amankwah-Amoah et al., 2019).

Analysing the literature, it is possible to extrapolate that there are at least four green capabilities, and to accomplish our framework (see Fig. 1) we had in mind all these four dimensions and the work of Martins et al. (2024). The four dimensions above mentioned are:

1) Green supplier development, which refers to the ability of the firm to identify and collaborate with their suppliers to achieve greater green outcomes (Blome et al., 2014);
2) Green marketing, which refers to the ability of the firm to effectively market its green products and processes to its customers and stakeholders (Shang et al., 2010);
3) Green manufacturing and packaging, which refers to the ability of firms to alter their manufacturing and packaging processes in a way to reflect green concerns (Shang et al., 2010);
4) Environmental participation, which refers to the ability of the firm to bring together its internal stakeholders (staff and departments) to work together in the implementation of green initiatives (Bae, 2017).

In conclusion, green capability enables companies to acquire ecological skills and competencies that allow them to improve their overall performance (Li et al., 2024). Thus, the green capability maximises the company's dividends while minimising the costs associated with environmental risk. Since the business' goal is to attain economic growth, the green capability brings positive changes in organisational ecological activities, thus contributing to achieving high business performance (Davies et al., 2023).

5 Green Consumption Values, Beliefs, and Attitudes

There is a general acceptance among researchers that environmental issues can be lessened or even solved through the correction of individual behaviour in the direction of greater environmental sustainability (Han, 2021). Nevertheless, there is still a lack of research on how consumers' cultural values and ethical ideologies lead to green consumption (Halder et al., 2020). However, the literature has a very sound theoretical basis on pro-environmental consumer behaviour in the hospitality sector.

Making changes regarding the individual's consumer behaviour when approaching, buying, and consuming products in an environmentally friendly manner is widely advocated as a core requirement for environmental sustainability (White et al., 2019; Halder et al., 2020; Wang et al., 2020). This statement stems from the fact that environmentally friendly consumption is nowadays and worldwide a decisive trend. Individuals have interiorised that environmentally friendly consumption can be an effective contribution to achieve global sustainable development goals (Zhao et al., 2022).

But despite that, research in psychology stumble upon an apparent attitude-behaviour gap (also described in the literature as the value-action gap) when scrutinising the interrelations between environmental attitudes and behaviour and between environmental behaviours and the confirmed ecological consequences of actions (described as the behaviour-impact gap) (Halder et al., 2020).

There is no consensus among academics regarding the definition of environmental attitudes as a unidimensional construct. However, some are more embraced than others. Environmental attitudes have been defined in various ways, from a more conventional perspective as a psychological tendency to positively or negatively evaluate the natural environment and the factors that influence its quality (Milfont, 2012) to a broader perspective as the set of beliefs, affects, and behavioural intentions regarding environmental activities or issues (Schultz et al., 2004). For example, the one of Gifford and Sussman (2012, p. 65) that define environmental attitudes as a "concern for the environment or caring about environmental issues". But in general, studies focus on the multidimensional nature of environmental attitudes (affective, cognitive, conative, and other) (Hadler et al., 2022).

Research pointed out the fact that values are a key antecedent of a profusion of pro-environmental behaviours (Steg et al., 2015). It is also demonstrated in the literature that the most important factors that explicate people's pro-environmental behaviour are 'biospheric values' (concern for nature and environmental protection), 'altruistic values' (concern with others well-being), 'egoistic values' (looking after and promoting personal interests), and 'hedonic values' (looking for pleasure effortless) (De Groot & Steg, 2009; Steg et al., 2014; Bouman & Steg, 2020; Wang et al., 2021). However, current research is now focusing on how self-control can explain the adoption of pro-environmental behaviour and not how perceived behavioural control can explain it (e.g., Gómez-Olmedo et al., 2021; Wyss et al., 2022). The reason for this change in focus resides in the fact (besides the identified gaps in the epigraph) that reality has shown that individual adoption of pro-environmental behaviour per se is currently insufficient to meet the United Nations' sustainable development goals (Fuso Nerini et al., 2018). Generally speaking, individuals need self-control when standing before a conflict between two mutually exclusive motives: a more significant motive that one expects to give rise to a proximal prize, and one less powerful motive that comes with a greater long-term prize (Duckworth et al., 2019). In such situations, self-control capacity can aid individuals to bring their behaviour in line with their long-term objectives (Hofmann et al., 2012; Milyavskaya & Inzlicht, 2017). Hence, a growing ability of people to sustain focus on their own behaviours and resist both hedonistic and selfish temptations would permit people to maintain their self-pro-environmental intentions, even when performing in favour of the environment requires going beyond habits or sacrificing short-term individual rewards (Colombo et al., 2023). This is broadly in line with those predictions made with the use of choice architecture tools such as the one of nudge.

Summarising, we can say that managers in the hospitality sector must understand the dynamics behind consumer behaviour and how their values and beliefs influence them if they are to develop and implement effective and efficient policies and programmes that can benefit the environment. Unfortunately, much is still to be studied regarding how consumers' cultural values and ethical ideologies may guide them into green consumption (Halder et al., 2020).

6 Green Marketing in the Hospitality Sector

A Global Consumer Insights Project Report (GlobeScan, 2019) states that issues related to the environment due to climate change and other factors are among humanity's most significant concerns in present times. Environmental sustainability is now a central matter of discussion in societies worldwide. Consequently, brands are growingly embracing green marketing strategies. Nonetheless, some studies demonstrate that there is still a gap between what people intend regarding sustainability and their actual behaviour (GlobeScan, 2019). For example, a study conducted by Šagovnović and Stamenković (2022) reveals that the eco-brand positively influences tourists' sustainable attitudes and behaviour, with attitudes being a significant predictor of this behaviour. Another study from Fauzi et al. (2024)

confirmed that tourists' decisions to stay at a green hotel are directly influenced by their subjective norms and perceived behavioural control.

Research on the purchase of green products and green marketing started in the 1970s. But real progress on this thematic only occurred in the 1980s and early 1990s, encompassing already the idea of eco-labelling (Horne, 2009; Smith & Brower, 2012; Braimah, 2015). Besides this, it is commonly accepted that the first definition of green marketing as green marketing (before authors spoke of ecological marketing or sustainable marketing) was coined by Mintu and Lozada (1993, p. 17)—they argued that "green marketing refers to the application of marketing concepts and tools to facilitate exchanges that satisfy organisational and individual goals in such a way that the preservation, protection, and conservation of the physical environment is upheld". Peattie (2001) argued that green marketing should not be seen merely as a variation on traditional marketing, suggesting some specific manners to lean upon it in way to avoid potential implementation failures. Over time and with the growing recognition of the planet's environmental issues and the acceptance of the environment, society, and economy interdependence, marketing begins to direct its efforts not only in terms of environmental damage reduction but also in terms of attaining sustainability widely (Dangelico & Vocalelli, 2017). Some of the most cited definitions of green marketing, besides the initial one, are those of Prakash (2002, p. 285), who says that green marketing refers to "strategies to promote products by employing environmental claims either about their attributes or about the systems, policies, and processes of the firms that manufacture or sell them"; of Mishra and Sharma (2012, p. 35) who express that "green marketing refers to holistic marketing concept wherein the production, marketing, consumption, and disposal of products and services happen in a manner that is less detrimental to the environment, with growing awareness about—i.e., the implications of—global warming, nonbiodegradable solid waste, the harmful impact of pollutants, etc."; or of the American Marketing Association (AMA, 2017), which states that it "refers to the development and promotion of products that are presumed to be environmentally safe (i.e., designed to minimize negative effects on the physical environment or to improve its quality). This term may also be used to describe efforts to produce, promote, package, and reclaim products in a manner that is sensitive or responsive to ecological concerns" and also "green marketing is the practice of actively associating a product offering with ideas relevant to environmental sustainability" (AMA, 2023).

Green marketing is then based on the triple-bottom-line principle (Elkington, 1994), which perspectives green marketing as the marketing of products and services that are presumed to be environmentally friendly and environmentally undamaged (Witjes et al., 2017). It is also an instrument for the resources, places and destinations promotion, allowing at the same time that businesses be placed in the market as organisations committed to nature (Shi et al., 2022). Using green marketing in these dimensions may provide a chance to positively influence the tourists consuming practices (Prieto-Sandoval et al., 2022) while promoting more friendly and sound marketing strategies (Apaza-Panca et al., 2024).

Thus, we may constate that in the hospitality business sector green marketing is being implemented to engender new needs in their customers, seeking to achieve a competitive advantage in the market (Salas Canales, 2018). For example, a study conducted by Patwary et al. (2022) showed that after using green marketing tools (e.g., eco-label, eco-brand and environmental advertising) in the Malaysian hospitality sector, the result was an improvement in green attitudes and behavioural intention to acquire green products. One other study conducted by Akram et al. (2024) examined the impact of green marketing instruments and motivations on green behavioural intentions in China and India. They pinpointed the influence of green marketing instruments on brand image and trust, besides revealing the noteworthy moderating role of brand experience. Finally, and through several studies, Wong and Lai (2024) explored the green certification logos influence and text on environmental consciousness and hotel guests' behaviour regarding water-savings. Their study shows that an obvious green message was more effective than an indirect one when aiming to reduce hotel shower time.

7 The Nudge Effectiveness in the Hospitality Sector

Since it was coined, academics and practitioners alike have embraced 'nudge' as an invigorating instrument for behaviour change hence it enables the possibility of exerting influence over individuals' decisions through basic changes in what is understood as choice architecture (Martins et al., 2024). This behaviour change plays a huge role in the hospitality business sector since many studies alert to the fact that day-to-day pro-environmental behaviours with direct impact on climate change (e.g., reduction of energy, waste disposal, meat consumption) are put aside almost entirely when engaging in tourism activities (Kim et al., 2020).

Due to its novelty (coined in 2008), studies proving nudging's effectiveness in the hospitality sector are still scarce. In spite of this, and having the purpose of this study in mind, we can find at least the ones by Goldstein et al. (2008), when nudging was still being born, by Kallbekken and Sælen (2013), by Chang et al. (2016), by Kim et al. (2020), and by Visser-Amundson (2020). In their research, Goldstein et al. (2008) confirmed that the majority of guests who participated in the towel reuse program followed the message "join your fellow guests in helping to save the environment"; inversely, a much smaller number followed the message "help save the environment". According to the study, 75% of the guests in the first sample complied with the program, but in reality, only 44.1% indeed followed the program to the letter. In the second sample, only 35% of the guests also adhered to the reuse of their towels. They believe that the difference has to do with social norms—those rules that guide group behaviour. Kallbekken and Sælen (2013) revealed that it was possible to diminish the food leftovers quantity in hotel restaurants by an average of 20% with two non-intrusive nudges. The researchers reduced the plate size and provided social cues in the study. Given the contribution of edible consumption and waste to climate change, this research should have become a beacon for the hospitality business sector and for academics to follow and apply. Chang et al. (2016) on their part conducted a research that attested the effectiveness of nudges in electricity saving. They proved that hotels may achieve a guest-driven electricity efficiency with properly design of nudges. The results were an indicative that guests are more inclined to involve themselves in electricity saving practices when they are informed in real time with respect to how much others are consuming. The study conducted by Kim et al. (2020) on the implementation of architectural choice nudges showed, through diverse experiments, that the majority of consumers who had to choose between an environmentally friendly hotel and a hotel with regular services chose unavoidable the environmentally friendly hotel due to a feeling of guilt resulting from the choices and perceived social constraints. It is also important to detach the research conducted by Visser-Amundson (2020) on a wide multi-stakeholder partnership to fight against food waste in the hospitality sector in the Netherlands. In it, 172 restaurants participated, and using nudges resulted in a 21% reduction in food waste. Nudging in the hospitality's food sector can be very important as one third of all edibles produced for human consumption worldwide are never consumed (Murphy et al., 2019).

The importance of applying the architecture choice tools resides in the fact that the most recent analysis on the scale of CO2 tourism industry emissions, as far as we know, is the one provided by Lenzen et al. (2018), and it points out a total contribution of about 8%. It is important to note that this estimate does not contemplate tourism's indirect contribution to global emissions. For example, the consumption of meat in the tourism sector is not contemplated, and according to Crippa et al. (2021), food systems are responsible for a third of global anthropogenic GHG emissions. Therefore, it seems clear the importance of nudging to help reduce at least the hospitality business sector's overall contribution.

8 Aiming Towards Green Capability in the Hospitality Sector Through Nudge

From what has been aforementioned, nudge can play an important role when a hospitality business has the goal to develop its green capability. As Lehner et al. (2016, p. 166) pointed out "there is a growing recognition that supply-side policies need to be complemented by demand side strategies that could help individuals make more sustainable decisions". Hence, research like the one by Khan et al. (2022) shows that green capability can be pivotal in adopting green practices.

Our framework (see Fig. 1) is based on the idea that applying resource-based view theory (RBT) will enable hospitality businesses to gain and/or maintain competitive advantage through the use of critical resources and, at the same time, creating new capabilities through skills acquisition and accumulation of their tangible and intangible resources. These resources include R&D, advanced technology adoption, brand and effective cost management, and knowledgeable management that companies integrate to develop/readapt their product/services or to create new products or services for the market (Adner & Helfat, 2003; Alkaraan et al., 2024). In this case, the focus is on using nudge as a green marketing tool to enhance hospitality businesses' green capability. In doing so, hospitality businesses will achieve greater performance and gain competitive advantage by creating a new capability—green capability—and at the same time acquiring skills and knowledge that will enable them to be more effective regarding cost effectiveness and at the knowledge and development levels.

The application of nudge according to our framework (Fig. 1) must follow what Bovens (2009) and Alemanno (2012) defended, i.e., nudges must satisfy the following parameters:

1) Intervention must not restrict individual choice;
2) It must be in the interest of the individual(s) being nudged;
3) It involves a change in the architecture choice;
4) It implies the cognitive use of cognitive biases;
5) The intended action does not result from a totally autonomous choice.

It is crucial to bear in mind that this framework foresees that nudging should only be considered after the development and implementation of a broader green marketing strategic plan, thus acting as a complement to a much broader, greener policy. Throughout this study, several constructs were identified due to the specifics of the context of our own research. Therefore, in this framework on tourist's green behaviour change through nudge, three core dimensions were identified:

1. Decision-making (e.g., knowledge, skills, judgement);
2. Individual behaviour and practices (e.g., energy conservation, water savings, avoiding food waste, etc.);
3. Organisational engagement with greener governance (e.g., institutional change, implementing green processes, eco-labelling, nudge, etc.).

In this framework (see Fig. 1), hospitality business green capability refers to the ability to bring into effective action resources and to achieve specific objectives (Nkrumah et al., 2021). Thus, when addressing the green capabilities of the hospitality business, we are speaking about a notion that was defined by Lee and Klassen (2008), i.e., all the assets, technologies, and expertise that businesses can explore to administrate the multiple environmental needs of customers and other stakeholders.

For addressing green capability, a sound green marketing strategy is needed. In that sense, "wherein the production, marketing consumption and disposal of products and services must happen in a manner that is less detrimental to the environment" (Mishra & Sharma, 2012, p. 35), and at the same time design a strategy that raises awareness about the implications of global warming, nonbiodegradable solid waste, harmful impact of pollutants, etc., among their consumers and internal stakeholders, besides designing and using nudging to reinforce that same awareness aiming towards a greener behaviour. As a result of attentive monitoring of nudge results, moreover, and if necessary, (re)adaptations must be executed timely. Guest consumption practices depend on their values and beliefs but also on their self-control.

Thus, self-control capacity can aid individuals in bringing their behaviour in line with their long-term objectives (Hofmann et al., 2012; Milyavskaya & Inzlicht, 2017). Hence, a growing ability of people to sustain focus on their own behaviours and resist both hedonistic and egoistic temptations would permit people to maintain their self-pro-environmental intentions, even when performing in favour of the environment asks them to go beyond habits or sacrifice short-term individual rewards (Colombo et al., 2023). Nudging allows guests to maintain their self-control and even incentivises it because nudging "helps individuals to make what, on reflection, they themselves would recognise as the better choice" (Infante et al., 2016, p. 4).

When addressing the internal environmental strategy, it will be crucial to environmental participation, which refers to the ability of the firm to bring together its internal stakeholders (staff and departments) to work together in the implementation of green initiatives (Bae, 2017). Therefore, green marketing managers must design a strategy that, by using nudging, reinforces environmental awareness among internal stakeholders, aiming towards not only greener behaviour but also towards the proposal, design, and supply of greener products and services.

Regarding the external environment strategy: wanting and becoming more environment-friendly by developing the business's green capability will imply alterations at the external environment level, i.e., at the supply chain level. Negotiating with the external stakeholders' greener products and services and even managing to achieve green/ecological/sustainable certifications will be imperative. Furthermore, the reduction of energy, water, and meat supply due to the successful application of nudging will also have an impact on the supply chain. Summarising, there is a need for the hospitality business to identify and collaborate effectively with suppliers for greater green results (Blome et al., 2014).

In what concerns environmental business performance, we can say that green capability development will allow businesses to embrace ecological skills and competencies that will enable hospitality businesses to improve their overall performance (Li et al., 2024). Thus, the green capability maximises the company's dividends while minimising the cost associated with environmental risk. As the business' goal is to attain economic growth, the green capability brings positive change in organisational ecological activities, thus aiding in achieving high business performance (Davies et al., 2023).

In what regards to business competitive advantage – is possible to say, based in the resource-based view theory (RBT) on the business and on the dynamic capabilities (including green capabilities), and in Nkrumah et al. (2021), that the green capabilities and resources that the hospitality business control will enable it a superior performance. When developing green capability, and through it implementing green practices, that as a whole will enable hospitality businesses to gain and/or maintain competitive advantage through the use of critical resources, and at the same time creating new capabilities through skills acquisition and accumulation of their tangible and intangible resources. These resources include R&D, advanced technology adoption, brand and effective cost management, and knowledgeable management that companies will integrate to develop/readapt their product/services or to create new products or services for the market (Adner & Helfat, 2003; Alkaraan et al., 2024).

It is important to observe that this framework also envisages that the use of nudging should only be considered after implementing a green marketing strategy, and it will act as a complement in a much broader green policy—one whose primary objective is to develop the hospitality business's green capabilities.

9 Conclusion

The unprecedented effects of climate change are increasingly impacting the hospitality sector, which is one of the greatest contributors to this state of things globally. Therefore, the sector is implementing strategies aiming to lessen its effects. Hence, marketers are choosing to adopt more sustainable processes, products, services and complementary offers (Rust, 2020). Thus, businesses are increasing their internal efforts and resources to become greener and achieve sustainable development because sustainability practices can improve their performance (Song et al., 2018).

Studies have consistently demonstrated a gap between manifested values and practical actions (White et al., 2019). Our literature review also points out the fact that the use of messages encouraging social comparability usually has a much greater effect on hospitality than messages that do not (Goldstein et al,. 2008; Gächter et al., 2013; Visser-Amundson, 2020). Thus, although calling to individual responsibility and moral responsibility produces some results, social comparability seems to result in a higher level of effectiveness.

The framework in epigraph detaches the role played by nudge, but also by green marketing in the pursuit of green capability in the hospitality business sector. The development and implementation of green marketing strategies should predate nudging—nudge should be an important variable but never be seen as a panacea. It is then our belief that this framework here presented provides clear answers to our research question, i.e. how can we promote 'greener' consumer behaviour in the hospitality sector through nudging and thereby reduce the impact of the sector in climate change. Based on the literature, we can presuppose that our green capability framework provides a critical understanding of the following:

- The role that guests and social practices may play in reducing water, food and energy waste, littering and carbon emissions;
- What is possible to achieve through the guests' actions;
- Which activities require organisational action and change;
- Nudge implementation;
- Green marketing management;
- The role that the social context plays in guest consumption;
- How to develop green capability within the hospitality business sector.

It is our understanding that this study will contribute to the literature on nudging by showing how hospitality organisations can put strategies into practice in order to achieve greater green effectiveness. Furthermore, it contributes to the literature with a novel framework, one that will aid in strengthening the debate on what is needed to implement nudge strategies with greater effectiveness. Our study is also important because, as Khan et al. (2022) clearly expressed, green capabilities play a key role in the adoption of green purchasing practices and may have a positive effect on economic and environmental performance.

Although, through this chapter we had enhanced the positive aspects of implementing a choice architecture based on nudge. The truth is that nudge is not an infallible tool, and perhaps one of its most renowned failures was in the fight against smoking. “The strength of the images used in the advertising campaigns of various brands weakens public policy efforts” (Sunstein, 2017, p. 20). Perhaps not by chance, Bang et al. (2020), in a study aiming to understand the role of perceived effectiveness on the acceptability of choice architecture, ‘discovered’ that it is crucial to disclose the design decision because that does not lessen its effectiveness. Furthermore, individuals perceive that certain designs are more effective with others than with themselves. Hence, disclosing the design decision not only does not reduce its effectiveness as understanding the intended objectives increases its acceptability but more importantly, it may facilitate social comparability.

Regarding future research avenues we must express that when considering that nudging and green marketing are closely linked to what is often referred to as ‘new governance’, further research must be made to explore that same interconnection. Besides, the majority of studies predict that in the future, there is a chance to reshape and promote new behaviours for a new trajectory for sustainable tourism and travel (Moyle et al., 2021), and thus, businesses that understand how to adequately explore this “new” concept, nudge, will be able to gain a competitive advantage and improve the profitability of hospitality services. However, are these predictions correct? Because of its novelty, much more research must be done on nudge to understand to what extent the hospitality business will or is already gaining a competitive advantage by just applying these tools. One step forward must also be to find partners in the hospitality sector that may be interested in implementing the framework. That would allow us to understand if the elements in the framework can be successfully applied, the extent to which they can be applied, and/or if some should be added, amended, or somehow ameliorated.

It is important to refer that our study has the limitations inherent to all sorts of conceptual studies. These limitations consist in the fact that conceptual studies are very dependent of the authors’ unique experiences and perspectives, consequently different authors will likely yield different insights and interpretations. Furthermore, it is not possible to us, in this stage, to know if the framework can be applied successfully or to understand whether the variables in the framework apply, the extent to which they apply, or whether some should be added or amended (Passey, 2020).

Competing Interests The authors have no conflicts of interest to declare relevant to this chapter’s content.

References

Adner, R., & Helfat, C. E. (2003). Corporate effects and dynamic managerial capabilities. *Strategic Management Journal, 24*(10), 1011–1025. https://doi.org/10.1002/smj.331

Akram, U., Lavuri, R., Bilal, M., Hameed, I., & Byun, J. (2024). Exploring the roles of green marketing tools and green motives on green purchase intention in sustainable tourism destinations: A cross-cultural study. *Journal of Travel & Tourism Marketing, 41*(4), 453–471. https://doi.org/10.1080/10548408.2023.2293022

Alemanno, A. (2012). Nudging smokers – The behavioral turn of tobacco risk regulation. *European Journal of Risk Regulation, 3*(1), 32–42. http://www.jstor.org/stable/24323124

Alharahsheh, H. H., & Pius, A. (2020). A review of key paradigms: Positivism vs interpretivism. *Global Academic Journal of Humanities and Social Sciences, 2*(3), 39–43. https://gajrc.com/media/articles/GAJHSS_23_39-43_VMGJbOK.pdf

Alkaraan, F., Elmarzouky, M., Hussainey, K., Venkatesh, V. G., Shi, Y., & Gulko, N. (2024). Reinforcing green business strategies with Industry 4.0 and governance towards sustainability: Natural-resource-based view and dynamic capability. *Business Strategy and the Environment, 33*(4), 3588–3606. https://doi.org/10.1002/bsc.3665

Amankwah-Amoah, J., Danso, A., & Adomako, S. (2019). Entrepreneurial orientation, environmental sustainability and new venture performance: Does stakeholder integration matter? *Business Strategy and the Environment, 28*(1), 79–87. https://doi.org/10.1002/bse.2191

American Marketing Association. (2017). Definitions of Marketing. https://www.ama.org/the-definition-of-marketing-what-is-marketing/#:~:text=Green%20Marketing,or%20to%20improve%20its%20quality)

American Marketing Association. (2023). Green marketing. Universal Marketing Dictionary Project. https://marketing-dictionary.org/g/green-marketing/

Apaza-Panca, C. M., Flores Quevedo, L. A., & Carranza Reyes, L. M. (2024). Green marketing to promote the natural protected area. *Sustainable Technology and Entrepreneurship, 3*(3), 100067. https://doi.org/10.1016/j.stae.2023.100067

Bae, H. S. (2017). The effect of environmental capabilities on environmental strategy and environmental performance of Korean exporters for green supply chain management. *The Asian Journal of Shipping and Logistics, 33*(3), 167–176. https://doi.org/10.1016/j.ajsl.2017.09.006

Bang, H., Shu, S., & Weber, E. (2020). The role of perceived effectiveness on the acceptability of choice architecture. *Behavioral Public Policy, 4*(1), 50–70. https://doi.org/10.1017/bpp.2018.1

Baron, J. (2010). Review of the book nudge: Improving decisions about health, wealth, and happiness, by R. H. Thaler & C. Sunstein. *Journal of Behavioral Decision Making, 23*(2), 224–226. https://doi.org/10.1002/bdm.652

Blome, C., Hollos, D., & Paulraj, A. (2014). Green procurement and green supplier development: Antecedents and effects on supplier performance. *International Journal of Production Research, 52*(1), 32–49. https://doi.org/10.1080/00207543.2013.825748

Bouman, T., & Steg, L. (2020). Engaging city residents in climate action: Addressing the personal and group value-base behind residents' climate actions. *Urbanisation, 7*(1), 26–41. https://doi.org/10.1177/2455747120965197

Bovens, L. (2009). The ethics of nudge. In G. Yanoff & O. Hansson (Eds.), *Preference change: Approaches from philosophy, economics and psychology* (pp. 207–219). Springer. https://doi.org/10.1007/978-90-481-2593-7_10

Braimah, M. (2015). Green brand awareness and customer purchase intention. *Management Science Letters, 5*, 895–902. https://doi.org/10.5267/j.msl.2015.8.007

Bryan, C. J., Tipton, E., & Yeager, D. S. (2021). Behavioural science is unlikely to change the world without a heterogeneity revolution. *Nature Human Behavior, 5*, 980–989. https://doi.org/10.1038/s41562-021-01143-3

Chang, H. "Sean", Huh, C., & Lee, M. J. (2016). Would an energy conservation nudge in hotels encourage hotel guests to conserve? *Cornell Hospitality Quarterly, 57*(2), 172–183. https://doi.org/10.1177/1938965515588132

Colombo, S. L., Chiarella, S. G., Raffone, A., & Simione, L. (2023). Understanding the environmental attitude-behaviour gap: The moderating role of dispositional mindfulness. *Sustainability, 15*(9), 7285. https://doi.org/10.3390/su15097285

Crippa, M., Solazzo, E., Guizzardi, D., Monforti-Ferrario, F., & Leip, A. (2021). Food systems are responsible for a third of global anthropogenic GHG emissions. *Nature Food, 2*, 198–209. https://doi.org/10.1038/s43016-021-00225-9

Dangelico, R. M., & Vocalelli, D. (2017). "Green marketing": An analysis of definitions, strategy steps, and tools through a systematic review of the literature. *Journal of Cleaner Production, 165*, 1263–1279. https://doi.org/10.1016/j.jclepro.2017.07.184

Davies, P., Bustinza, O. F., Parry, G., & G., & Jovanovic, M. (2023). Unpacking the relationship between digital capabilities, services capabilities, and firm financial performance: A moderated mediation model. *Industrial Marketing Management, 115*, 1–10. https://doi.org/10.1016/j.indmarman.2023.09.005

De Groot, J. I. M., & Steg, L. (2009). Mean or green: Which values can promote stable pro-environmental behavior? *Conservation Letters, 2*(2), 61–66. https://doi.org/10.1111/j.1755-263X.2009.00048.x

Duckworth, A. L., Milkman, K. L., & Laibson, D. (2019). Beyond willpower: Strategies for reducing failures of self-control. *Psychological Science in the Public Interest, 19*(3), 102–129. https://doi.org/10.1177/1529100618821893

Elkington, J. (1994). Enter the triple bottom line. In A. Enriques & J. Richardson (Eds.), *The triple bottom line: Does it all add up?* (pp. 1–16). Earthscan.

Fauzi, M. A., Hanafiah, M. H., & Kunjuraman, V. (2024). Tourists' intention to visit green hotels: Building on the theory of planned behaviour and the value-belief-norm theory. *Journal of Tourism Futures, 10*(2), 255–276. https://doi.org/10.1108/JTF-01-2022-0008

Fuso Nerini, F., Tomei, J., To, L. S., Bisaga, I., Parikh, P., Black, M., Borrion, A., Spataru, C., Castán Broto, V., Anandarajah, G., Milligan, B., & Mulugetta, Y. (2018). Mapping synergies and trade-offs between energy and the Sustainable Development Goals. *Nature Energy, 3*, 10–15. https://doi.org/10.1038/s41560-017-0036-5

Gächter, S., Nosenzo, D., & Sefton, M. (2013). Peer effects in pro-social behavior: Social norms or social preferences? *Journal of the European Economic Association, 11*, 548–573. https://doi.org/10.1111/jeea.12015

Gifford, R., & Sussman, R. (2012). Environmental attitudes. In S. D. Clayton (Ed.), *The Oxford handbook of environmental and conservation psychology* (pp. 65–80). Oxford University Press. https://doi.org/10.1093/oxfordhb/9780199733026.013.0004

GlobeScan. (2019). *Healthy & sustainable living – A global consumer insights project.* Highlights Report. https://www.ethicalmarkets.com/healthy-sustainable-living-a-global-consumer-insights-project

Goldstein, N. J., Cialdini, R. B., & Griskevicius, V. (2008). A room with a viewpoint: Using social norms to motivate environmental conservation in hotels. *Journal of Consumer Research, 35*(3), 472–482. https://doi.org/10.1086/586910

Gómez-Olmedo, A. M., Carrero Bosch, I., & Martínez, C. V. (2021). Volition to behave sustainably: An examination of the role of self-control. *Journal of Consumer Behaviour, 20*(3), 776–790. https://doi.org/10.1002/cb.1905

Gössling, S., & Scott, D. (2018). The decarbonisation impasse: Global tourism leaders' views on climate change mitigation. *Journal of Sustainable Tourism, 26*(12), 2071–2086. https://doi.org/10.1080/09669582.2018.1529770

Hadler, M., Klösch, B., Schwarzinger, S., Schweighart, M., Wardana, R., & Bird, D. N. (2022). Measuring environmental attitudes and behaviors. In *Surveying climate-relevant behavior* (pp. 15–35). Palgrave Macmillan. https://doi.org/10.1007/978-3-030-85796-7_2

Hage, J. (2019). Law and sustainability. In B. Akkermans & G. Van Dijck (Eds.), *Sustainability and private law* (pp. 3–16). Eleven International Publishing. ISBN:978-94-6236-986-3.

Halder, P., Hansen, E. N., Kangas, J., & Laukkanen, T. (2020). How national culture and ethics matter in consumers' green consumption. *Journal of Cleaner Production, 265*, 121754. https://doi.org/10.1016/j.jclepro.2020.121754

Han, H. (2021). Consumer behavior and environmental sustainability in tourism and hospitality: A review of theories, concepts, and latest research. *Journal of Sustainable Tourism, 29*(7), 1021–1042. https://doi.org/10.1080/09669582.2021.1903019

Hansen, P. G. (2016). The definition of nudge and libertarian paternalism: Does the hand fit the glove? *European Journal of Risk Regulation, 7*(1), 155–174. https://doi.org/10.1017/S1867299X00005468

Hausman, D. M., & Welch, B. (2010). Debate: To nudge or not to nudge. *Journal of Political Philosophy, 18*(1), 123–136. https://doi.org/10.1111/j.1467-9760.2009.00351.x

Hofmann, W., Baumeister, R. F., Förster, G., & Vohs, K. D. (2012). Everyday temptations: An experience sampling study of desire, conflict, and self-control. *Journal of Personality and Social Psychology, 102*(6), 1318–1335. https://doi.org/10.1037/a0026545

Horne, R. E. (2009). Limits to labels: The role of eco-labels in the assessment of product sustainability and routes to sustainable consumption. *International Journal of Consumer Studies, 33*(2), 175–182. https://doi.org/10.1111/j.1470-6431.2009.00752.x

Imenda, S. (2014). Is there a conceptual difference between theoretical and conceptual frameworks? *Journal of Social Sciences, 38*(2), 185–195. https://doi.org/10.1080/09718923.2014.11893249

Infante, G., Lecouteux, G., & Sugden, R. (2016). Preference purification and the inner rational agent: A critique of the conventional wisdom of behavioral welfare economics. *Journal of Economic Methodology, 23*(1), 1–25. https://doi.org/10.1080/1350178X.2015.1070527

Juvan, E., & Dolnicar, S. (2017). Drivers of pro-environmental tourist behaviors are not universal. *Journal of Cleaner Production, 166*, 879–890. https://doi.org/10.1016/j.jclepro.2017.08.087

Kahneman, D. (2011). *Thinking, fast and slow*. Farrar.

Kallbekken, S., & Sælen, H. (2013). Nudging' hotel guests to reduce food waste as a win–win environmental measure. *Economics Letters, 119*(3), 325–327. https://doi.org/10.1016/j.econlet.2013.03.019

Khan, S. A. R., Yu, Z., Umar, M., & Tanveer, M. (2022). Green capabilities and green purchasing practices: A strategy striving towards sustainable operations. *Business Strategy and the Environment, 31*(4), 1719–1729. https://doi.org/10.1002/bse.2979

Kim, J., Kim, S. (Sam), Lee, J.-S., Kim, P. B., & Cui, Y. (Gina) (2020). Influence of choice architecture on the preference for a pro-environmental hotel. *Journal of Travel Research*, 59(3), 512–527. https://doi.org/10.1177/0047287519841718

Korteling, J. E. H., Paradies, G. L., & Sassen-van Meer, J. P. (2023). Cognitive bias and how to improve sustainable decision making. *Frontiers in Psychology, 14*, 1129835. https://doi.org/10.3389/fpsyg.2023.1129835

Kraus, S., Rehman, S.-U., & García, F. J. S. (2020). Corporate social responsibility and environmental performance: The mediating role of environmental strategy and green innovation. *Technological Forecasting and Social Change, 160*, 120262. https://doi.org/10.1016/j.techfore.2020.120262

Kuehnhanss, C. R. (2019). The challenges of behavioural insights for effective policy design. *Policy and Society, 38*(1), 14–40. https://doi.org/10.1080/14494035.2018.1511188

Lee, S.-Y., & Klassen, R. D. (2008). Drivers and enablers that foster environmental management capabilities in small- and medium-sized suppliers in supply chains. *Production and Operations Management, 17*, 573–586. https://doi.org/10.3401/poms.1080.0063

Lehner, M., Mont, O., & Heiskanen, E. (2016). Nudging: A promising tool for sustainable consumption behavior? *Journal of Cleaner Production, 134*(A), 166–177. https://doi.org/10.1016/j.jclepro.2015.11.086

Lenzen, M., Sun, Y. Y., Faturay, F., Ting, Y.-P., Geschke, A., & Malik, A. (2018). The carbon footprint of global tourism. *Nature Climate Change, 8*, 522–528. https://doi.org/10.1038/s41558-018-0141-x

Li, Z., Rasool, S., Cavus, M. F., & Shahid, W. (2024). Sustaining the future: How green capabilities and digitalization drive sustainability in modern business. *Heliyon, 10*(1), e24158. https://doi.org/10.1016/j.heliyon.2024.e24158

Lim, W. M. (2023). Philosophy of science and research paradigm for business research in the transformative age of automation, digitalization, hyperconnectivity, obligations, globalization and sustainability. *Journal of Trade Science, 11*(2–3), 3–30. https://doi.org/10.1108/JTS-07-2023-0015

Martins, M., Guerra, R. J., Santos, L., & Lopes, L. (2023). The exploration of the metaverse by destination management organisations towards sustainability. In Z. Tučková, S. Dey, H. H. Thai, & S. Hoang (Eds.), *Impact of industry 4.0 on sustainable tourism* (pp. 105–117). Emerald Publishing Limited. https://doi.org/10.1108/978-1-80455-157-820231007

Martins, M., Guerra, R. J., Santos, L., & Lopes, L. (2024). Aiming towards green capability in the hospitality sector using nudge strategies. In C. Martins & M. Martins (Eds.), *The image and sustainability of tourism destinations* (pp. 254–265). XV International Tourism Congress. Sciendo. https://doi.org/10.2478/9788367405287-027

Milfont, T. L. (2012). The psychology of environmental attitudes: Conceptual and empirical insights from New Zealand. *Ecopsychology, 4*(4), 269–276. https://doi.org/10.1089/eco.2012.0058

Milyavskaya, M., & Inzlicht, M. (2017). What's so great about self-control? Examining the importance of effortful self-control and temptation in predicting real-life depletion and goal attainment. *Social Psychological and Personality Science, 8*(6), 603–611. https://doi.org/10.1177/1948550616679237

Mintu, A. T., & Lozada, H. R. (1993). Green marketing education: A call for action. *Marketing Education Review, 3*(Fall), 17–23. https://doi.org/10.1080/10528008.1993.11488420

Mishra, P., & Sharma, P. (2012). Green marketing: Challenges and opportunities for businesses. *Journal of Marketing & Communication, 8*, 35–41.

Mollick, E. (2014). The dynamics of crowdfunding: An exploratory study. *Journal of Business Venturing, 29*(1), 1–16. https://doi.org/10.1016/j.jbusvent.2013.06.005

Moyle, B., Moyle, C., Ruhanen, L., Weaver, D., & Hadinejad, A. (2021). Are we really progressing sustainable tourism research? A bibliometric analysis. *Journal of Sustainable Tourism, 29*(1), 106–122. https://doi.org/10.1080/09669582.2020.1817048

Münscher, R., Vetter, M., & Scheurle, T. (2016). A review and taxonomy of choice architecture techniques. *Journal of Behavioral Decision Making, 29*(5), 511–524. https://doi.org/10.1002/bdm.1897

Murphy, J., Gretzel, U., Pesonen, J., & Elorinne, A. L. (2019). Wicked problem: Reducing food waste by tourist households. *Journal of Gastronomy and Tourism, 3*(4), 247–260. https://doi.org/10.3727/216929719X15506764047162

Nkrumah, S. K., Asamoah, D., Annan, J., & Agyei-Owusu, B. (2021). Examining green capabilities as drivers of green supply chain management adoption. *Management Research Review, 44*(1), 94–111. https://doi.org/10.1108/MRR-01-2020-0015

OECD – Organisation for Economic Cooperation and Development. (2017). *Behavioural insights and public policy: Lessons from around the world.* OECD Publishing. https://doi.org/10.1787/9789264270480

Passey, D. (2020). Theories, theoretical and conceptual frameworks, models and constructs: Limiting research outcomes through misconceptions and misunderstandings. *Studies in Technology Enhanced Learning, 1*(1). https://doi.org/10.21428/8c225f6e.56810a1a

Patwary, A. K., Mohamed, M., Rabiul, M. K., Mehmood, W., Ashraf, M. U., & Adamu, A. A. (2022). Green purchasing behavior of international tourists in Malaysia using green marketing tools: Theory of planned behavior perspective. *Nankai Business Review International, 13*(2), 246–265. https://doi.org/10.1108/NBRI-06-2021-0044

Peattie, K. (2001). Golden Goose or Wild Goose? The Hunt for the Green Consumer. *Business Strategy and the Environment, 10*, 187–199. http://doi.org/10.1002/bse.292

Prakash, A. (2002). Green marketing, public policy and managerial strategies. *Business Strategy and the Environment, 11*, 285–297. https://doi.org/10.1002/bse.338

Prieto-Sandoval, V., Torres-Guevara, L. E., & García-Díaz, C. (2022). Green marketing innovation: Opportunities from an environmental education analysis in young consumers. *Journal of Cleaner Production, 363*, 132509. https://doi.org/10.1016/j.jclepro.2022.132509

Qiu, L., Jie, X., Wang, Y., & Zhao, M. (2020). Green product innovation, green dynamic capability, and competitive advantage: Evidence from Chinese manufacturing enterprises. *Corporate Social Responsibility and Environmental Management, 27*(1), 146–165. https://doi.org/10.1002/csr.1780

Ramos, J. (2019). *Analysing the role of cognitive biases in the decision-making process*. IGI Global. https://doi.org/10.4018/978-1-5225-2978-1

Ravitch, S. M., & Riggan, M. (2016). *Reason and rigor. How conceptual framework guide research* (2nd ed.). Sage.

Reisch, L., & Gwozdz, W. (2013). Smart defaults and soft nudges: How insights from behavioral economics can inform effective nutrition policy. In J. Scholderer & K. Brunsø (Eds.), *Marketing, food and the consumer: Festschrift in Honour of Klaus G. Grunert* (pp. 189–200). Pearson Education Ltd..

Rust, R. T. (2020). The future of marketing. *International Journal of Research in Marketing, 37*(1), 15–26. https://doi.org/10.1016/j.ijresmar.2019.08.002

Šagovnović, I., & Stamenković, I. (2022). Investigating values of green marketing tools in predicting tourists' eco-friendly attitudes and behavior. *Journal of Ecotourism, 22*(4), 479–501. https://doi.org/10.1080/14724049.2022.2075003

Salas Canales, H. J. (2018). Marketing ecológico: La creciente preocupación empresarial por la protección del medio ambiente. *Fides et Ratio - Revista de Difusión cultural y científica de la Universidad La Salle en Bolivia, 15*(15), 151–170. http://www.scielo.org.bo/scielo.php?pid=S2071-081X2018000100010&script=sci_abstract

Schmidt, R. (2024). A model for choice infrastructure: Looking beyond choice architecture in behavioral public policy. *Behavioural Public Policy, 8*(3), 415–440. https://doi.org/10.1017/bpp.2021.44

Schultz, P. W., Shriver, C., Tabanico, J. J., & Khazian, A. M. (2004). Implicit connections with nature. *Journal of Environmental Psychology, 24*(1), 31–42. https://doi.org/10.1016/S0272-4944(03)00022-7

Shang, K.-C., Lu, C.-S., & Li, S. (2010). A taxonomy of green supply chain management capability among electronics-related manufacturing firms in Taiwan. *Journal of Environmental Management, 91*(5), 1218–1226. https://doi.org/10.1016/j.jenvman.2010.01.016

Shi, J., Yang, D., Zheng, Z., & Zhu, Y. (2022). Strategic investment for green product development and green marketing in a supply chain. *Journal of Cleaner Production, 366*, 132868. https://doi.org/10.1016/j.jclepro.2022.132868

Smith, K. T., & Brower, T. R. (2012). Longitudinal study of green marketing strategies that influence millennials. *Journal of Strategic Marketing, 20*(6), 535–551. https://doi.org/10.1080/0965254X.2012.711345

Snyder, H. (2019). Literature review as a research methodology: An overview and guidelines. *Journal of Business Research, 104*, 333–339. https://doi.org/10.1016/j.jbusres.2019.07.039

Song, W., Ren, S., & Yu, J. (2018). Bridging the gap between corporate social responsibility and new green product success: The role of green organizational identity. *Business Strategy and the Environment, 28*(1), 88–97. https://doi.org/10.1002/bse.2205

Steg, L., Perlaviciute, G., & Van der Werff, E. (2015). Understanding the human dimensions of a sustainable energy transition. *Frontiers in Psychology, 6*, 805. https://doi.org/10.3389/fpsyg.2015.00805

Steg, L., Perlaviciute, G., Van der Werff, E., & Lurvink, J. (2014). The significance of hedonic values for environmentally relevant attitudes, preferences, and actions. *Environment and Behavior, 46*(2), 163–192. https://doi.org/10.1177/0013916512454730

Sunstein, C. (2017). Nudges that fail. *Behavioural Public Policy, 1*(1), 4–25. https://doi.org/10.1017/bpp.2016.3

Tagliabue, M., Squatrito, V., & Presti, G. (2019). Models of cognition and their application in behavioral economics: A conceptual framework for nudging derived from behavior analysis and relational frame theory. *Frontiers in Psychology, 10*, 2418. https://doi.org/10.3389/fpsyg.2019.02418

Thaler, R. H., & Sunstein, C. R. (2008). *Nudge: Improving decisions about health, wealth, and happiness*. Yale University Press.

Thaler, R. H., Sunstein, C. R., & Balz, J. P. (2013). *Choice architecture*. In E. Shafir (Ed.), *Foundations of public policy* (pp. 428–439). Princeton University Press. https://doi.org/10.13140/2.1.4195.2321

UNEP – United Nations Environment Programme. (2023). *Emissions Gap Report* 2023. https://doi.org/10.59117/20.500.11822/43922

Visser-Amundson, A. (2020). A multi-stakeholder partnership to fight food waste in the hospitality industry: A contribution to the United Nations Sustainable Development Goals 12 and 17. *Journal of Sustainable Tourism, 30*(10), 2448–2475. https://doi.org/10.1080/09669582.2020.1849232

Wang, S., Wang, J., Li, J., & Yang, F. (2020). Do motivations contribute to local residents' engagement in pro-environmental behaviors? Resident-destination relationship and pro-environmental climate perspective. *Journal of Sustainable Tourism, 28*(6), 834–852. https://doi.org/10.1080/09669582.2019.1707215

Wang, X., Van der Werff, E., Bouman, T., Harder, M. K., & Steg, L. (2021). I am vs. we are: How biospheric values and environmental identity of individuals and groups can influence pro-environmental behavior. *Frontiers in Psychology, 12*, 618956. https://doi.org/10.3389/fpsyg.2021.618956

White, K., Habib, R., & Hardisty, D. J. (2019). How to SHIFT consumer behaviors to be more sustainable: A literature review and guiding framework. *Journal of Marketing, 83*(3), 22–49. https://doi.org/10.1177/0022242919825649

Witjes, S., Vermeulen, W. J. V., & Cramer, J. M. (2017). Assessing corporate sustainability integration for corporate self-reflection. *Resources, Conservation and Recycling, 127*, 132–147. https://doi.org/10.1016/j.resconrec.2017.08.026

Wong, J. W. C., & Lai, I. K. W. (2024). Influence of the green certification logo and text on environmental awareness and hotel guests' water-saving behaviour. *Journal of Travel & Tourism Marketing, 41*(4), 491–507. https://doi.org/10.1080/10548408.2024.2310763

Wyss, A. M., Knoch, D., & Berger, S. (2022). When and how pro-environmental attitudes turn into behavior: The role of costs, benefits, and self-control. *Journal of Environmental Psychology, 79*, 101748. https://doi.org/10.1016/j.jenvp.2021.101748

Zhao, G., Geng, Y., Sun, H., Tian, X., Chen, W., & Wu, D. (2022). Mapping the knowledge of green consumption: A meta-analysis. *Environmental Science and Pollution Research, 27*, 44937–44950. https://doi.org/10.1007/s11356-020-11029-y

Taxation on Non-Recoverable Packaging as a Sustainability Measure in Tourist-Intensive Regions

Susana Cristina Rodrigues Aldeia

Abstract The main objective of this paper is to examine the legislative measures the Portuguese government took regarding single-use packaging and how these fiscal instruments constitute a sustainability measure in tourist-intensive regions. The research methodology includes a review of legislative documents. The application of the polluter pays principle necessarily leads to environmental taxation as an instrument for preventing, restoring, and combating pollution. Taxes are used to raise funds for environmental protection. From this perspective, the tax principle requires that the costs of pollution be borne by those responsible. In an extra-fiscal way, it encourages a conservative operating environment and combats polluting behaviour. In conclusion, this study highlights the significant intersection between tourism, single-use packaging regulation, and climate change mitigation, suggesting that effective legislative measures can promote sustainable tourism practices.

1 Introduction

The use of plastic leads to pollution, harms marine life, and contributes to the climate crisis (Albuquerque, 2019). Frequently plastics are made from fossil fuels, and the industry's impact is evident. While renewable sources like wood fibres and algae can be used, they are not widely used in industry. Most plastic materials can be recycled, but most are not due to challenges such as contamination, harmful additives, and lack of economic incentives (Kamalakkannan et al., 2022). This delays sending materials to landfills or incinerators, contributing to the accumulation of plastic waste and greenhouse gas emissions, ultimately impacting the climate crisis and ecosystems (Kim, 2004).

S. C. R. Aldeia (✉)
Universidade Portucalense | REMIT – Research on Economics, Management and Information Technologies, Porto, Portugal
e-mail: saldeia@upt.pt

J. Marques et al. (eds.), *Tourism and Climate Change*, Tourism, Hospitality & Event Management, https://doi.org/10.1007/978-3-031-99128-8_7

Plastic's widespread use in everyday life, including short-life products often not designed for effective reuse or recycling, negatively impacts the environment, economy, and health (da Silva et al., 2022). The European Commission (EU) has highlighted the need for a specific legal regime to address these issues through the "Closing the Loop—An EU Action Plan for the Circular Economy" (European Commission, 2015).

The European Green Deal has launched a strategy to achieve climate neutrality by 2050 (Fetting, 2020). The Circular Economy Action Plan aims to reduce waste, implement sustainable services, product-as-service models, and digital solutions, and ensure a well-functioning internal market for high-quality secondary raw materials (European Commission, 2020a). The EU aims to tackle plastic pollution by introducing mandatory requirements for recycled content and waste reduction measures in crucial products and capturing microplastics throughout product lifecycles (European Commission, 2015). Therefore, the Commission is working to guarantee the timely implementation of the Single-Use Plastic Products Directive (European Commission, 2019). The focus is on the harmonised interpretation of covered products, product labelling, and rules on measuring recycled content. The aim is to prevent plastic littering and pollution while safeguarding the single market (European Commission, 2020).

The Commission is creating a policy framework to tackle sustainability issues related to bio-based and biodegradable plastics. The focus is on assessing environmental benefits and regulating product labelling to prevent confusion and pollution. The Directive on Single-Use Plastic Products has been implemented to harmonise interpretation, classification, and rules for measuring recycled content. The goal is to prevent plastic pollution, protect the single market, and promote sustainable consumption and production patterns (European Commission, 2019).

Therefore, the main objective of this document is to analyse the methods through which Portuguese policymakers have implemented the European Union's directives regarding the taxation of plastic materials within the country. Specifically, the focus is on examining the legislative measures the Portuguese government took in relation to single-use packaging. The study applies the legal research methodology by studying Portuguese regulation.

The results show that legislator has introduced some legal measures to promote regulation. The matter has been regulated and clarified, particularly by the approval of Law no. 75-B/2020 of December 31 (State Budget for 2021) and by Ordinance no. 331-E/2021, of December 31. Subsequently, the Tax Authority issued an Ordinance no. 35.189, of January 3, 2023. More recently, Law 75-B/2020 was repealed by the State Budget Law for 2024. Articles 278 and 279 add to Law no. 82-D/2014, of December 31, Chapter VI, "Contribution on single-use packaging". The chapter includes Articles 49-A to 49-P.

Law no. 75-B/2020, approved on December 31, 2020, passed the State Budget for 2021 and introduced a contribution to single-use packaging of plastic, aluminium, or multi-material, that includes plastic or aluminium, that is purchased in ready-to-eat meals, takeaway, or home delivery schemes. Article 320 of the law outlines the rules and general principles of application of this Contribution. On

December 31, 2021, Ordinance No. 331-E/2021 was issued to regulate this Contribution. This measure aims to promote national environmental policy objectives in the transition towards a circular economy (European Commission, 2020a). Its goal is to encourage a sustained decrease in the consumption of single-use packaging and the resulting reduction in packaging waste (Neto, 2023). Additionally, it seeks to introduce more environmentally friendly reuse systems (da Costa, 2021).

Starting from July 1, 2022, a contribution of 30 cents per package, plus Value-Added Tax (VAT), applies to plastic or multi-material packaging with plastic that has been introduced into consumption since that date. Similarly, from January 1, 2023, the Contribution applies to aluminium or multi-material packaging with aluminium.

To meet legal requirements, the invoice must provide a detailed breakdown of the levy along the entire supply chain, from packaging production to end-user consumption (Carlos et al., 2024). It includes specifying the product as "single-use packaging", indicating the number of units sold or provided, and detailing the pricing, including the applicable Contribution.

When considering eateries that offer pre-made meals and disposable packaging, it is up to the establishment whether or not to include the cost of the packaging in the meal price. This decision is part of their overall pricing strategy. However, the cost must be passed on to the customer if they provide single-use packaging.

According to recent clarifications from the Tax Authority, published on the Finance Portal, the delivery of plastic or multi-material packaging with plastic is now subject to a 30 cents contribution as of July 1, even if no value is charged for the packaging. Both the value of the packaging, when charged, and the value of the Contribution are part of the taxable value of meals sold on a ready-to-eat or take-away basis or with home delivery. Therefore, on the invoice to be issued for the sale of ready-to-eat meals, the intermediate VAT rate of 13 % must be applied to the meal itself, the value of the packaging (if charged) and the value of the Contribution (which must be charged), and these amounts must be itemised.

Clearly, the amounts corresponding to disposable packaging and the respective Contribution are part of the taxable value of meals sold on a ready-to-eat or take-away or home delivery basis, and the VAT rate of 13% applies to them. It should be noted that this Contribution does not apply to the Autonomous Regions of Azores and Madeira.

Since single-use packaging and the respective Contribution are passed on outside the scope of the supply of ready-to-eat meals, under the ready-to-eat or home delivery systems, namely between producers, wholesalers, and retailers, they are taxed at the standard VAT rate of 23 %. In a restaurant, the taxable amount subject to VAT then corresponds to the value of the meal itself, plus the value of the packaging, plus a contribution, plus VAT at the intermediate rate. Thus, the tax framework for single-use packaging highlights important tax principles, such as the polluter pays principle.

2 Plastics and Climate Change

The impact of plastic pollution is a significant environmental concern, as it often leads to marine litter that poses serious threats to marine life. Many marine animals are at risk of suffocation or injury due to the ingestion of plastic bags and bottles, highlighting the urgent need for effective waste management and reduction strategies (van Leeuwen & Surya, 2024). However, rubbish in the oceans is not the only consequence of the use of plastics; the production of these products has a strong environmental impact (Azman et al., 2024). The plastics industry is a source of greenhouse gases and therefore has a strong impact on the climate crisis (Zhang et al., 2024).

Plastics are made up of a group of materials with different characteristics. However, most plastics are made from fossil fuels, such as oil, coal, or gas. In this context, the relationship between the fossil fuel industry and plastic production is clear. This is why several fossil fuel companies invest in plastic production industries. However, renewable sources can be used in the production of plastics, such as wood fibres or algae. Worldwide, this industry does not have a significant representation.

Although the majority of plastic materials can be recycled, most are not (Kuroda et al., 2023). There are several challenges to effective recycling, for example: the contamination of plastic waste, the existence of harmful chemical additives (Jang et al., 2024), and the lack of economic incentives. As a result, a significant part of the recycling that takes place today is not real recycling. There is a delay in sending materials to landfills or incinerators, contributing to the accumulation of plastic waste (Riber et al., 2009). In this way, plastics contribute significantly to the climate crisis, because due to the greenhouse gas emissions from its entire production process, plastic harms ecosystems.

The functionality and low cost of plastic mean that it is frequently used in everyday life (Tran et al., 2024). Although plastic plays an important role in the economy, with applications in many sectors, its increasing use in short-life products is a reality (Albuquerque, 2019). Often, short-use products are not designed to be reused or recycled effectively.

The impact on the environment, the economy, and health of some plastic products requires the implementation of a specific legal regime to promote the effective reduction of these negative effects (Kamalakkannan et al., 2022).

In this context, the European Commission's Communication of December 2 2015, entitled "Closing the loop—An EU action plan for the circular economy", determines the need for a European strategy for plastics. This action stems from the action plan for the circular economy.

On January 16, 2018, the European Commission (EC) issued a new communication entitled "European Strategy for Plastics in a Circular Economy". This action promotes increasing the circular life cycle of plastics. It is important to find a solution to the constant increase in the production of plastic waste and the dispersion of plastic waste in the environment, particularly in the marine environment (Zuo et al.,

2024). The European strategy for plastics is a step towards creating a circular economy in which the design and production of plastics and plastic products fully respect the need to reuse, repair, and recycle, and in which more sustainable materials are developed and promoted. The extension of the circular economy to the different economic agents in general contributes to achieving climate neutrality.

The issue of marine litter is recognised as a growing global problem. The United Nations (UN) Sustainable Development Goal 14 (SDG) promotes the reduction of litter at sea by sustainably conserving the oceans and marine resources (Borras & Rivera, 2022).

The EU therefore has an important role to play in moving towards a regenerative growth model, in which resource consumption does not exceed the limits of the planet itself. We often see the "extract, manufacture, use, and throw away" pattern that is characteristic of the linear economy (European Commission, 2020b). The production process requires more materials to be used in packaging. At the same time, there is an excessive consumption of plastic, estimated to double in the next 20 years (European Commission, 2020, 2020a).

In this context, the European Union is essential in preventing and combating pollution. The European bloc has been working with its partners (G20, G7, and UN) to develop concerted action. The EU must demonstrate a commitment to ensuring that its exports of plastic waste do not increase marine litter in other regions of the planet.

3 Green Taxation Agenda in the European Union

The European Union recognises that climate change and ecological degradation pose significant existential threats to Europe and the world. This understanding is reflected in the information on the European Commission's website, highlighting the profound concerns and commitment to addressing these critical issues.[1] It is imperative for the European Union to undergo a comprehensive transformation to evolve into a modern economy that can effectively mitigate and adapt to the pressing environmental challenges of our time (Mia et al., 2021).

The EU's ambitious European Green Deal seeks to transform the European Union into a modern, efficient, and competitive economy (Sun et al., 2011). The primary objectives include achieving net-zero greenhouse gas emissions by 2050, promoting economic growth independent of resource consumption, and ensuring that all individuals and regions benefit from these efforts (Zicari & Renouard, 2018). To meet these goals, the European Commission has proposed reducing net greenhouse gas emissions by at least 55% by 2030, compared to 1990 levels, across critical sectors such as climate, energy, transport, and taxation (European Commission). The EU's ultimate aim is to lead the world as the first climate-neutral continent while simultaneously staying on course to meet its climate ambitions by 2030 in an

[1] https://climate-pact.europa.eu/index_en

equitable, cost-effective, and competitive manner. The EU is actively pursuing various initiatives and strategies to realise this vision.

The Objective 55 package comprises a set of measures aimed at reviewing and revising existing EU legislation, as well as introducing new initiatives. The goal is to ensure that EU policies align with the climate objectives set by the Council and the European Parliament. These proposals establish a comprehensive and equitable framework for attaining the EU's climate objectives. Specifically, the package aims to (1) facilitate a fair and socially equitable transition, (2) bolster innovation and competitiveness in EU industry while ensuring a level playing field with economic entities from non-EU countries, and (3) support the EU's leadership role in the global effort to combat climate change. The name "Objective 55" reflects the EU's target of reducing greenhouse gas emissions by at least 55% by 2030 (Klotz et al., 2022). The proposed package seeks to harmonise EU legislation with the 2030 target.

From a financial perspective, the European Union has implemented several measures to advance an ambitious green fiscal agenda. These initiatives include (i) the plastics tax, (ii) the Energy Tax Directive (ETD), (iii) the Carbon Border Adjustment Mechanism (CBAM), (iv) the new Emissions Trading Scheme (ETS 2), and the existing Emissions Trading Scheme (ETS 1). For example, the plastics tax involves collecting revenue based on the consumption of non-recycled plastic. Each member state has the option to introduce its own plastics tax. In the United Kingdom, starting from April 1, 2022, any packaging primarily made of plastic with less than 30% recycled content is subject to a tax of £0.21083 per kilogramme. Similarly, in Spain, non-reusable plastic packaging incurs a tax of €0.45 per kilogramme, a measure that took effect on January 1, 2023.

4 The Taxation of Plastics in the European Union

On June 5 2019, the European Parliament and the Council of the European Union introduced Directive (EU) 2019/904. This regulation addresses the issue of reducing the impact of plastic products on the environment. In particular, it promotes a circular approach that prioritises sustainable products and reuse systems. It, therefore, encourages a reduction in the use of single-use products and, consequently, the waste generated (European Commission, 2019). In this sense, this Directive contributes to the realisation of Sustainable Development Goal 12. SDG 12 aims to ensure sustainable consumption and production patterns in which the value of products and materials is preserved for as long as possible (Di Vaio et al., 2024). This practice makes it possible to generate less waste and meets the objectives set out in the 2030 Agenda for Sustainable Development (Omeje, 2021).

The issue of pollution in the seas is transboundary and a growing global problem. In this sense, the UN has also recognised the importance of the issue and has determined that reducing marine litter is a fundamental action for implementing the Sustainable Development Goal. The European Union has worked with international partners (G20, G7, and UN) to take on a standard-setting role capable of combating and preventing litter.

Several documents[2] emphasise that member states must promote efficient waste management. This action is fundamental to preventing and reducing marine litter from the sea and the land. According to the European Commission, 80 per cent of marine litter is made up of plastic, but single-use plastic items account for 50 per cent (European Commission, 2019). This category of plastics includes many everyday consumer products discarded after being used only once. Single-use products are rarely recycled and tend to become rubbish. Therefore, this type of product is a severe problem in marine litter, as it poses a risk to biodiversity, marine ecosystems, and human health. Consequently, it causes damage to the tourism sector.

Therefore, proper waste management is essential to prevent all types of rubbish, including marine litter. Regulations at the European Union level have already provided some solutions to the issue of marine litter.[3] However, the impact of these measures is not yet sufficient, and there are differences in the adoption of these measures between member states. It justifies the need to promote measures that cover the use of single-use plastic products. Identifying these products is part of the criteria for the most commonly found products on beaches in the European Union. European Union estimates indicate that the single-use plastic products covered by measures under the Directive represent around 86 % of the single-use plastic products found on beaches in the EU (European Commission, 2019).

The Directive determines the concept of "single-use plastic product", the standard clarifies that products produced to make multiple journeys or rotations in their life cycle are excluded. Single-use plastic products are intended to be used only once (or for a short period).

According to the European Directive, examples of single-use products are (1) food containers, (2) fast food containers, (3) meal boxes for sandwiches, wraps and salads, with hot or cold food, (4) containers for fresh or processed food that requires no further preparation, such as fruit, vegetables, or desserts. The standard explains that food containers that contain (i) dry food or food sold cold that requires further preparation, (ii) containers that hold more than one individual serving of food, or (iii) food containers that are sized for one individual serving and are sold in more than one unit are not considered single-use plastic products.

According to paragraph 13 of the Directive, single-use plastic products should be subject to one or more measures. In particular, consideration should be given to (i)

[2] The United Nations Convention on the Law of the Sea of 10 December 1982 (UNCLOS) (5), the Convention on the Prevention of Marine Pollution by Dumping of Wastes and Other Matter of 29 December 1972 ('London Convention') and its 1996 Protocol ('London Protocol'), Annex V to the International Convention for the Prevention of Pollution from Ships 1973 (MARPOL), as modified by the Protocol of 1978 relating thereto, and the Basel Convention on the control of transboundary movements of hazardous wastes and their disposal of 22 March 1989 (6) and with Union waste legislation, namely Directive 2008/98/EC and Directive 2000/59/EC of the European Parliament and of the Council.

[3] Directives 2008/98/EC, 2000/59/EC, 2000/60/EC and 2008/56/EC and Council Regulation (EC) No 1224/2009.

the availability of suitable and more sustainable alternatives, (ii) the feasibility of changing consumption patterns, and (iii) the extent to which they are already covered by existing EU legislation.

In this context, Directive (EU) 2019/904 requires Member States to take the necessary measures to achieve a significant and sustained decrease in the consumption of single-use plastic products. Part A of the Annex to the Directive immunises the products to be considered in this context. This type of product includes drinking glasses (including lids) and food containers such as food boxes (with or without lids). It includes food containers used in fast food restaurants.

5 The Portuguese Single-Use Packaging Tax Regime

In response to the European Union Directives on Packaging and Packaging Waste (Directive 94/62/EC and Directive 2019/904, both of the European Parliament and of the Council), the Portuguese legislator introduced the single-use packaging levy. These regulations aim to implement tax policies that promote the reduction of pollution caused by plastic packaging waste. Therefore, they aim to reduce the consumption of single-use packaging and encourage the use of reusable packaging from 2022 onwards in the catering sector (Neto, 2023).

This Contribution covers single-use plastic or aluminium packaging or multi-material packaging (with plastic or aluminium). The Contribution has an environmental purpose: to promote the reuse of plastic or aluminium packaging in the catering sector, provided it is purchased in ready-to-eat meals.

At the legislative level, Law no. 75-B/2020, of December 31 (State Budget for 2021), created this Contribution to single-use plastic or aluminium packaging for ready meals. This law aims to incentivise the reduction of the environmental impact of specific plastic products. Ordinance 331-E/2021, of December 31, regulated this Contribution, amended by Ordinance 312-C/2022, of December 30. Recently, the Tax Authority issued Circular Letter no. 35.189 of January 3, 2023. This letter seeks to clarify key aspects of the regime. In the meantime, Order no. 270/2023, of August 29, 2023, made the second amendment to Order no. 331 E/2021, of December 31. Initially, the Contribution was scheduled to start applying on January 1, 2023, and was then postponed to September 1, 2023. Finally, given the constraints expressed by various economic agents, as well as the need to extend the scope of application of the ordinance to other materials, the legislator decided, in the meantime, to extend the application of the Contribution to aluminium packaging or multi-material with aluminium to January 1, 2024. As such, the State Budget for 2024 repealed Article 320 of Law 75-B/2020 of December 31 and Ordinance 331-E/2021 of December 31. In this sense, the Contribution on single-use packaging used in ready-to-eat meals is now included in Chapter VI of Law no. 82-D/2014 (Green Tax Law).

In this context, this Contribution applies to primary packaging, including single-use (non-reusable) service packaging for food and drink. The products are made from plastic, aluminium, or multi-material (with plastic or aluminium), and they are purchased in ready-to-eat meals, takeaway, or home delivery schemes. The law stipulates that if the non-reusable packaging consists of more than one part (where the parts are placed on the market separately), the Contribution applies to the main component, i.e. the container.

This Contribution also covers single-use packaging for ready-to-eat meals. This measure applies even if the meals have not been cooked to the final consumer at the point of sale.

Producers and importers (with their head office or permanent establishment in mainland Portugal) are liable for the tax. Purchasers of single-use packaging from suppliers with registered offices or permanent establishments in the Autonomous Regions of the Azores and Madeira or another EU Member State are also liable.

The Portuguese legislator excludes from this tax single-use packaging that contains ready-to-eat meals, but which (i) have not been packaged at the point of sale to the final consumer, (ii) those made available in the context of non-sedentary catering or drinking activities, and (iii) those made available through vending machines for the supply of ready-to-eat meals.

Exporting single-use packaging is also exempt from this tax. In particular, packaging that is dispatched or transported to another Member State of the European Union or the Autonomous Regions of the Azores and Madeira. In addition, packaging used in a social or humanitarian context is exempt from the tax. Packaging used in the social distribution of food (or to combat food waste) by charities or other organisations when donating meals is also exempt.

It should be noted that this Contribution applies to any plastic. The origin of the material, whether biological or fossil, does not matter. In the same way that its recyclability is irrelevant, it is essential that it is used only for the purposes described above.

The tax on single-use packaging is set at €0.30 (thirty cents). This tax is subject to value-added tax and is charged to the final consumer. Therefore, the economic agents in the commercial chain must pass on the economic burden of the tax to the purchaser.

The amount of the tax must be itemised on the invoice throughout the entire economic chain. This economic circuit runs to the final consumer, who is the purchaser of the packaging. To this end, the invoice must include "single-use packaging". Other information must be included on the invoice, such as the number of units sold (or made available), the amount charged as a price, and VAT.

The release for consumption operation is formalised through the electronic submission of the "declaration of release for consumption" ("e-DIC") or, upon import, through the respective customs declaration. The e-DIC must be processed quarterly by the 5th of the month following the end of each quarter in which the releases for consumption occurred. Taxpayers are then notified by the 15th of the same month to make their payment, which must be made by the 15th of the second month following the quarter.

Receipt and storage of this packaging can only be formalised through a tax warehouse, so taxpayers must apply for authorised warehouse status with the competent customs authority. They must also set up a tax warehouse for production and storage if they do not comply with these procedures.

6 The Tax on Non-Returnable Packaging in the Food Sector and Tourism's Sustainability

Many economies around the world are dependent on tourism, and this dependence has positive and negative economic, fiscal, and social ramifications (Aguero et al., 2006). The sustainability question of basic industry is particularly pertinent, for example in Clark County, Nevada, 30% of the region's workforce is directly employed in the leisure and hospitality industry and that sector represents the lion's share of state and local tax collections (Aguero et al., 2006).

Moreover, the importance of sustainable tourism is unquestionable, tourism and hotel companies around the world are moving towards more sustainable product offerings (Tiago et al., 2021). The United Nations World Tourism Organization (UNWTO) defines sustainable tourism as "tourism that takes full account of its current and future economic, social, and environmental impacts, and addresses the needs of visitors, the industry, the environment, and host communities" (Sarigiannidis et al., 2023). Therefore, there is a need to balance environmental protection with the growth of the global tourism industry (Suhardono et al., 2024).

Sustainability is itself becoming an element of differentiation and competitiveness in tourism markets (Durán-Román et al., 2022). However, the transition to sustainable tourism is a difficult task that requires time, money, and the will of the stakeholders (Durán-Román et al., 2022). For example, the incorrect management of municipal solid waste (MSW) leads to the contamination of water, soil, and air, with negative impacts on human health (Molinos-Senante et al., 2022). The lack of treatment of residual waste has huge implications for the fulfilment of circular economy objectives (Molinos-Senante et al., 2022). In Chile, the cost of reducing residual waste was 81.10 US dollars per kilo (Molinos-Senante et al., 2022). Therefore, sustainable consumption focuses on the behaviour of individuals and the characteristics of products (Melo et al., 2022).

Environmental taxes are an important tool for governments to adjust the relative prices of goods and services (OECD, 2017). Green taxes are an instrument for the development of many destinations where excessive tourism generates different externalities, helping to mitigate them and create sustainable tourism (Ortega-Rodríguez et al., 2024). The aim of green taxation is to raise awareness of environmental policy, as well as to carry out a public tax reform (Kim & Hyun, 2021). The funds obtained through green taxes can be used to finance conservation, environmental recovery, and sustainable development initiatives (Ortega-Rodríguez et al., 2024).

The maturity of tourist destinations, coupled with the growing competitiveness of the sector and the evolution of tourist habits, requires the implementation of a series of strategies to increase the sustainability of these destinations and improve the tourist experience (Durán-Román et al., 2021). Thus, imposing taxes and/or levies on different tourist activities has become a viable option for financing these policies (Durán-Román et al., 2021).

Tourism taxation is one of the tools that can make an effective contribution to obtaining resources that favour the development of policies to improve sustainability and the tourist experience in the destination (Cárdenas-García et al., 2022). However, these taxes are often unknown to tourists visiting a city and can generate distrust and even discomfort when they are forced to pay them (Ortega-Rodríguez et al., 2024). In terms of management implications for destinations, green taxes should be seen and passed on as a means of achieving the economic and environmental sustainability of destinations and not as yet another tax to be borne by tourists (Ortega-Rodríguez et al., 2024).

Additionally, the polluter pays principle is a concept that directly applies in this scenario. This principle asserts that the party responsible for pollution should bear the costs of managing it to prevent harm to the environment or human health. In the context of consumer goods, this translates to consumers paying a tax when they purchase products that contribute to environmental pollution. This tax serves as a form of accountability, placing an additional financial burden on those who consume non-recyclable or environmentally harmful products. In contrast, consumers who opt for recyclable packaging are not subjected to this additional financial responsibility, as they are not contributing to the same level of pollution.

The polluter pays principle is a widely accepted environmental policy approach that aims to hold those responsible for pollution financially accountable (Aldeia, 2023). This principle leads to implementing environmental taxation as a means of preventing and combating pollution while aiding in its recovery (Rahman & Edwards, 2004). Environmental taxation is carried out through both fiscal and extra-fiscal means (Glazyrina et al., 2006). In fiscal approaches, tax values are adjusted to generate funds specifically earmarked for environmental protection (Fishman et al., 2012). It ensures that the costs of pollution are incurred by those responsible for it, aligning with the fundamental principle of the polluter pays approach (Baldock, 1992).

Moreover, extra-fiscal methods of environmental taxation help create a regulatory framework that promotes a conservative operational environment and discourages polluting behaviour (Milon, 2019). The literature emphasises that taxpayers need to comprehend the financial implications of taxation, reinforcing the importance of understanding the effects of environmental taxation (Chamizo-Gonzalez et al., 2018). It is well-documented that company managers consider the tax burden associated with their decisions, highlighting the significant impact of environmental taxation on business operations and financial considerations (Schmidtchen et al., 2021).

In the light of the polluter pays principle, the implementation of the single-use packaging levy requires tourists to honour the consumption of this type of packaging. Tourists who consume more of this type of packaging pay more environmental tax. On the other hand, tourists who are more aware of the consumption of non-reusable plastic/aluminium packaging are subject to less taxation.

Negative aspects of implementing the contribution in the catering sector could include an increase in the cost of meals, with a potential impact on the revenue of economic operators and the state. A wide range of contributions of this nature have proliferated in the Portuguese legal system, which has led to the creation of a (parallel) tax system made up of coercive measures of the same nature as taxes that are not called taxes.

7 Conclusion

This paper aims to understand how Portuguese legislators have been incorporating the European Union's guidelines on the taxation of plastics at home. In particular, it examines what measures the Portuguese legislature has adopted regarding single-use packaging in the tourism sustainability context.

As Portugal is a member of the European Union, it is only natural to follow EU guidelines. In this case, the Portuguese legislator introduced a levy on single-use plastic products for restaurants.

The implementation of the polluter pays principle inherently involves the use of environmental taxation as a vital tool for preventing, restoring, and addressing pollution issues and of the sustainable tourism. Through these taxes, governments can generate significant funding dedicated to protecting and maintaining the environment. This principle of taxation asserts that those who contribute to environmental degradation should bear the financial responsibility for the costs associated with that damage.

Additionally, it promotes a more sustainable and responsible approach to resource management by discouraging harmful practices and incentivising more environmentally friendly behaviours. In this way, the polluter pays principle not only holds polluters accountable but also fosters a more conservative and mindful approach to environmental stewardship.

Therefore, there is a straightforward application of the polluter pays principle, as the consumer pays the tax if they buy this type of product. It places an additional tax on those who consume the most, which is not borne by those who use recyclable packaging.

References

Aguero, J., Conway, S., & Navis, I. (2006, September 06–08). Community monitoring in a tourism-dependent economy. *Wit transactions on ecology and the Environment* [Sustainable Tourism II]. In 2nd International Conference on Sustainable Tourism, Bologna, ITALY.

Albuquerque, S. (2019). *Contributo para a redução dos plásticos descartáveis provenientes da praça de alimentação (foodcourt) do LoureShopping [Contribution to the reduction of single-use plastics from LoureShopping's foodcourt]*. ISEG - Universidade de Lisboa.

Aldeia, S. C. R. (2023). Environmentally friendly tax measures in the Portuguese jurisdiction. Enforcement of the polluter pays tax principle [Article; Early Access]. *International Journal of Law and Management, 65*(6), 511–522. https://doi.org/10.1108/ijlma-02-2023-0018

Azman, D. Q., Daud, W., Patah, M. F. A., Amir, Z., & Saw, P. A. (2024, May). Plastic waste management through liquefaction in hydrogen donating solvents: A review [Article]. *Journal of Environmental Management, 359*, 18, Article 120961. https://doi.org/10.1016/j.jenvman.2024.120961

Baldock, D. (1992). The polluter pays principle and its relevance to agricultural policy in European Countries [Article]. *Sociologia Ruralis, 32*(1), 49–65. https://doi.org/10.1111/j.1467-9523.1992.tb00918.x

Borras, M. F., & Rivera, V. S. (2022, May–June). Implementing SDG 14 under principles, ethical values, and environmental and economic justice for small-scale artisanal fishermen in Costa Rica [Editorial Material]. *Marine Technology Society Journal, 56*(3), 25–30.. <Go to ISI>://WOS:000811597500006

Cárdenas-García, P. J., Pulido-Fernández, J. I., Durán-Román, J. L., & Carrillo-Hidalgo, I. (2022, July). Tourist taxation as a sustainability financing mechanism for mass tourism destinations [Article]. *International Journal of Tourism Research, 24*(4), 577–592. https://doi.org/10.1002/jtr.2523

Carlos, A. B., Abreu, I. A., Durão, J. R., & Pimenta, M. E. (2024). *Guia dos impostos em Portugal [Guide to taxes in Portugal]*. Quid Juris.

Chamizo-Gonzalez, J., Cano-Montero, E. I., & Munoz-Colomina, C. I. (2018, May). Does funding of waste services follow the polluter pays principle? The case of Spain [Article]. *Journal of Cleaner Production, 183*, 1054–1063. https://doi.org/10.1016/j.jclepro.2018.02.225

da Costa, M. R. (2021). *Alternativas aos plásticos de utilização única, foco em soluções comestíveis [Alternatives to single-use plastics, focus on edible solutions]*. Universidade Nova de Lisboa.

da Silva, D. L. B., Gil, J., do Nascimento, E. P., Costa, H. A., & Paixao, R. (2022). Plastic pollution in the Brazilian coast: Perceptions from hotel managers on the consumption of disposables [Article]. *Revista Brasileira De Pesquisa Em Turismo, 16*, 18. https://doi.org/10.7784/rbtur.v16.2481

Di Vaio, A., Dell'Amura, G., Chhabra, M., & Garofalo, A. (2024, May). Circular economy and waste production models for sustainable development goals 12 and 14: Evidence from cruise sustainability reporting [Article; Early Access]. *Sustainable Development, 17*. https://doi.org/10.1002/sd.3051

Durán-Román, J. L., Cárdenas-García, P. J., & Pulido-Fernández, J. I. (2021, January). Tourist tax to improve sustainability and the experience in mass tourism destinations: The Case of Andalusia (Spain) [Article]. *Sustainability, 13*(1), 20, Article 42. https://doi.org/10.3390/su13010042

Durán-Román, J. L., Pulido-Fernández, J. I., Rey-Carmona, F. J., & Núñez-Tabales, J. M. (2022, September). Willingness to pay by tourist companies for improving sustainability and competitiveness in a mature destination [Article; Early Access]. *Leisure Sciences, 22*. https://doi.org/10.1080/01490400.2022.2123072

European Commission. (2015). *Closing the loop - An EU action plan for the Circular Economy*.

European Commission. (2019). *Directive (EU) 2019/904 of the European Parliament and of the Council of 5 June 2019 on the reduction of the impact of certain plastic products on the environment (Text with EEA relevance)*.

European Commission. (2020). *Delivering the European Green Deal*. https://commission.europa.eu/strategy-and-policy/priorities-2019-2024/european-green-deal/delivering-european-green deal_cn

European Commission. (2020a). *Circular economy action plan* (978-92-76-19107-0).

European Commission. (2020b). *A new circular economy action plan for a cleaner and more competitive Europe. For a cleaner and more competitive Europe*

Fetting, C. (2020). *The European Green Deal*. ESDN Office.

Fishman, Y., Becker, N., & Shechter, M. (2012). The Polluter Pays Principle as a policy tool in an externality model for nitrogen fertilizer pollution [Article]. *Water Policy, 14*(3), 470–489. https://doi.org/10.2166/wp.2011.139

Glazyrina, I., Glazyrin, V., & Vinnichenko, S. (2006, September). The polluter pays principle and potential conflicts in society [Article]. *Ecological Economics, 59*(3), 324–330. https://doi.org/10.1016/j.ecolecon.2005.10.020

Jang, M., Shim, W. J., Cho, Y., Han, G. M., Ha, S. Y., & Hong, S. H. (2024, February). Hazardous chemical additives within marine plastic debris and fishing gear: Occurrence and implications [Article]. *Journal of Cleaner Production, 442*, 10, Article 141115. https://doi.org/10.1016/j.jclepro.2024.141115

Kamalakkannan, S., Abeynayaka, A., Kulatunga, A. K., Singh, R. K., Tatsuno, M., & Gamaralalage, P. J. D. (2022, November). Life cycle assessment of selected single-use plastic products towards evidence-based policy recommendations in Sri Lanka [Article]. *Sustainability, 14*(21), 25, Article 14170. https://doi.org/10.3390/su142114170

Kim, H. L., & Hyun, S. S. (2021, April). The anchoring effect of aviation green tax for sustainable tourism, based on the nudge theory [Article]. *Journal of Sustainable Tourism, 29*(7), 1082–1097. https://doi.org/10.1080/09669582.2020.1820017

Kim, J. H. (2004). Sustainable urban waste management system in Metropolitan Seoul, South Korea. *Advances in Architecture Series* [Sustainable city iii: Urban regeneration and sustainability]. In 3rd *International Conference on Urban Regeneration and Sustainability*, Siena, ITALY.

Klotz, S., Probst, S. L., & Hulliger, O. (2022). The Carbon border adjustment mechanism – A tax in disguise? [Article]. *European Taxation, 2022*(8), 350–353. https://www.scopus.com/inward/record.uri?eid=2-s2.0-85148635362&partnerID=40&md5=2aebaf5e08bc60311cec8a3b30180cbe

Kuroda, H., Amasawa, E., Nakatani, J., & Hirao, M. (2023, November). Linear programming approach to design bio-based plastics strategies for Japan: Integration of material characteristics, product applications, and end of life options [Article]. *Resources Conservation and Recycling, 198*, 10, Article 107137. https://doi.org/10.1016/j.resconrec.2023.107137

Melo, S. R. S., Silva, M. E., Melo, F. V. S., & Vo-Thanh, T. (2022, September). The practice of (un)sustainable tourism in a National Park: An empirical study focusing on structural elements [Article]. *Journal of Outdoor Recreation and Tourism-Research Planning and Management, 39*, 8, Article 100548. https://doi.org/10.1016/j.jort.2022.100548

Mia, P., Rana, T., & Ferdous, L. T. (2021). Government reform, regulatory change and carbon disclosure: Evidence from Australia [Article]. *Sustainability (Switzerland), 13*(23), Article 13282. https://doi.org/10.3390/su132313282

Milon, J. W. (2019, March). The polluter pays principle and Everglades restoration [Article]. *Journal of Environmental Studies and Sciences, 9*(1), 67–81. https://doi.org/10.1007/s13412-018-0529-y

Molinos-Senante, M., Maziotis, A., Sala-Garrido, R., & Mocholi-Arce, M. (2022, October). How much does it cost to collect recyclable and residual waste in medium-income countries? A case study in the Chilean waste sector [Article]. *Journal of the Air & Waste Management Association, 72*(10), 1083–1094. https://doi.org/10.1080/10962247.2022.2083722

Neto, P. S. (2023). *Embalagens reutilizáveis: O impacto numa cadeia de restauração rápida e no setor de QSR [Reusable packaging: The impact on a fast food chain and the QSR sector].* ISCTE - Instituto Universitário de Lisboa].

OECD. (2017). *Environmental tax.* https://data.oecd.org/envpolicy/environmen-tal-tax.htm

Omeje, K. (2021). Natural resources and rentier capitalism [Book Chapter]. *International Political Economy Series*, pp. 85–116. https://doi.org/10.1007/978-3-030-75170-8_4

Ortega-Rodríguez, C., Vena-Oya, J., Barreal, J., & Józefowicz, B. (2024). How to finance sustainable tourism: Factors influencing the attitude and willingness to pay green taxes among university students [Article]. *Green Finance, 6*(4), 649–665. https://doi.org/10.3934/gf.2024025

Rahman, A. M., & Edwards, C. A. (2004, January). Electricity: Taxes on emission liabilities. An examination of the economic effectiveness of Polluter Pays Principles [Article]. *Energy Policy, 32*(2), 221–235, Article Pii s0301-4215(02)00285-9. https://doi.org/10.1016/s0301-4215(02)00285-9

Riber, C., Petersen, C., & Christensen, T. H. (2009, April). Chemical composition of material fractions in Danish household waste [Article]. *Waste Management, 29*(4), 1251–1257. https://doi.org/10.1016/j.wasman.2008.09.013

Sarigiannidis, C., Halkiopoulos, C., & Boutsinas, B. (2023, August 29–31). An environmental management accounting system for sustainable tourism based on business intelligence. *Springer Proceedings in Business and Economics* [Recent advancements in tourism business, technology and social sciences, vol 1, iacudit 2023]. In 10th International Conference Recent Advancements in Tourism Business, Technology and Social Sciences (IACuDiT), Crete, GREECE.

Schmidtchen, D., Helstroffer, J., & Koboldt, C. (2021, January). Regulatory failure and the polluter pays principle: Why regulatory impact assessment dominates the polluter pays principle [Article]. *Environmental Economics and Policy Studies, 23*(1), 109–144. https://doi.org/10.1007/s10018-020-00285-4

Suhardono, S., Fitria, L., Suryawan, I. W. K., Septiariva, I. Y., Mulyana, R., Sari, M. M., Ulhasanah, N., & Prayogo, W. (2024, March). Human activities and forest fires in Indonesia: An analysis of the Bromo incident and implications for conservation tourism [Article]. *Trees Forests and People, 15*, 12, Article 100509. https://doi.org/10.1016/j.tfp.2024.100509

Sun, Z. Y., Dong, J. J., & IEEE. (2011, March 25–28). Research on Green taxation policy for promoting low-carbon economy.*Asia-Pacific Power and Energy Engineering Conference* [2011 Asia-Pacific Power and Energy Engineering Conference (APPEEC)]. In Asia-Pacific Power and Energy Engineering Conference (APPEEC), Wuhan, PEOPLES R CHINA.

Tiago, F., Gil, A., Stemberger, S., & Borges-Tiago, T. (2021, January–March). Digital sustainability communication in tourism [Article]. *Journal of Innovation & Knowledge, 6*(1), 27–34. https://doi.org/10.1016/j.jik.2019.12.002

Tran, V. C. M., Le, P. C., & Quyen, H. H. (2024, July). A study on plastic waste generation and disposal habits in riverside and coastal households towards the promotions of reducing plastic leakage into the ocean in Da Nang City, Vietnam [Article; Early Access]. *Journal of Material Cycles and Waste Management*, 13. https://doi.org/10.1007/s10163-024-02024-3

van Leeuwen, J., & Surya, I. R. F. (2024, September). Network power and exclusion of informal waste pickers when plastic flows change: A case study of community waste banks in Klaten Municipality in Indonesia [Article]. *Marine Policy, 167*, 9, Article 106285. https://doi.org/10.1016/j.marpol.2024.106285

Zhang, Y. M., Cao, Y. Q., Chen, B., Dong, G. H., Zhao, Y. Y., & Zhang, B. Y. (2024, March). Marine biodegradation of plastic films by<i> Alcanivorax</i> under various ambient temperatures: Bacterial enrichment, morphology alteration, and release of degradation products [Article]. *Science of the Total Environment, 917*, 12, Article 170527. https://doi.org/10.1016/j.scitotenv.2024.170527

Zicari, A. P., & Renouard, C. (2018). A forgotten issue: Fiscal responsibility in the CSR debate [Book Chapter]. *Critical Studies on Corporate Responsibility, Governance and Sustainability, 12*, 243–259. https://doi.org/10.1108/S2043-905920180000012012

Zuo, Z. C., Zhang, L., Ni, J., Zhang, X. Y., Lang, X. P., He, Z., & Yang, G. P. (2024, June). Occurrence of halogenated organic contaminants in surface sediments of the Yangtze River estuary and its adjacent marine area [Article]. *Environmental Research, 251*, 10, Article 118579. https://doi.org/10.1016/j.envres.2024.118579

Capacity-Building Initiatives to Foster Sustainable Tourism in a Climate Change Context in Bolivia

Georgina Chávez, Vidfa Garvizu, Freddy Soria, and Omar Salinas

Abstract Tourism in Bolivia generates more jobs than the mining and natural gas industry, therefore, measures to mitigate climate change in tourist places are vital not just for the preservation of nature and ecosystems in general, but also for the country's economic development. Considered a megadiverse country, Bolivia, with its protected areas and national parks with ecosystems of mountains, forests, and savannas, Andean-Chaco-Amazonian tropics positions the country as an ideal actor for sustainable tourism initiatives. Bolivia's geography, characterized by high-altitude landscapes and tropical rain forests, makes it particularly vulnerable to climate change impacts. Rapid glacial retreat in the Andean region threatens the viability of adventure tourism activities such as glacier trekking and mountaineering. Moreover, shifts in precipitation patterns and increasing occurrences of extreme weather events pose risks to tourism infrastructure and visitor safety, exacerbating challenges for sustainable tourism development. This chapter examines sustainable tourism practices and challenges in Bolivia, with the main emphasis on the role of capacity-building programs and education initiatives within a climate change context to foster long-term sustainability in the tourism sector, especially in those areas with an emerging tourism industry outside the cities.

1 Introduction

Sustainable tourism development is increasingly recognized as a vital pathway for economic growth (León-Gómez et al., 2021), social equity (Wu, 2024), and environmental preservation in countries as well as the development of quality products and meeting the needs of tourists (Chan, 2023). For the case of low- and mid-income

G. Chávez (✉) · V. Garvizu · F. Soria · O. Salinas
Centro de Investigación en Agua, Energía y Sostenibilidad (CINAES) Universidad Católica Boliviana "San Pablo", La Paz, Bolivia
e-mail: gchavez@ucb.edu.bo; vgarvizu@ucb.edu.bo; fsoria@ucb.edu.bo; osalinas@ucb.edu.bo

J. Marques et al. (eds.), *Tourism and Climate Change*, Tourism, Hospitality & Event Management, https://doi.org/10.1007/978-3-031-99128-8_8

economies, as these nations leverage their unique cultural and natural assets to attract tourists, the importance of capacity building is documented principally with a focus on women, rural communities, smallholders, indigenous and in general unattended populations as a form to provide a voice to locals and to promote gender equality and empowerment (Carrera Villacrés et al., 2021; Higgins-Desbiolles et al., 2022; Syafruddin et al., 2023; Torres et al., 2023; Yu et al., 2023). Capacity building refers to the process of enhancing the skills, knowledge, and resources of individuals and communities, enabling them to effectively engage in and manage tourism initiatives and promote development. This approach was defined as an emerging challenge a couple of decades ago as critical for fostering sustainable tourism that benefits local populations while preserving their heritage and environment (European Travel Commission, 2021; Moscardo, 2008), and more recently has been linked as an actor to promote acceleration to achieve the 2030 Agenda in the framework of the Sustainable Development Goals, finding that there is limited research to respond to the remaining research questions in the topic (Khizar et al., 2023).

Tourism development in Latin America has historically been driven by external investors and transnationals, following a concept of development that surged in Classic Greece and captured global acceptance in the middle of the last century. The success of the tourism sector is often measured based on the number of tourist arrivals, investments, employment, the income of foreign currency, and its contribution to the Gross Domestic Product. However, these variables do not necessarily translate into social well-being as shown by the case studies reviewed in Latin America (Gliemmo & Moscoso, 2022). The capacity-building concept in tourism is sustained by the possibility that it can bring balance and empowerment to local communities, as part of a strategy leading to an increase of local input and control. Capacity-building initiatives aim at equipping community members with the skills and knowledge needed to actively participate in tourism planning and decision-making processes, under the expectation that when communities take ownership of their tourism initiatives, they are more likely to invest in their success and work collaboratively toward sustainable development. In practice, results are still under evaluation and the sustainability of the initiatives struggles to survive the COVID-19 pandemic impacts. The case studies reviewed later in this document better describe and analyze that situation.

The objective of this chapter is to provide a general context of Bolivia regarding tourism and explore concepts, initiatives, and possible future directions driven by case studies regarding the role of capacity building in the tourism industry in Bolivia.

The literature review was structured according to the criteria and recommendations of Page et al. (2021). The literature search was carried out in scientific databases such as Scopus, Mendeley, and Research4Life, using specific keywords, combining terms such as "capacity-building tourism," "sustainable tourism," "tourism developing countries" and words associated with the geographical scope of the review. Inclusion and exclusion criteria were defined to select relevant studies, such as study type (case study, systematic reviews), publication date, and language. The review and evaluation of the quality of the information followed the relevance criteria under the question: Does the study directly address the research question? and the transparency (are the study cases and results presented?).

2 General Context

2.1 *Climate Change in Bolivia and Its Effects on Tourism*

The first scientific study on geographic indicators related to climate change in Bolivia can be traced to the tropical Andes. Glacier melt was first warned in the early 1990s by Francou et al. (1995) under concerns based on the ENSO (El Niño—Southern Oscillation) impacts on glacier mass balance in the Zongo glacier, a solid ice and snow mass in the Cordillera Real nearby La Paz. Climate change was not a term used often by scientists during those years, but it was clear that something complex was occurring in the tropics and subtropics. In the 2000s, concerns were reassured highlighting the association between air temperature rise and the complex changes occurring in the physicochemical processes over the highlands of the Andes (Coudrain et al., 2005). Today evidence is strong and there is high confidence in the causing factors, the consequences, and the need to define mitigation and adaptation measures to face the current and coming challenges further enhanced by serious economic challenges ahead in a country with an economy thoroughly dependent on the exploitation of natural resources, which accounts for more than 40% of economic activity and almost 80% of exports.

The loss of small glacier caps is a climate change indicator in the Central Andes significantly affecting tourist attractions. The decadence in tourism in the Bolivian Andes began with the reduction of skiing activities in the Chacaltaya Peak resort in 1987, which found an endpoint when the glacier disappeared in 2009 (Kaenzig et al., 2016). The place is now a lonely mountain with rusty vestiges of a tourist attraction that used to be in full glow in the twentieth century, leaving behind an empty diner and a stone-made cabin that is currently falling apart. The road that used to take visitors to the attraction is no longer maintained and the current users are only scientists who visit a research site for cosmos observation in a mountain where since 1939 was established the highest skiing station in the world at 5421 meters above sea level (m.a.s.l.).

However, Chacaltaya is not the only attraction in the zone, and it has the potential to attract tourists who seek adventures in trekking, mountaineering, and hiking over ancient ice walls. The east branch of the Andean Cordillera crossing Bolivia has a geography that provides a wonderful view of tropical glaciers struggling against climate change in La Paz and El Alto. Contrary to what occurs in the Northern Hemisphere, the sensitivity of tropical glaciers to climatic variations is due to the simultaneous occurrence of snow precipitation and ablation processes during the rainy summer season, and the year-round occurrence of glacier melt, providing short periods for the glaciers to recover under a scenario with sustained trends of rising air temperatures (Soria & Kazama, 2011). The consequent impacts are varied, among which the most concerning are the loss of water resources affecting drinking water, irrigation, water security in general, and the overall ecosystem needs for survival in an area where the weather is arid and semiarid (Kinouchi et al., 2019).

Downstream the Cordillera and to the west, the harsh living conditions of the Altiplano turn into fertile land fed by Lake Titicaca. The Titicaca, in the Central Andes, is the world's highest navigable freshwater lake at 3810 m.a.s.l., with a diverse ecosystem and rich native wetlands that have taken the site to be cataloged as part of a world RAMSAR site shared between Peru and Bolivia. Lake Titicaca is a popular tourist spot; it has been relevant for life for centuries despite its natural richness and it has faced important challenges since the beginning of the CE (Common Era). In the ancient past, extreme climate fluctuations were responsible for the decline and disappearance of wealthy and powerful societies due to the stress caused on agrosystems, and a similar scenario has been hypothesized for the ancient Tiwanaku civilization, which developed approximately 1500 years ago in the southern basin of Lake Titicaca. For 700 years, Tiwanaku was the dominant sociopolitical power in the region, relying on camelid herding and intensive farming, and connecting a mix of cultures throughout the south-central Andes (Janusek, 2004).

From a touristic perspective, the Andes is a hiking paradise and a site for adventure tourism due to small anthropogenic interventions in vast areas and the lack of roads and other complementary infrastructure. Yet, the Eastern Bolivian Andes is still a remote region, with limitations in infrastructure and little support and guidance provided to visitors. With little information available to scheme a sound projection of climate change impacts in the tourism industry, the region lacks a strategy to adapt and mitigate the changes suffered by the environment and the people. The region has faced climate stress since the beginning of the Common Era and locals have learned to develop resilient measures to survive, but regarding tourism, the perceptions of locals and stakeholders differ when the topic of climate change is brought into discussion. To the best of our knowledge, no initiatives exist to attract tourists to witness environmental changes, neither from government authorities nor from travel agencies, despite the common awareness of the impacts of environmental changes in the landscape, because that perspective belongs to experts or members of the Andean Club and may not be shared by travel agencies (Kaenzig et al., 2016).

Climate change impacts are visible in glaciers and in water bodies of the Altiplano natural system at elevations greater than 3659 m.a.s.l., attractive as well to visitors from all over the world as part of the Peru-Bolivia-Chile touristic circuit advertised in most international travel fairs. The dynamic history of the Bolivian Altiplano, a plateau formed during the uplift of the Andes Mountains, is recorded in the salt flats, among which the Uyuni Salt Flat with about 8000 km^2 is perhaps the most renowned tourist location in Bolivia. Uyuni Salt Flats, the largest Salt Lake in the world, is in a region with no drainage outlets. Sediments in the salt flats record fluctuations in water levels that occurred as the lakes evaporated. These sediments provide valuable paleoclimate records for the region. Bolivia's salt flats, colored lagoons, and flamingo sanctuaries attract thousands of visitors every year, who may take tours covering neighboring areas in Chile. However, the changing intensity of climatic events such as the no longer irregular ENSO events, intensified by climate change, has been threatening the life and subsistence of local inhabitants and the subsistence of the endangered flora and fauna.

The Andes vanish to the east of Bolivia along an Amazonian landscape. The climate crisis and the intensive influence of human activities including the seasonal human-ignited Amazonian Forest fires have also been negatively shaping life and in consequence tourism (Cochrane & Laurance, 2002; Lugo-Carvajal et al., 2023; Singh et al., 2022). The Amazon comprises a large region where Brazil is the most significant player, but Venezuela, Colombia, Peru, Ecuador, and Bolivia are also relevant actors. In a region where tourism is internal, the impacts of climate change on tourism are expected to arise from the transformation from a heterogeneous and humid landscape to a deforested and degraded land composed of a unique species, economically profitable for local farmers, but deadly for the ecosystem due to the poisoning impacts in soil and water by intensive use of agrochemicals, which in turn increase plagues and disease outbreaks (MacDonald & Mordecai, 2019).

On track to the East, the Amazonian lowlands provide worldwide nature wonders such as the Pantanal, the world's largest tropical wetland. The Pantanal is largely located in Brazil, but with a significant area inside Bolivia. Scientists warn of a serious risk of the collapse of the Pantanal due to climate change, intensified droughts human deforestation, and ecosystem endangerment. The large swamp is currently being shaped by intensive human actions which are turning the beauty into a savanna shaped by fire, cattle, and water, as well as the intensive interference of Brazil's large infrastructure projects whose main argument is an urgent need for economic development of one of the largest economies in the planet. Smaller countries such as Bolivia and Paraguay, which comprise 18% and 4% of the total of 185,000 km^2 of Pantanal, have little to say against a giant like Brazil. On the same path, the enormous concerns of scientists may have little to say, but the evidence is growing on the possible tragedy ahead (Tortato et al., 2022).

Bolivia was situated in place 10 among the countries more affected by climate change (Eckstein et al., 2021), indicating that the possibility of a natural disaster in a country affected by several threats is latent.

2.2 *Relevance of Tourism in the Country's Economy*

The tourism sector is transversal to other economic activities with direct multiplier effects, as stated in the 2030 Agenda indicated in Sustainable Development Goal 8 "Decent Work and Economic Growth" and Goal 8.9 referring to: "develop and implement policies aimed at promoting sustainable tourism that creates jobs and promotes culture and local products" (United Nations, 2015). Such is a reason that countries worldwide strive to generate specific policies that promote the development of this important economic sector.

In Bolivia, activities related to tourism constitute one of the most important economic sectors, together with the export of hydrocarbons, and agricultural production focused on soybeans and their derivatives, quinoa, and other products. Other activities, such as transportation, lodging, travel agencies, and community-based entrepreneurship, had as well played an important role in the development of the sector by 2019.

Before the pandemic, the tourism industry constituted the fourth most important factor in the economy of the country. For instance, during 2018 tourism in Bolivia generated more jobs than mining and hydrocarbons combined (Fig. 1) (Análisis Real –Latinoamérica, 2018). However, public investment was several times lower than that in the oil, mining, and agriculture sectors (Andersen et al., 2023).

The positive economic impact that tourism activity has generated in Bolivia has been evident over recent years, being of great importance compared to other economic sectors in the country. Data from the National Institute of Statistics show that the economic income provided by this activity was $837 million in 2019 from inbound tourism. It represented a contribution of 2.23% to the Gross Domestic Product (GDP). Regarding the generation of employment in tourism in Bolivia, the International Labor Organization (ILO) reported for the year 2019 a contribution of 7.5% as shown in Table 1. Therefore, tourism is a fundamental pillar of the national economy and the generation of employment, infrastructure works, investment in accommodation and gastronomy, and transportation, among others.

Tourism in Bolivia has the potential to engine sustainable development. The road ahead still demands public policies to support it as well as capacity-building actions to enhance the role of stakeholders and locals in the journey.

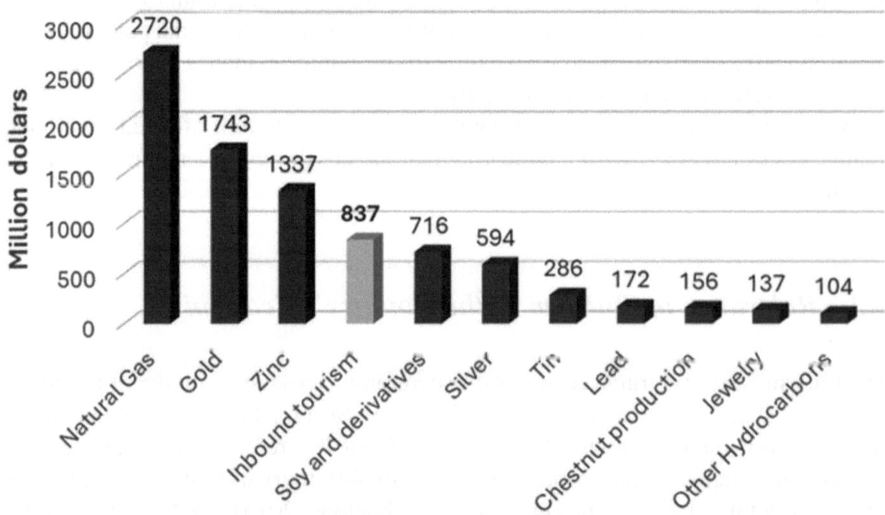

Fig. 1 Inbound tourism revenues in relation to other export sectors 2019 (in millions of dollars). Source: National Institute of Statistics (2019)

Table 1 Contribution of the tourism sector to employment (2019)

Service type	Percentage of contribution to employment (%)
Hotels	0.4
Restaurants	7
Travel agencies	0.1
Total	7.5

Source: Based on data from the Sectoral Plan for Integral Development to Live Well Tourism Sector (2021–2025)

Due to the importance of the sector in the Bolivian economy, state policies are established to promote the sustainable development of tourism. The legal framework that guides the climate change approach taken by Bolivia will be discussed later in this chapter.

2.3 *Legal Context to Foster the Tourism Industry*

The Law 292, "Bolivia waits for you" establishes the general policies and tourism regime of the Plurinational State of Bolivia, aimed at the development, dissemination, and promotion of the productive activity of public, private, and community-based tourism-related sectors, through adaptation to the existing management models, strengthening the community-based tourism model, within the framework of the exclusive powers assigned to the central level of the State by the Political Constitution of the State. That Law constitutes an essential instrument aimed at guiding a search for well-being, enhancing the rational and balanced use of natural and cultural resources. Thus, it also establishes the harmonious and sustainable development of tourism (Article 12), pointing out that the tourist activity carried out by the forms of economic, community, state, and private organizations will be carried out within the framework of a harmonious and sustainable development, of way that has long-term economic sustainability and promotes its growth based on rational and responsible management in environmental, cultural, social, and economical matters.

Based on Law 292, the Sectoral Plan for Integral Development to Live Well—Tourism Sector (2021–2025) collects and implements all the technical criteria, methodological guidelines, philosophical foundations, and current political mandate to become a tool that allows orienting and operationalizing specific programs, projects, and actions toward 2025 in the tourism sector, intending to reactivate, articulate, and strengthen the different actors in this sector of the plural economy and promote harmony with Mother Earth.

Bolivia has a very progressive approach to climate change and the environment in general. "Bolivia's 2009 Constitution recognizes the importance of protecting nature. The rights of nature, Pachamama or Mother Earth in the Bolivian indigenous tradition, and its protection are set out in two statutes: (a) Law N° 071 of December 21, 2010 Mother Earth Rights Law (Ley de Derechos de la Madre Tierra)—the Mother Earth Law—which establishes specific rights to which Mother Earth is entitled; and (b) Law N° 300 of October 15, 2012, Framework Law of Mother Earth and Integral Development for Living Well (Ley Marco de la Madre Tierra y Desarrollo Integral para Vivir Bien)—Mother Earth Framework Law—which seeks to put into operation the rights of Mother Earth set out in the former law, in the context of Integral Development (Desarrollo Integral) for Living Well (Vivir Bien)" (Muñoz, 2023).

By the look of these laws, Mother Earth should be kept well protected, and ecosystems should be conserved. That is not the case, since, as mentioned before, the biggest part of the economy relies still upon the use of natural resources and the application of these statutes is not as expected. Thus, Indigenous people and rural communities, where some of the most promising scenarios to develop sustainable tourism, are not conserved and a need for safekeeping and restoring ecosystems is imperative.

Anyhow, let's go back to the National Policy for Climate Change, there is not a specific mention of how to tackle the climate crisis for the tourism sector, but discusses other interesting topics regarding the conservation of the environment by local communities and pathways for capacity building, which should be the spearhead to generate projects and program to support sustainable initiatives. It is mentioned that academic institutions are expected to not only generate knowledge to fight the climate crisis but also academic programs to better cope with future scenarios.

Tourism contributes to the construction of pillar 3 "Food security with sovereignty, promotion of exports with added value and tourism development" of the Economic and Social Development Plan (PDES) 2021–2025 of Bolivia, considering its strategic line which is entitled to promote the country's tourism development based on natural and cultural diversity, promoting entrepreneurship, private initiative, and strategic alliances.

Within that strategic framework, the following goals are outlined:

- To implement a national tourism development and competitiveness policy.
- To establish strategies for articulating local tourism products, routes, and community-based tourism networks.
- To implement a strategy to promote domestic tourism.
- The international promotion of touristic destinations in Bolivia.
- To implement tourist complexes at the national level to promote community-based, gastronomic, adventure, accessible, heritage, agrotourism, ecotourism, rural experiential, ancestral health, etc.

2.4 *Key Capacity-Building Concepts*

Considering that capacity building is essential to establish strategies in any context, which allow the strengthening of sustainable tourism in a climate change scenario, it is important to consider the following aspects to be able to establish it:

2.4.1 Sustainable Practices in Tourism

The concept of sustainable tourism has been proposed as a response to the impacts of tourism on local ecosystems. As such, sustainable tourism was defined by the World Tourism Organization in 1993 (updated in 1998, (WTO, 2017)) as a concept that implies that the natural, historical, and cultural resources for tourism are conserved for continuous use in the future as well as the present. Therefore, sustainable tourism also means that tourism development does not generate serious environmental or socio-cultural problems, tourist satisfaction is maintained so that tourist markets are retained, and the benefits of tourism are spread throughout society (WTO, 2017).

Capacity building is expected to play a crucial role in promoting sustainable practices within the tourism sector. Training programs focused on environmental conservation, sustainable resource management, and cultural heritage protection should empower local stakeholders to minimize the ecological impact of tourism. For example, communities can learn how to implement waste management systems, engage in eco-friendly business practices, and promote responsible visitor behavior. Such initiatives not only protect local ecosystems and cultural sites but also enhance the overall tourist experience. By adopting sustainable practices, communities can attract environmentally conscious travelers, creating a niche market that benefits both the local economy and the environment.

Enhancing Economic Resilience

Latin American economies often face economic vulnerabilities, making capacity building essential for enhancing resilience. By providing training in entrepreneurship and business management, the expectation is that capacity building can help local communities diversify their tourism offerings. Communities may develop new attractions, promote local cultural festivals, or create unique experiential tours that cater to specific interests (Guerra et al., 2024; Ticse et al., 2023). By broadening their tourism products, communities could secure more stable income streams, reduce dependence on a single revenue source, or provide alternatives to reduce migration to cities (García & Manquillo, 2023).

2.4.2 Improving Service Quality

In the competitive tourism market, service quality is paramount. Capacity-building initiatives often focus on hospitality training, customer service, and professional development. The purpose is to improve products and services to offer values that are perceived by consumers (Zhang et al., 2022). The Latin America and Caribbean (LAC) region has nearly 150 World Heritage sites (WHS) and 90 Intangible Cultural Heritage (ICH) elements, because of which sustainable tourism is key. Under a WHS context, the perceived value can be important to attract tourists, but it may face challenges given that it may be more difficult to make adaptations because of the nature and characteristics of the site's history (Moreno-Manzo et al., 2022). At WHS sites in LAC, the destination management could increase the perceived value by providing tourists with affordable costs, but also improving the quality of the service and the standard of accessibility in cultural tourist sites and essential locations for their visits (Moreno-Manzo et al., 2022). However, the perceptions of the value of visitors do not only depend on functional and tangible aspects of the destination but also the emotional and experiential characteristics (Perez Galvez et al., 2023). Thus, capacity building has multidimensional relevance, not only focused on the front desk but also on the other actors in the destination, becoming a complex key factor associated with sustainable tourism.

2.4.3 Strengthening Governance and Collaboration

Effective governance is essential for sustainable tourism development, and capacity building can enhance local governance structures. Training in governance practices and stakeholder collaboration helps local institutions manage tourism resources more effectively. Adverse and critical events such as those faced by most touristic destinations during the COVID-19 pandemic had tested the resilience of most systems. Public–private partnership, smart information systems, local and international certification, assertive touristic promotion campaigns, as well as reactivation protocols, had demonstrated to be components of efficient strategies to foster dialogue among government agencies, private sector actors, and local communities, where capacity building creates cohesive strategies that align with broader community goals (Lacruhy Enríquez & Morandin Ahuerma, 2023). Capacity building promotes strong partnerships, as an essential tool for maximizing tourism benefits, ultimately leading to a more sustainable and inclusive tourism sector, but the economy and political structures of countries are not always ready to face global challenges. Capacity building around work groups promoted among the different levels of government, the private sector, and academia launches collaborative approaches imperative during times of crisis to make decisions and draw up action plans (Farfán-Pacheco et al., 2024). By fostering dialogue among government agencies, private sector actors, and local communities, capacity building creates social cohesion (Iglesias Alonso, 2021), but challenges are still present given the gaps in State-society relationships, bureaucratic government structures, and other transversal problems (De Castro et al., 2020).

2.4.4 Ensuring Long-Term Sustainability

The goal of capacity building in tourism development is to ensure long-term sustainability. By investing in the skills and knowledge of local communities, developing countries can create tourism systems that are economically viable, socially inclusive, and environmentally responsible. This holistic approach not only meets the immediate needs of current generations but also protects the interests of future generations, in concordance with the definition of sustainability itself.

Capacity building is a cornerstone that is expected to promote sustainable tourism development. To achieve that, strategies have considered approaches such as sustainability international (Bastidas-Goyes et al., 2021) certifications, which by 2022 were well spread in Brazil, Peru, and Mexico, with some experiences in Colombia (Araya & Correa, 2023). Although the literature draws attention mostly to rural and community-driven tourist experiences in Latin America, it is relevant to the interest that medical tourism has captured. While it is hardly a new trend, it has been growing steadily over the past decades under the rising trends of medical treatment in countries such as the USA. Although few sources that account for medical tourism estimations in Latin America have been found over the WEB and literature search made for the current book chapter, some estimations suggest revenues of 6.75 billion USD in Mexico (Corpart, 2024). Older data cited in Flores (2021) estimate that by 2012, the medical tourism market in Brazil and Mexico was evaluated at 722.4 and 10.3 million USD, respectively. The same source estimates that the first exporter of health services is the United States with 4097 million dollars and in the fifth place Mexico is positioned for 347 million dollars. Although ethical concerns remain regarding the possible country-specific laws and regulations, questions remain on the actual benefits that medical tourism brings to local economies and tourism stakeholders (Bastidas-Goyes et al., 2021; Cajiao, 2023).

By empowering local communities, promoting sustainable practices, enhancing economic resilience, and improving service quality, capacity building lays the foundation for a tourism sector that should benefit stakeholders. With capitalism driven by developed nations being a relevant player in the tourism industry, the path to sustainable tourism is expected to lie in the hands of local communities, making capacity building an important factor in the strategy to promote the sustainability of tourism. Some success stories are documented in the literature, where they have been shown the limitations still faced and possible alternatives as well to define a path to guide stakeholders and policymakers.

3 Capacity Building Toward Sustainable Tourism in Bolivia

The lack of a solid structure to support the promotion and communication of the potential of tourism is an important obstacle to overcome, and it is often attributed to the governmental levels, even as it was detailed earlier, it appears to be a proper legal framework to do so. However, there are several actions that capacity building could foster. It is also important to identify individuals among the different stakeholders who can act as the driving force, serving as leaders and trainers of trainers to expand the number of trained individuals.

At the national level, the most important governmental actor in tourism in Bolivia is the Vice Ministry of Tourism, and at the local level are the municipalities and regional governments. The government of Bolivia invests economic resources in capacity building referred to as the social capital in tourism; however, there is still a gap to reaching its full potential as the structure needed for its development is still under construction (Ministry of Productive Development and Plural Economy, 2023).

Thus, the need to implement a process of development and strengthening of the skills, capacities, processes, and resources of all people involved in Bolivia's tourism activities is identified.

According to the characteristics and particularities of the Bolivian context, two important areas that promote capacity building in sustainable tourism can be identified: (1) Formal education and (2) Practical capacitation.

Formal education is directed at schools and universities. The goal is to create awareness among the students about the impact of climate change in natural areas, while capacity building in universities is directed to the formation of future tourist operators and other professions related to tourism.

On the other hand, practical education has been financed mainly for international initiatives, and its focus is working with tourism entrepreneurs who are already working in natural areas and/or capacitation to those interested in making tourism more sustainable and resilient to climate change.

3.1 Formal Education

Education is a fundamental pillar framed in the Sustainable Development Goals SDG 4. According to Vera (2018), the tourism sector comprises multidisciplinary activities that affect the study of the relationships and interactions between tourists, visitors, hikers, and the tourist destination. The tourism sector considers aspects such as the behavior of tourists, the environmental impact, and the impact generated by the activity, highlighting the need to generate spaces for capacity-building strategies aimed at strengthening that sector of the economy. Moreover, sustainable tourism is intrinsically related to climate change effects, especially in natural tourist destinations, for instance, in most cases climate change is a topic included in sustainable tourism subjects.

In Bolivia, capacity building in sustainable tourism is considered at different levels of education: primary, secondary, and higher education in both public and private institutions, without specific focus on the relevance of the sector for the country.

At the primary and secondary education level, there is no specific space in tourism capacity building; however, Educational Law 070 "Avelino Siñani-Elizardo

Table 2 Tourism Education Institutions recognized by the CEUB

Institution	Program name
Universidad Católica Boliviana San Pablo	Tourist Administration
Universidad Mayor de San Andrés	Tourism Career
Universidad Pública de El Alto	Tourism and Hotel Management
Universidad Mayor de San Simón	Tourism Career
Universidad San Francisco Xavier	Tourism Career
Universidad Autónoma Gabriel René Moreno	Tourism Management Career
Universidad Autónoma Tomás Frías	Tourism Career
Universidad Técnica de Oruro	Tourism Technician
Universidad Autónoma del Beni José Ballivián	Tourism Career
Universidad Amazónica de Pando	Sustainable Tourism Career

Source: Own elaboration, based on CEUB Data (2024)

Pérez" promotes the development of education in coexistence with nature and community health, as an articulator axis. It has been implemented in various educational experiences through the Complementary Training Program (PROFOCOM) for teachers and practicing educators. In this context, children are generally trained to respect and preserve natural resources, often with a focus on sustainable tourism.

At the undergraduate and higher technical level, within the universities recognized by the Executive Committee of the Bolivian University (CEUB), the syllabus of a career is determined to establish some equivalences between universities.

The program names and the Institutions that offer tourism careers are shown in Table 2.

Universidad Amazónica de Pando is the only institution that offers an entire career focused on sustainable tourism, including a postgraduate course officially registered at CEUB.

Other universities, such as Universidad Mayor de San Simón and Universidad San Francisco Xavier, offer courses with a focus on sustainable development and sustainable tourism, incorporating subjects like the environment and sustainable development into their curricula. However, according to the latest sectoral report on tourism careers (CEUB, 2015), no courses specifically addressing climate change have been included. Universidad Católica Boliviana San Pablo is the only university offering two courses that focus separately on climate change and sustainable tourism: Sustainable Tourism Development and Environmental Heritage, and Climate Change.

Bolivia is historically a country with catholic roots. In this sense, catholic educational institutions at primary, secondary, and higher education levels use the encyclical Laudato SI dedicated to the care of the "Common Home" as a guideline for the preservation of the environment (Pope Francisco, 2015). The content of this document is often referred to in religion and environmental subjects.

In different regions of the country, other private universities do not belong to the Bolivian University System, which also offers training in Tourism with a Sustainability focus.

At the level of continuous training, such as short courses, other types of institutions carry out training in different programs related to sustainable tourism such as the training of guides in Ecotourism, training of park rangers in protected areas, etc.

3.2 *Practical Initiatives*

Bolivia, being a developing country, receives financing from different sources for the execution of various projects that allow the progress of its population. In recent years, tourism has become a prioritized sector for the execution of various initiatives that promote, above all, sustainable tourism, as a strategy that allows the preservation of natural resources and at the same time an economic livelihood.

Among the notable efforts to promote sustainable tourism and inclusive development are several key initiatives, each addressing critical challenges through unique approaches.

3.2.1 Sustainable Development Solutions Network (SDSN)

The SDSN is a network that contributes to the implementation of efficient solutions and strategies to advance toward the Sustainable Development Goals in Bolivia. It has developed specific capacity-building material related to sustainable tourism; the material varies from blogs to books where it is possible to find information about the relation of tourism in the frame of SDGs including the index of risk related to climate change effects as a forest fire, frosts, droughts, and floods (Andersen et al., 2023). These materials can be used to improve the skills of decision-makers involved in tourism and to establish baselines useful to define priority actions.

3.2.2 Project: Tourism as an Engine of Gender-Inclusive and Sustainable Development in Bolivia

More than 75% of the population employed in tourism are women, the jobs in which women are employed are more precarious than those of men, and women are underrepresented in tourism enterprise leadership. The COVID-19 pandemic struck the tourism sector severely. This project proposes to foster sustainable tourism for a low-carbon transition, women's economic empowerment, and the involvement of local communities in COVID-19 recovery in Bolivia. In this sense, the project set out to investigate whether tourism could become an engine of sustainable development and female empowerment in Bolivia.

3.2.3 Orbita Observatory

It is an initiative for the formation of a research and sustainable tourism observatory, which was implemented in Bolivia in 2022, under the leadership of the Fundación Innovación en Empresariado Social (IES, a private non-profit organization) and the United Nations Sustainable Development Solutions Network (SDSN). The Orbita Project is supported by Canada's International Development Research Center (IDRC), which is an initiative that champions and funds research and innovation within and alongside developing regions to drive global change (IDRC, 2024). The purpose of this observatory is to compile current information about tourism practices and develop research; moreover, its action scope includes capacity building for tourism companies and providing technical advice to those communities with tourism potential.

According to what is suggested by IES, Orbita seeks to establish collaboration ties between private, public, and non-governmental institutions for the common implementation of dissemination, training, and research activities that benefit tourism. Orbita's intervention model is based on decision-making based on evidence that allows sustainable actions to be taken, in such a way as to generate the connection of tourism with other large production and service chains in Bolivia. This model is based on three components: Intelligence, connection, and action.

Intelligence: Digitized and democratized information to learn (capacity building) and share to strengthen business, public, and community decisions. Basing the efforts on promoting and improving the tourist experience through the creative and technological industries.

Action: Promote changes that seek to improve economic opportunities, generating financial and social stability through a precise understanding of the tourism industry. Promoting protection and harmony with the environment and the responsible use of resources and production factors.

Connection: Focus on fostering networks among tourism companies, territories, governments (at all levels in Bolivia), and the supporting industries. The aim is to position tourism as a transformative axis, moving away from a natural resource-dependent model while strengthening opportunities for social and personal development for both women and men.

3.2.4 Project: Sustainable Community Tourism in the Motacucito Nuevo Community

This project was funded by the Small Grants Programme in Bolivia belonging to the United Nations Development Programme (UNDP), which aims to contribute to the conservation, use, and sustainable exploitation of biodiversity resources and to the improvement of the livelihoods of the Motacucito Nuevo community through the development of community tourism in harmony with the priorities of the Otuquis National Park and Integrated Management Natural Area (PN ANMI) (Fundesoc, 2023).

In capacity building, one of the main results of the project was to generate spaces for reflection/training on gender, leadership, teamwork, rights, sustainable tourism, etc. Mainly strengthening the capacities of women for guiding, interpreting, and serving visitors, trying to enable them to identify a means of support for their families.

3.2.5 The Chalalán Eco-lodge

The Madidi National Park hosts the Chalalán Eco-lodge, which is a successful example of sustainable tourism led by local guides, including activities that involve close contact of the visitor with nature. The immersion starts with the arrival of the visitor who is taken to the lodge in a boat. Once in the place, different tours are offered making tourists gain a deeper appreciation of the area. However, what makes the initiative to be considered sustainable is that their activities are services carried out by locals, for instance, the incomes from the lodge and other activities generate benefits for indigenous families from Quechua and Tacana cultures from San José de Uchupiamonas. The sense of conservation and the inclusion of the community in the whole concept of Chalalán Eco-lodge have been the main factors to be awarded for its responsible and Indigenous tourism.

The Chalalán eco-lodge constitutes a benchmark for sustainable tourism in Bolivia, which has maintained stability over time and has survived crises such as COVID-19. An important key to its success came before its opening in 1998, and it was based on capacity building. The topics covered were basic accounting, lodge management, cooking, natural history, sales, basic marketing, among others. A total of 20 people from the communities were part of a six-month training program, along which some of them even took a complementary training course that involved visiting capacity-building centers in cities in Bolivia to receive specialized training in some other topics (Peaty & Portillo, 2009).

3.2.6 The Family Community Integral Tourism (TICOF)

The Amazonian region is wide and diverse. It does not only hold humid and wild forests but also semiarid climates. In such an environment lies The Chiquitania.

In that area, the Family Community Integral Tourism (TICOF) project was established and was implemented by the Wildlife Conservation Society (WCS) in close collaboration with local communities. Educational programs offered by TICOF include workshops on sustainable resource management, guided tours exploring nature, and hands-on experiences in agroforestry and wildlife monitoring. By integrating capacity building, environmental education, and conservation into tourism activities, TICOF fosters environmental stewardship among visitors and local residents alike, contributing to the long-term sustainability of the region. This project has been recognized nationally for its positive impact on the communities where they work.

Although both Chalalán and TICOF promote community tourism through social and economic entrepreneurship, the capacity-building component in these communities has also contributed to promoting gender equality. The TICOF initiative empowers women from the Yampara culture, where leadership roles have been fostered to highlight their roles in their communities. Thus, while Chalalán focuses more on the conservation of nature, the TICOF initiative prioritizes cultural heritage (Condori & Ateaga, 2023).

3.2.7 The Senda Verde Eco-lodge

Another example of the role of capacity building is the Senda Verde eco-lodge. Senda Verde is engaged with nature preservation. Their mission is to rescue injured wild animals captured as pets by locals and foreign hunters. Senda Verde also rescues animals that migrated from the forests destroyed by humans or affected by climate change, such as jaguars and bears, which appeared and were captured by locals when sighted in nearby populated areas. Senda Verde is in Yolosa, a community in the province Yungas of La Paz, in the transition from the Andes to the Amazon, where currently sheltered an average of 800 species (monkeys, felines, reptiles among others). In Senda Verde, capacity-building activities are offered to visitors and locals, to promote a better understanding between humans and wildlife.

3.2.8 Community Environmental Tourism Development Project "Opening Our Doors"

This initiative is developed by the foundation "PORVENIR VOCES LIBRES" which is an independent and non-profit organization, made up mainly of professionals in the field of the Environment and social community education, which aims to promote the active participation of the different sectors of society to build a more just, equitable, and supportive world (Fundación Porvenir, 2024).

"Opening Our Doors" is a project that promotes the development of the communities of the Valle Alto, Bajo, and Chapare, in the Cochabamba Department, to promote eco-sustainable tourism.

It aims to support and promote the community's own potential, promoting technical education in young people and the defense and promotion of the right to quality education for all communities.

Through this initiative, a training and support program was developed for community leaders and young entrepreneurs for the revaluation and conservation of the environment, cultural heritage, and gastronomy. This is done to promote the relevance and knowledge of the cultural-historical, archaeological, artisanal, and traditional heritage, reinforce identity, and generate tourist awareness, for the generation of economic initiatives to improve their quality of life.

3.2.9 Other Initiatives

On a smaller scale, there are also some initiatives carried out by local universities and international cooperation. Among them, it has showcased an initiative of capacity building promoted by the Japan International Cooperation Agency (JICA). The activities were referred to as solid waste management in the Uyuni Salt Flats. The purpose is to help locals and visitors to adequately dispose of solid waste in an area where municipal staff is insufficient. The training also reached local tourist operators on topics such as waste disposal, reduction-reuse-recycling, and its relevance in sustainable tourism. The initiative was supported by volunteers from universities such as the ECO-Tomodachi group, which helped in the implementation of a system to improve the sanitary conditions in the area.

After COVID-19, specific courses to reactivate the tourism industry have been given through the aid of the Business Scout for Development and the Chamber of Commerce and Industry German Bolivian (AHK Bolivia), this capacitation has been directed to tourism operators; however, these courses are not permanent. Although these capacity-building initiatives contribute to building a more sustainable tourism ecosystem, there is still a lack of articulation among all the tourism stakeholders, and that is maybe the reason some initiatives are successful while others do not last over time.

4 Community Resilience and Adaptation Strategies to Foster Tourism

Considering the different ecological floors in Bolivia, it is natural to have a great variety of indigenous people and communities scattered throughout the country. Many of these initiatives have started to enhance tourism dynamics in the territory with mixed results. Although some good practices have been described previously, in general terms, they lack a management background and are lately being affected by variations in climate patterns (Shane, 2023).

There are better conditions to tackle these issues in urban areas, with access to technology and facilities through capacity-building programs and implementing good practices (Sustacha et al., 2023). Distances are shorter and the availability of natural resources in the urban areas is constant, even though the rural areas are the ones providing them. It is contradictory that although water resources tend to be scarce in urban areas, which could be even tougher as the effects of the climate crisis are felt stronger over the following years (Caretta et al., 2022), it is generally safe for human activities (drinking, cooking, among others). In rural areas, particularly in the Amazon basin, there is water everywhere, but it is not safe to consume and needs some basic treatment before it can be used. This influences the visitors since they are recommended to carry their water or at least some disinfectant tablets to use in the water before drinking it.

Some basic resources and services are not constantly available, such as electricity and communication, which may give an interesting feeling of adventure and excitement (Pomfret et al., 2023), but also jeopardize the appeal of these places to a broader public. Therefore, the focus will be on rural areas and their capacity to tackle these obstacles.

In general terms, capacity building in tourism is challenging since there are not always appropriate conditions (lack of academic or practical programs, trainers, and limited technological capabilities) to address it (Moscardo, 2008). This is enhanced by a deficiency of basic learning knowledge in rural areas due to a poor educational system. Add to this the effects that climate change is having on the Communities' dynamics and local practices that were effective before but have been compromised and do not offer the same results, then you will end up with great challenges with little support to address them.

On the other hand, the government and society have started several initiatives to support the public with climate change through mitigation and adaptation strategies to build resilience. In 2023, the central government updated the National Policy for Climate Change (APMT, 2023), which establishes how Bolivia is preparing and improving mechanisms to deal with the "climate crisis," a term that is suggested should replace climate change, which does not demonstrate the urgency in which the actions should be implemented.

Additionally, capacity building is included in several strategic axes of Bolivian policy, emphasizing that processes should strengthen the women's and youngsters' capacities to generate more climate-resilient communities (Bryan et al., 2024).

So, how should the adaptation strategies for the tourism sector be prepared? First, they need to be aligned with the Climate policy, to seek governmental support and/or financing, and if this cannot be obtained from a state source, look elsewhere. Second, they must rely on local people and their communities since they are the protagonists to make these strategies successful. Third, these protagonists need to increase their skills and managerial capacities, for which agreements with academic institutions, but also with NGOs or others that can teach basic knowledge in tourism and climate change, are imperative.

4.1 *Initiatives of Sustainable Tourism in Bolivia*

The development of the tourism industry has not had a planned and organized growth, instead arising because of the opportunity of the moment, with the concomitant result of heterogeneous and unfinished landscapes which frequently do not harmonize with the geography of the place. Nowadays, tourism has been included in the vision of the indigenous towns who see an opportunity to build a sustainable model based on planning and proper management. For almost 30 years, Bolivia has been using tourism to integrate indigenous communities with other economic and productive activities (Cox, 2009). Until 2019, 138 initiatives of sustainable tourism

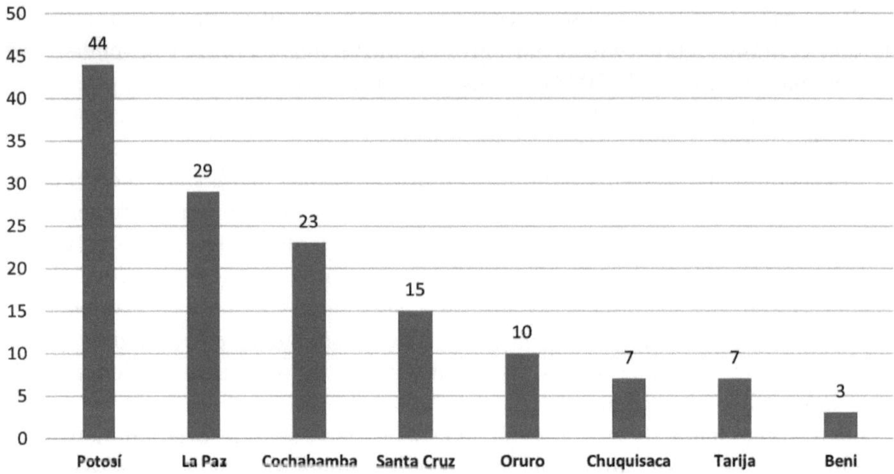

Fig. 2 Number of communitarian tourism entrepreneurships by department, 2019. Source: San Martin (2024)

were working actively in Bolivia, the distribution of these initiatives in the different departments is seen in Fig. 2 (San Martin et al., 2024).

Tourism is involved in all 17 SDG (Pacheco et al., 2024) and according to the 2030 Agenda for Sustainable Development and the Sustainable Development Goals, a major effort must include a change in policies, business practices, and consumer behavior to reach: (1) Sustainable economic growth; (2) Social inclusiveness, employment, and poverty reduction; (3) Resource efficiency, environmental protection, and climate change; (4) Cultural values, diversity, and heritage; and (5) Mutual understanding, peace, and security (World Tourism Organization and United Nations Development Programme, 2017).

The examples described in the following lines are within the SDG framework.

The Amazonian area of the country has an interesting potential to promote sustainable tourism activities given the remoteness of the region and the possibility of offering visitors close encounters with the Amazonian ecosystems.

Also located in the Amazonian region of the country, the Tsimane Project has been implemented. It started in 2000 and is in the indigenous territory of the National Park Isiboro Secure (TIPNIS), in an area of transition between the Amazon basin and the Andes mountains. Such transition provides biodiversity richness, sensitive to the impacts of climate change due to the altered rainfall patterns resulting from climate change, the expansion of agricultural activities based on aggressive foreign species, and the uncontrolled use of agrochemicals. The Project hires local park rangers and it has been a source of economic income to the indigenous communities. Their approach is to work with small groups of tourists, assuring in this way a minimal impact on the environment will occur.

The overall goal of sustainable tourism in protected areas, especially in those vulnerable to climate change effects, is managing resources and visitors today so that tomorrow's visitors can also experience quality sites and the conservation values that these places represent (Eagles et al., 2002). However, sustainability can also be achieved in urban areas located in the arid and semiarid regions of the Andes. An example of such sustainable entrepreneurship is promoted by Tourism El Alto and Colibri Eco-lodge, at locations between 4000 m.a.s.l. and 3500 m.a.s.l., along the most populated cities in the country. Although these initiatives are in more impacted areas and have a more modest scope, they are great examples of the inclusion of sustainable practices.

Tourism El Alto is an operator dedicated to provide tours where participants explore the biodiversity of birds and insects in urban and suburban contexts. The target audience of this project is not the typical international or national tourist; instead, the approach is focused on educating locals about the biodiversity richness that can be found in the city. Colibri Eco-lodge is in the Valley of Flowers in the southern part of La Paz. This private entrepreneurship has a variety of thematic cabins that harmonize with the landscape. Being close to the city, engagement with the community is sought through several tour packages, where the most popular is a walk aimed to show the challenges that women may face in a city, which may include cooking lessons using fresh products from nearby communities.

Another initiative inside the city is Tourism in Auquisamaña, supported by the Direction of Protected Areas, Forests, and Climate Change of the Municipality of La Paz, where young volunteers educate the visitors about the importance of maintaining the natural resources.

The initiatives described are just a glance at the actions taken in Bolivia to promote tourism as a sustainable industry. However, there are still several barriers to overcome, among which the support from an underdeveloped governmental structure and the robustness of mechanisms to take legal frameworks to promote efficient actions seem to be among the most important challenges to overcome. Some aspects to be considered are: (i) the need to establish profitable relations cost/benefit to take individual initiatives to last in time; (ii) to reinforce the role of government–private partnerships as an active component in the formulation of agreements, permanent monitoring; (iii) the establishment of sanctions, the development of proper mechanism for conflict resolution (Choque, 2023). In addition, the lack of an adequate and national marketing strategy has a direct impact on the lack of visibility; inside the country, few know about the activities that the country has to offer because they are not published unless they are showcased by local influencers through social networks (Facebook, Instagram, and TikTok); the situation is worst considering the image of the country abroad since activities and sustainable tourism offers are not easily found in booking commercial platforms which made them less competitive than other similar touristic experiences in neighboring countries. Capacity building could then be a way to advance.

4.2 *Challenges in Sustainable Tourism: Addressing Climate Change in Bolivia*

The pathways to address climate change impacts in tourist spots will depend on the damage and risk they are exposed to and whether the purpose is to mitigate or to take actions to adapt. Mitigation demands responses from higher governmental levels, where tourism in Bolivia is still developing. Then, perhaps the most appropriate strategies are on adaptation, at the local-individual level, which explains why most projects focus on.

Approaches suggested in the literature regarding climate change risk evaluation in tourism (Roselló-Nadal, 2014) would start with an assessment of changes in geographical conditions, health impacts as well as infrastructure essentials to tourism, followed using climate indexes to measure the attractiveness of tourist destinations; modeling tourism demand together with climate factors would provide a holistic approach when including socio-economic factors.

To assess the changes in geographical conditions, the Intergovernmental Panel on Climate Change (IPCC) has promoted in the last decade significant advances in climate change science, the physical underlying concepts, impacts, mitigation and adaptation measures, reducing the geographical disparities regarding underrepresentation of views and research from developing countries in the literature; as a result, advances, understanding, and public awareness have risen at a level that would have been impossible to reach otherwise. The use of climate risk factors such as those developed by the Sustainable Development Solutions Network (SDSN) constitutes a step ahead, based on which they guided specific strategies to define actions in the country's region. For Bolivia, the challenge ahead is to generate specific data to address issues in the tourism sector, from an integral perspective rather than from the efforts of individuals.

Concerning limitations in infrastructure associated with the potential threats of climate change and the impacts on the tourism industry, the advances from the IPCC developed with limited contributions from low- and mid-income countries but with significant findings provided by the work from developed economies, could, in turn, be extrapolated to areas with scarce knowledge. Unfortunately, it is an important obstacle that such a case would not represent the situation regarding health in low- and mid-income countries, where extrapolations would induce important uncertainties, suggesting a need for specific knowledge referred to the conditions and challenges.

Also referred to as infrastructure, the geographical location of Bolivia as a landlocked country and the limited aerial international connections increase costs and traveling times, discouraging travelers from visiting the country. Considering that the tourism industry has a big contribution to greenhouse gas emissions, Bolivia needs to take action to decrease emissions by 50% by 2030. The first action will be to have a baseline of CO_2 print with data on transportation, accommodation, food, and residue generation. Since transportation is the factor that produces the largest amount of CO_2, we must encourage the use of electric vehicles, especially in areas where it is not possible to walk or to take biking tours, such as in the Uyuni Salt Flats.

When discussing capacity building, establishing programs that include education for local and external tourists and training for operators would raise awareness of climate change effects in tourist areas. A certification of accomplishment in these courses awarded from a university or another education institution should be mandatory, especially for those operators working in preserved areas.

Finally, it is important to consider that we must be prepared for an eventual disaster and prepare and develop plans and infrastructure to deal with extreme weather events and other climate-related impacts. A strong investment in research will also be helpful to detect possible risks and to provide solutions to act promptly in case of an eventual disaster.

All the proposed strategies will demand political engagement at national and regional levels. Policies' development on emissions reduction, waste management, and conservation of natural resources need to be socialized between different stakeholders. As an incentive to those who adopt sustainable tourism practices, a tax discount could be a great incentive.

The active participation of the stakeholders involved in sustainable tourism will be key to implementing these strategies, making it possible for Bolivia to mitigate the impacts of climate change and position itself as a leading destination for sustainable tourism, benefiting both the environment and local communities.

5 Conclusions

This chapter discussed a general context of the tourism industry in Bolivia, opportunities for capacity-building activities, and initiatives to foster sustainable tourism while considering the effects of climate change.

Several examples of the effects of climate change on different ecosystems in the country are addressed, highlighting the impact that has already affected some areas. Complementary, economic, and legal context is presented to understand how tourism has a role in Bolivia.

Later, while discussing formal education, it was established that higher education institutions have several academic programs in tourism in different parts of the country, some of which have adjusted the curricula on sustainable tourism.

There have been several practical initiatives carried out in the country over the past few years, with mixed results, but it is important to create an entrepreneurship culture involving local people. Most of them have centered views and activities in sustainable tourism, where there are actions to foster capacity building for the locals, as well as for the visitors. However, some of them are struggling due to the effects of climate change and need to reinvent themselves to continue to be an alternative for the future.

With a closer look, important factors to make these initiatives feasible are not always available, particularly in rural areas, where some basic services such as electricity or drinking water can jeopardize their sustainability. Therefore, when capacity building is discussed, it must also include methodologies and technologies that can support the service provided.

Finally, there are important challenges to tackle to incorporate the impacts of climate change in the tourism sector in Bolivia. Lack of planning and infrastructure are components that must be addressed to cope with the tasks required to adapt the current initiatives and promote future ones. Mitigation of extreme climate events is a must, but at the same time, needs support from governmental institutions.

One of the responses to the issues at hand is capacity building. A set of programs that can train the people to not only host guests, but manage natural resources is a clear pathway to increase and improve opportunities for initiatives all over the country, taking advantage of the variety and beauty of the ecosystems in Bolivia to, as a final goal, improve the quality of life of people while caring for their environment.

References

Análisis Real -Latinoamérica. (2018). *The economic system of local systems: The potential of the 339 municipalities of Bolivia*. Real-Latin America Analysis and Jubilee Foundation.

Andersen, L., Medinaceli, A., Pacheco, E., García, L., Choque, S., Paz, F., & Ramos, I. (2023). *Índice Municipal de Potencial Turístico*. Universidad Privada Boliviana, SDSN Bolivia.

Araya, N., & Correa, F. (2023). *Certificaciones empresariales de sostenibilidad en América Latina y el Caribe*. Repository Cepal.

Autoridad Plurinacional de la Madre Tierra - APMT. (2023). *Política Plurinacional de Cambio Climático*. La Paz Bolivia. www.madretierra.gob.bo

Bastidas-Goyes, L. G., Méndez-Castillo, E., & Bonilla-Estévez, C. M. (2021). Ethical and legal analysis of medical tourism in Colombia. Risks in communicative instrumentalization. *Revista de Bioetica y Derecho, 52*, 121–138. https://doi.org/10.1344/RBD2021.52.34224

Bryan, E., Alvi, M., Huyer, S., & Ringler, C. (2024). Addressing gender inequalities and strengthening women's agency to create more climate-resilient and sustainable food systems. *Global Food Security, 40*, 100731–100745. https://doi.org/10.1016/j.gfs.2023.100731

Cajiao, X. (2023). Colombia and medical tourism. *Voices in Bioethics, 9*. https://journals.library.columbia.edu/index.php/bioethics/article/view/11941

Caretta, M. A., Mukherji, M., Arfanuzzaman, R., Betts, A., Gelfan, Y., Hirabayashi, T., Lissner, J., Liu, E., Lopez, R., Morgan, S., & Supratid, S. (2022). Water. In H.-O. Pörtner, D., Roberts, M., Tignor, E., Poloczanska, K., Mintenbeck, A., Alegría, M., Craig, S. Langsdorf, S., Löschke, V., Möller, A., Okem, B., & Rama (Eds.), *Climate change 2022: Impacts, adaptation and vulnerability. Contribution of Working Group II to the Sixth Assessment Report of the Intergovernmental Panel on Climate Change*. Cambridge University Press (pp. 551–712). https://doi.org/10.1017/9781009325844.006

Carrera Villacrés, F., Vernaza Quiñones, L., & Vicente Da Silva, E. (2021). Strengthening community tourism through landscape diagnosis in Bunche. *FIPCAEC, 6*(3), 367–392. https://doi.org/10.23857/fipcaec.v6i3.403

Catholic Church. Pope (2013- : Francis). (2015). *Laudato SI: On care for our common home: Encyclical letter*. Huntington.

Chan, J. (2023). Sustainable rural tourism practices from the local tourism stakeholders' perspectives. *Global Business & Finance Review, 28*, 136–144. https://doi.org/10.17549/gbfr.2023.28.3.136

Choque Ticona, O. B. (2023). *El manejo de los recursos de uso común para el turismo comunitario: Un estudio de caso de la comunidad de Caripe en el PN y ANMI Sajama—Oruro*. Tesis de maestría, Universidad Mayor de San Andrés. Cides-UMSA.

Cochrane, M. A., & Laurance, W. F. (2002). Fire as a large-scale edge effect in Amazonian forests. *Journal of Tropical Ecology, 18*(3), 311–325. https://doi.org/10.1017/S0266467402002263

Comité Ejecutivo de la Universidad Boliviana. (2015). *Informe de la reunión sectorial de Carreras de Turismo*. Retrieved from https://ceub.edu.bo/doc/sna/sectoriales/138_Turismo_UGRM_2015.pdf

Condori, C., & Ateaga, M. (2023). Community based tourism in Bolivia: Empowering SDG through TICOF and Chalalán Ecological Hostel [Poster presentation]. *11th international conference of sustainable development*. https://doi.org/10.13140/RG.2.2.25657.95845

Corpart, G. (2024). *A closer look at medical tourism in Latin America*. Global Health Intelligence. https://globalhealthintelligence.com/ghi-analysis/a-closer-look-at-medical-tourism-in-latin-america-2/

Coudrain, A., Francou, B., & Kundzewicz, Z. W. (2005). Glacier shrinkage in the Andes and consequences for water resources - Editorial. *Hydrological Sciences Journal, 50*(2). https://doi.org/10.1623/hysj.2005.50.6.925

Cox, R. (2009). *Turismo indígena y comunitario en Bolivia: Un instrumento para el Desarrollo socio-económico e intercultural*. Plural Ed. La Paz, Bolivia.

De Castro, S. L., Guerrero-Rodríguez, F. A., Tobón, G. J., & Nina-Baltazar, E. A. (2020). Juntas de Acción Comunal y gobernanza rural: retos para la participación y organización comunitaria en seis territorios de Nariño, Colombia. *OPERA, 28*, 239. https://doi.org/10.18601/16578651.n28.11

Eagles, P. F. J., McCool, S. F., & Haynes, C. D. A. (2002). *Sustainable tourism in protected areas: Guidelines for planning and management*. IUCN Gland, xv + 183 pp.

Eckstein, D., Künzel, V., & Schäfer, L. (2021). *Global climate risk index 2021. Who suffer most extreme weather events? Weather –related loss events in 2019 and 2000–2019*. Germanwatch e.V. https://www.germanwatch.org/en/cri

European Travel Commission. (2021) *Sustainable tourism implementation: Framework and toolkit*. Recovered from https://etc-corporate.org/reports/sustainable-tourism-implementation-toolkit/

Farfán-Pacheco, K., Alvarado-Vanegas, B., & Espinoza-Figueroa, F. (2024). Superando la adversidad: ¿resiliencia del turismo en la ruralidad durante la COVID-19? Comunidades en el sur de Ecuador como marco de estudio. *Investigaciones Turísticas, 28*, 103–126. https://rua.ua.es/dspace/bitstream/10045/144804/3/Investigaciones_Turisticas_28_06.pdf

Flores, J. (2021). Medical tourism business ecosystem in Mexico y Brazil. *El Periplo sustentable, 41*, 39–54. https://doi.org/10.36677/elperiplo.v0i41.15698

Francou, B., Ribstein, P., Saravia, R., & Tiriau, E. (1995). Monthly balance and water discharge of an inter-tropical glacier: Zongo Glacier, Cordillera Real, Bolivia, 16°S. *Journal of Glaciology, 41*(137), 61–67. https://doi.org/10.1017/S0022143000017767

Fundación para el Desarrollo Sostenible del Chaco (FUNDESOC). (2023). *Parque Nacional y Área Natural de Manejo Integrado Otuquis*. Retrieved from https://www.fundesoc.org.bo/index.php?Itemid=127&catid=9&id=135%3Aproyecto-turismo-comunitario-sostenible-en-la-comunidad-motacucito-nuevo&option=com_content&view=article&utm_source=chatgpt.com

Fundación Porvenir. (2024). *Iniciativas y proyectos para un futuro mejor en Bolivia*. Retrieved from https://fundacionporvenir.org.bo/

García, S., & Manquillo, C. (2023). *Elements for the formulation of a productive diversification and economic development program as a strategy to reduce youth emigration in the municipality of Nimaima, Cundinamarca*. Institutional Repository of Universidad Distrital Francisco José de Caldas. Retrieved from https://repository.udistrital.edu.co/items/fa082935-06bd-428b-bd17-b1bd2f780be3

Gliemmo, F., & Moscoso, F. V. (2022). *Geografía turística latinoamericana: Aportes conceptuales y estudios de caso para la comprensión de un escenario en transición*. Universidad Nacional de La Plata. https://sedici.unlp.edu.ar/handle/10915/148712

Guerra, D. N., García Hurtado, D., Acosta, D. R., & Rodríguez Torres, E. (2024). Work stress and its influence on female performance: A systematic review. *Health Leadership and Quality of Life, 3*(1), 546. Retrieved from https://hl.ageditor.ar/index.php/hl/article/view/546

Higgins-Desbiolles, F., Scheyvens, R. A., & Bhatia, B. (2022). Decolonising tourism and development: From orphanage tourism to community empowerment in Cambodia. *Journal of Sustainable Tourism, 31*(12), 2788–2808. https://doi.org/10.1080/09669582.2022.2039678

Iglesias Alonso, Á. (2021). La gobernanza local del turismo rural como respuesta a los efectos de la Covid-19. Barataria. *Revista Castellano-Manchega de Ciencias Sociales, 30*, 86–98. https://doi.org/10.20932/barataria.v0i30.614

Instituto Nacional de Estadística. (2019). Retrieved from https://www.ine.gob.bo/index.php/estadisticas-economicas/turismo/

International Development Research Centre (IDRC). (2024). *Orbita: An observatory designed to stimulate responsible tourism.* https://idrc-crdi.ca/en/stories/orbita-observatory-designed-stimulate-responsible-tourism

Janusek, J. W. (2004). *Tiwanaku: Portrait of an Andean civilization.* Routledge.

Kaenzig, R., Rebetez, M., & Serquet, G. (2016). Climate change adaptation of the tourism sector in the Bolivian Andes. *Tourism Geographies, 18*(2), 111–128. https://doi.org/10.1080/14616688.2016.1144642

Khizar, H., Younas, A., Kumar, S., Akbar, A., & Poulova, P. (2023). The progression of sustainable development goals in tourism: A systematic literature review of past achievements and future promises. *Journal of Innovation & Knowledge, 8*(4), 100442. https://doi.org/10.1016/j.jik.2023.100442

Kinouchi, T., Takeuchi, K., & Yamagata, Y. (2019). Water security in high mountain cities of the Andes under a growing population and climate change: A case study of La Paz and El Alto, Bolivia. *Water Security, 8*, 100025. https://doi.org/10.1016/j.wasec.2019.100025

Lacruhy Enríquez, C. C., & Morandin Ahuerma, I. (2023). Estrategias de competitividad en las empresas turísticas en un escenario de incertidumbre. Caso Los Cabos, México. *El Periplo Sustentable, 46*, 50–70. https://doi.org/10.36677/elperiplo.v0i46.20112

León-Gómez, A., Ruiz-Palomo, D., Fernández-Gámez, M. A., & García-Revilla, M. R. (2021). Sustainable tourism development and economic growth: Bibliometric review and analysis. *Sustainability, 13*, 2270. https://doi.org/10.3390/su13042270

Lugo-Carvajal, A., et al. (2023). Fish on fire: Shifts in Amazonian fish communities after floodplain forest fires. *Journal of Applied Ecology, 60*(5), 1851–1863. https://doi.org/10.1111/1365-2664.14165

MacDonald, A., & Mordecai, E. (2019). Amazon deforestation drives malaria transmission, and malaria burden reduces forest clearing. *Proceedings of the National Academy of Sciences, 116*(44), 22212–22218. https://doi.org/10.1073/pnas.1905315116

Ministry of Productive Development and Plural Economy. (2023). *Sectoral Plan for Integrated Development to Live Well in the Tourism Sector 2021–2025.* Retrieved from https://www.turismo.produccion.gob.bo/wp-content/uploads/2024/03/DIAGRAMADO-PSDI-TURISMO-09-08-2023-1_compressed.pdf

Moreno-Manzo, J., Gassiot-Melian, A., & Coromina, L. (2022). Perceived value in a UNESCO world heritage site: The case of Quito, Ecuador. *Journal of Cultural Heritage Management and Sustainable Development.* https://doi.org/10.1108/JCHMSD-03-2022-0049

Moscardo, G. (2008). Building community capacity for tourism development: Conclusions. *Building Community Capacity for Tourism Development.*, 172–179.

Muñoz, L. S.. (2023). Bolivia's mother earth Laws, is the ecocentric legislation misleading? *ReVista – Harvard Review of Latin America.* https://revista.drclas.harvard.edu/bolivias-mother-earth-laws-is-the-ecocentric-legislation-misleading/

Pacheco, E., Altamirano, A., & Andersen, L. (2024). *Índice Municipal de Potencial Turístico en Bolivia (IMPT).* sdsn Bolivia, upb y fes Bolivia. https://sdsnbolivi.org/impt

Page, M., McKenzie, J., Bossuyt, P., Boutron, I., Hoffmann, T., Mulrow, C., Shamseer, L., Tetzlaff, J. M., Akl, E. A., Brennan, S. E., Chou, R., Harris, J. R., Hartling, L., Karanicolas, M., Lalu, M., Li, T., Mavergames, C., Munn, Z., Puhan, M., Rehfuess, E., Rios, P., Salanti, G., Thomas, J., Tyndall, J., Veroniki, A., Wallace, B., & Wells, G. (2021). The PRISMA 2020 statement: An updated guideline for reporting systematic reviews. *PLoS Medicine, 18*(3), e1003583. https://doi.org/10.1371/journal.pmed.1003583

Peaty, D., & Portillo, A. (2009). Community-based tourism in Bolivia: Projects and perspectives. *Ritsumeikan Annual Review of International Studies, 21*(3), 111–120.

Perez Galvez, J. C., Fuentes Jimenez, P. A., Rodriguez Gutierrez, P., & Medina Viruel, M. J. (2023). Emotional perception and cultural motivation on loyalty to a World Heritage Sites destination. *Geojournal of Tourism and Geosite, 49*(3), 1165–1175. https://doi.org/10.30892/gtg.49331-1115

Pomfret, G., Sand, M., & May, C. (2023). Conceptualising the power of outdoor adventure activities for subjective well-being: A systematic literature review. *Journal of Outdoor Recreation and Tourism, 42*, 1–10. https://doi.org/10.1016/j.jort.2023.100641

Rosselló-Nadal, J. (2014). How to evaluate the effects of climate change on tourism. *Tourism Management, 42*, 334–340. https://doi.org/10.1016/j.tourman.2013.11.006

San Martin, R., Plata, M., Jemio, M., & Tamayo, A. (2024). *Turismo comunitario para el emponderamiento local y desarrollo sostenible de Bolivia.* Boletín de Inteligencia Turística No 17. Observatorio Boliviano para la Industria Turística Sostenible-ORBITA. https://orbita.bo/bit/bit-17/

Shane, I. (2023). Rising global temperatures are already affecting the tourism industry – here's how. *World Economic Forum.* http://www.weforum.org/stories/2023/08/temperatures-tourism-climate-impact/

Singh, S., Jain, S., Mishra, P., & Goll, M. (2022). Assessing the impact of land-use change on water resources in the Amazon Basin. *Physics and Chemistry of the Earth Parts A, B, C, 126*, 103053. https://doi.org/10.1016/j.pce.2021.103053

Soria, F., & Kazama, S. (2011). *Monte Carlo experiments for uncertainty investigation of glacier melt discharge predictions through surface energy balance analysis* (p. 346). IAHS-AISH Publication.

Sustacha, I., Baños-Pino, J., & Del Valle, E. (2023). The role of technology in enhancing the tourism experience in smart destinations: A meta-analysis. *Journal of Destination Marketing & Management, 30*, 100817–100829. https://doi.org/10.1016/j.jdmm.2023.100817

Syafruddin, S., Salahuddin, M., & Nurhilaliati. (2023). Accessibility of women's empowerment in tourism industry from Mandalika special economic zone. *Journal of Environmental Management and Tourism, 2*(66), 491–499.

Ticse, E., Zubieta R., Luque A., & Herrera, M. (2023). Stress and academic procrastination in students of a Private University Arequipa 2022. In *Proceedings of the Latin American and Caribbean conference for engineering and technology (LACCEI 2023)*, Buenos Aires, Argentina. Retrieved from https://laccei.org/LACCEI2023-BuenosAires/meta/FP1199.html

Torres, O., Álvarez, V., Espinoza, J., & Rodríguez, M. (2023). Metodología para mejoramiento comercial de la oferta de bares en Ciudades Creativas Gastronómicas: caso Portoviejo. *Ecuador. ECA Sinergia, 14*(3), 7–17. https://doi.org/10.33936/ecasinergia.v14i3.5695

Tortato, F., Tomas, W. M., Chiaravalloti, R. M., & Morato, R. (2022). Tragedy of the commons: How subtle, "legal" decisions are threatening one of the largest wetlands in the world. *Bioscience, 72*(7), 609. https://doi.org/10.1093/biosci/biac025

United Nations. (2015). Retrieved from https://www.un.org/es/about-us/

Vera, I. (2018). *Glosario turístico.* Retrieved from https://www.entornoturistico.com/es-el-turismo-una-ciencia/

World Tourism Organization (WTO) and United Nations Development Program. (2017). *Tourism and the sustainable development goals—Journey to 2030.* UNWTO.

Wu, D. (2024). City walk in a gap day: Potential and opportunities for tourism and leisure. *Tourism Review, 79*(9)1576–1581. https://doi.org/10.1108/TR-09-2023-0614

Yu, L., Zhao, P., Tang, J., & Pang, L. (2023). Changes in tourist mobility after COVID-19 outbreaks. *Annals of Tourism Research, 98*, 103522. https://doi.org/10.1016/j.annals.2022.103522

Zhang, L., Yang, S., Wang, D., & Ma, E. (2022). Perceived value of, and experience with, a World Heritage Site in China—The case of Kaiping Diaolou and villages in China. *Journal of Heritage Tourism, 17*(1), 35. https://doi.org/10.1080/1743873X.2020.1820014

Exploring the Ecomuseum Paradigm for Sustaining Siwa's Cultural Heritage Amid Climate Change

Shaimaa Matar

Abstract This chapter explores the potential of the ecomuseum model in enhancing community-based tourism and preserving Siwa's cultural heritage amid climate change adaptation. Using qualitative research methods, it examines Siwa's cultural resources, their historical development, and their role within the local community. The data collected were analyzed thematically to identify community perspectives on cultural heritage, sustainability, and economic opportunities. Findings indicate that ecomuseums can support regional business ownership, community engagement, and sustainable economic development. The study highlights potential synergies between ecomuseums and community-based tourism, emphasizing their role in enhancing visitor experiences, promoting environmental responsibility, and ensuring equitable benefit distribution. The chapter concludes with policy recommendations for integrating traditional ecological knowledge, sustainable resource management, and climate adaptation strategies into ecomuseum initiatives. It underscores the importance of stakeholder collaboration and community participation in developing a resilient and inclusive heritage tourism model.

1 Introduction

Siwa Oasis stands as a testament to the resilience and creativity of its local community, renowned for its distinctive cultural and natural heritage. However, this exceptional region faces substantial challenges, including the need for efficient resource management, the impacts of climate change, and the prevalence of

S. Matar (✉)
MMS, National Museum of Egyptian Civilization, Cairo, Egypt
e-mail: shaimaamatar.nmec@gmail.com

J. Marques et al. (eds.), *Tourism and Climate Change*, Tourism, Hospitality & Event Management, https://doi.org/10.1007/978-3-031-99128-8_9

unsustainable tourism practices. In this context, the concept of an ecomuseum emerges as a promising solution to empower the residents of Siwa and safeguard their heritage.

This chapter investigates the feasibility of establishing an ecomuseum in Siwa to integrate sustainable tourism practices with community engagement. By emphasizing the preservation of cultural identities and promoting environmentally conscious practices, an ecomuseum can serve as a critical platform for economic development, tourism, and education.

An ecomuseum, developed through collaboration with local stakeholders, can protect Siwa's resources while fostering community pride and ownership.

Furthermore, this chapter will explore how an ecomuseum can address the urgent challenges posed by climate change and advocate for adaptive strategies that enhance community resilience. By utilizing sustainable practices and traditional knowledge, the ecomuseum can help preserve Siwa as a vibrant and sustainable destination for future generations. Ultimately, we aim to provide a roadmap for a thriving Siwa Oasis by highlighting the potential synergies between sustainable tourism, community empowerment, and cultural heritage preservation as we delve into the intricacies of this model.

2 Literature Review

In the 1960s, environmental awareness grew, prompting questions about preserving and transmitting heritage. The "new museology" emphasizes engaging diverse audiences and recognizing museums as part of society, not isolated institutions. Ecomuseums emerged in the 1970s, aiming to rejuvenate communities through joint efforts of local leaders (Ohara, 1998).

de Varine (1996) begins by explaining the distinction between a traditional museum and an ecomuseum. In traditional museums, the focus is on a collection of objects housed within a building, with visitors coming to learn from and enjoy the exhibits, which are managed by professionals in the museum and antiquities field. In contrast, ecomuseums concern themselves with the heritage of a specific territory, including both tangible and intangible elements, and are managed by the community itself. The goal of ecomuseums is to develop and promote the heritage of the territory while involving the local population in the process (Fig. 1).

In 1980, George Henri Rivière wrote an evolutionary definition of the ecomuseum.

- An instrument conceived, fashioned, and operated jointly by a public authority and a local population.
- It serves as a mirror for the local population to discover its own image.
- It is an expression of man and nature. It situates man in his natural environment.
- An expression of time.
- An interpretation of space,

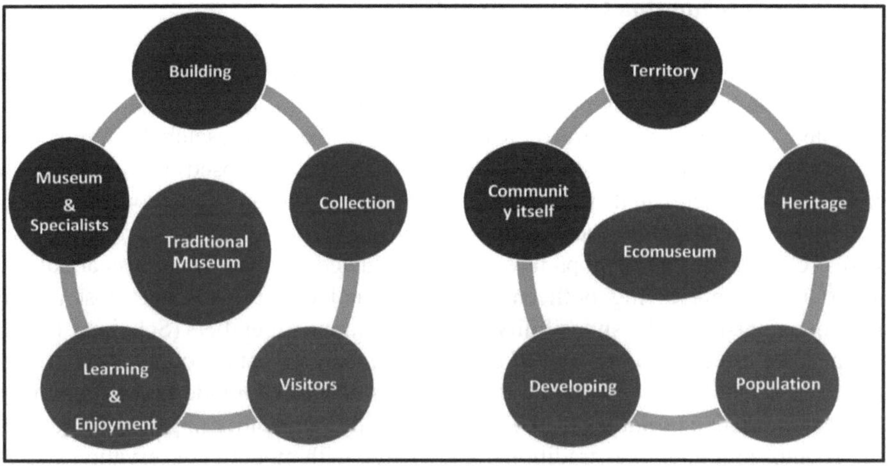

Fig. 1 Key differences between traditional museum and ecomuseum. Source, author

- It is a laboratory because it studies the past and present of the population and its environment and trains specialists in these fields in collaboration with outside research bodies.
- It is a conservation center insofar as it contributes to the preservation and development of the population's natural and cultural heritage.

It is a school, insofar as it involves the population in its work of study and protection and encourages it to have a clearer grasp of its own future (Rivière, 1985; Sabrina, 2010).

The principles of ecomuseums align with UNESCO's recommendations on *heritage interpretation and community engagement* (UNESCO, 2012).

2.1 Ecomuseum Applications Worldwide

Cultural heritage is a dynamic concept that evolves over time, encompassing both tangible and intangible elements. According to Ferretti et al. (2014), its definition continues to expand, moving beyond individual monuments such as architectural works, sculptures, and painted memorials to include groups of buildings, archaeological sites, regional systems, landscapes, itineraries, and intangible heritage. Similarly, the UNESCO World Heritage Convention (1972) defines cultural heritage as including monuments, buildings, landscapes, and traditions. Several studies underscore the effective application of the ecomuseum model in various cultural and geographical settings, offering valuable insights into how such models can be adapted to diverse socio-economic contexts and heritage preservation needs.

2.1.1 Ecomuseum Ilha Grande (Brazil)

The Ecomuseum Ilha Grande, founded in 1999 in Brazil, represents a significant example of cultural heritage preservation integrated with sustainable development. Affiliated with the University of Rio de Janeiro, the ecomuseum highlights the island's rich history and ecology, while addressing threats posed by rapid tourism expansion, environmental degradation, and cultural erosion. The ecomuseum has successfully engaged the local community in participatory decision-making processes. Community workshops have been conducted to raise awareness about the importance of preserving both environmental and cultural heritage, fostering a sense of ownership and responsibility among community members (Scheiner, 1970).

The Ecomuseum's "Girls and Women in Science" project encourages local girls to pursue scientific careers, while the "Ecomuseum Ilha Grande and Community Tourism" initiative promotes local tourism and cultural identity to generate income. The ecomuseum integrates traditional knowledge with modern conservation strategies. For example, it revives traditional fishing methods, adapting them to current environmental conditions to ensure ecological sustainability and cultural continuity (Mendon, 2015).

In 2011, the "Ecomuseum Recycle" project was launched to tackle waste accumulation during peak tourist seasons by establishing a waste collection system in Vila Dois Rios. This project promotes recycling and repurposing materials into crafts, culminating in a traveling exhibition to raise awareness. Using a participatory approach, the project involves community members in workshops and activities, fostering environmental responsibility. Ecomuseum Recycle has successfully reduced PET waste and serves as a model for other communities while supporting local artisans and promoting cultural heritage.

Educational initiatives aimed at both locals and visitors have been developed, focusing on the importance of biodiversity and the cultural significance of local practices. These include courses on museum and heritage management, environmental education, workshops, music, language instruction, firefighter training, and curatorial training for youth and professionals. By educating the community and visitors, the ecomuseum enhances understanding and appreciation for the local culture and environment.

The ecomuseum collaborates with researchers to monitor environmental changes and develop adaptive strategies. This partnership has led to the implementation of community-based conservation efforts that not only protect the environment but also preserve the cultural practices tied to those ecosystems. For instance, the Parque Botânico launched a PET bottle wall project in 2019, reusing around eight thousand bottles to reduce building costs and promote sustainability. It also emphasizes reusing local materials, such as old cinema roof beams for the Plant Nursery, training residents and creating jobs (De Almeida, 2022).

Angra dos Reis shares many characteristics with the Egyptian Siwa Oasis, as both are rich in cultural heritage and natural beauty, attracting tourists while facing similar environmental challenges. We aim to draw on the practices implemented at the Ecomuseum in Brazil to propose a project for the Siwa Ecomuseum.

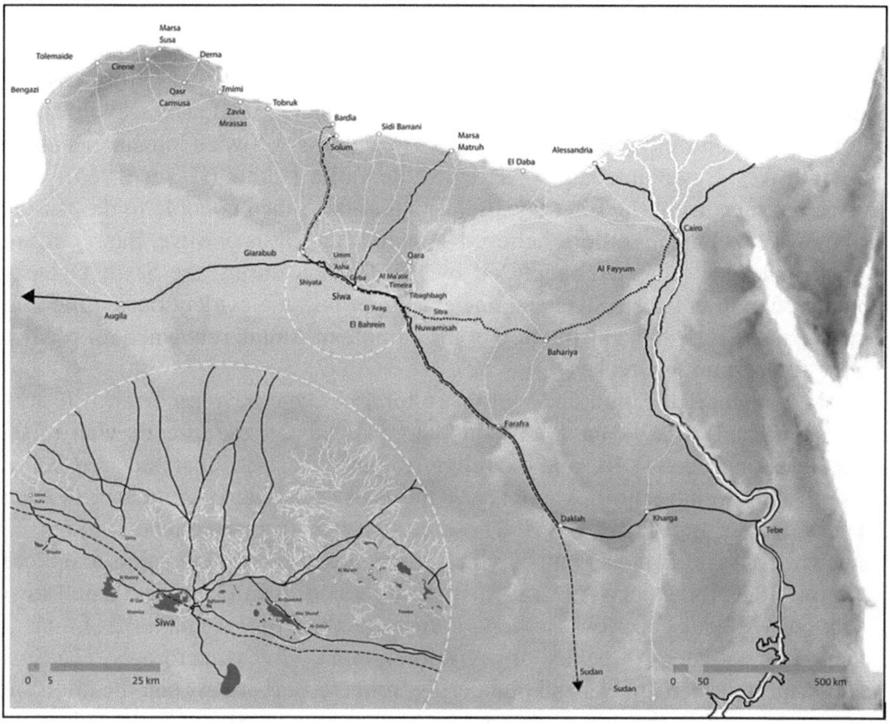

Fig. 2 Map of Siwa, a historical crossroads of ancient trade routes in the Western Desert (Source: Montalbano & Petruccioli, 2011)

2.2 *Siwa Oasis*

The Siwa Oasis is located at the crossroads of major ancient desert trade routes (Fig. 2). One such route, the Augila–Thebes route, stretched from Thebes in the Nile Valley to Augila in the Libyan desert, continuing westward across the Sahara. Along this route, key oases include Khargah, Dakhlah, Farafra, and Siwa in Egypt, as well as Augila and Jaghbub in Libya. Another important route, the Oasis route, emerged in the seventh century, connecting Solum on the Mediterranean coast with Sudan to the south (Montalbano & Petruccioli, 2011).

2.2.1 Climate in Siwa

Siwa, 18 meters below sea level, has a continental climate with cold winters and hot summers. Daytime temperatures reach 34–48 °C in summer and 4–20 °C in winter, with nighttime lows dropping below freezing (Weatherspark.com, n.d.).

2.2.2 Historical Background

Archaeological evidence, including flint tools resembling artifacts from Algeria, Morocco, Cyrenaica, Nubia, and Egypt, suggests that Siwa Oasis was inhabited during the Palaeolithic and Neolithic eras. The culture of Siwa's inhabitants aligns with both West Egyptian countries and the Nile Valley culture (Hayes, 1964).

Ancient texts from the First Dynasty mention the Tahenu people to the west of the Nile Valley. Tahenu settlers, believed to include the Oasis of Siwa, Bahriyah, and Barqah in Libya, were later replaced by the Temehu during the Sixth Dynasty. Communication between Siwa and ancient Egypt became evident during the 26th Dynasty, marked by the construction of the Temple of Amun, renowned for predicting the future among Greeks (الدميرى، ٢٠١٦).

Siwa's strategic location and vibrant cultural heritage have rendered it historically significant. Significant events, including Alexander the Great's visit to the Temple of Amun in 331 BC and the religious transformations that occurred during the sixth and seventh centuries, emphasize the oasis's historical significance (Vivian, 2008; Vale, 2014; Fakhry, 1944). These events gain further context through the framework of cultural preservation, which the proposed ecomuseum aims to uphold.

In the fifteenth century A.D., al-Maqrizi described Siwa Oasis as a small town inhabited by about 600 Siwi Berbers, known for their Siwit[1] language similar to Zanata. It had gardens of palms, olives, figures, and 20 freshwater pools. The Siwa manuscript[2] revealed attacks and plundering from the Berber between the ninth and twelfth centuries. In the thirteenth century, construction on Shali Castle revived the oasis (الدميرى، ٢٠١٦).

Siwa remained isolated until Musa bin Naseer's failed invasion in 708 AD resulted in heavy casualties during a prolonged siege. In the third century AH, Arabs accessed the oasis via the Bani Hilal tribes.

By 1820, internal conflicts led factions to seek help from Muhammad Ali Pasha, who sent a military unit under Hussein Bey Al-Shamma Sharji. Sharji ultimately annexed Siwa to Egypt with an annual tribute of 1000 riyals (Fakhry, 1944).

In the twentieth century, interest in Siwa's historical and archaeological resources grew, leading to its "Archaeological area" inscription on UNESCO's tentative World Heritage Sites list in 1994. That same year, Law No. 4 was enacted to protect the environment, and Siwa was recommended as one of 31 nature reserves (EEAA, 2002).

The Prime Minister's 2002 decision No. 1219 (SIS, 2023) declared the 7800-square kilometer Siwa Oasis region in Matrouh Governorate a natural reserve.

---

[1] Siwit is the first language spoken in homes and on the street, but they speak Arabic well and use it more after television entered homes, children learned Arabic early.

[2] The title of the Siwa Manuscript is "The Manuscript of the History of the Siwan and Their Most Significant Returns from the Islamic Conquest to 1327 AH," and it was written by Muslim Omar, the head of the Abu Muslim family and the religious judge of Siwa in the late nineteenth century. The viewing of this manuscript is restricted to a small number of researchers. It has not been released to the public yet.

One sector is on the western border of the Qattara Depression and includes Hattayat Sitra, Nawamisah, Al-Arj, and Bahrain, Tabbagh, Umm Al-Saghir, Umm Al-Ghazlan, Jerba, Shayata, and Al-Malfa with an area of about 1700 km^2.

2.2.3 The Origin of the Name Siwa

The name Siwa has evolved from "Tha" to the "Oasis of Jupiter-Ammon" in the Greco-Roman era, underscoring its historical significance (Fakhry, 1974; Vale, 2014). Al-Idrisi referred to it as "Santryah," home to Berbers and Bedouins in the eleventh and twelfth centuries. By the fifteenth century, Al-Maqrizi also called it "Santryah" and "Al-Waha Al-Aqsa." Ibn Khaldun named it "Tiswa," linking its people to the Zanata tribe, while Dr. Coleman noted that the ancient name in the Temple of the Oracle resembles "Santryah" as "sa-ntr" or "Sa-n-drw" (الدميرى، ٢٠١٦).

The historical events that have influenced Siwa throughout the centuries have played a crucial role in creating its distinct cultural heritage. To truly understand the importance of these developments, it is necessary to assess both the tangible and intangible cultural assets that exist today.[3]

Siwa Oasis exemplifies this expanded definition, as it is a culturally rich region in Egypt whose tangible and intangible heritage plays a vital role in shaping local identity and promoting sustainable tourism (Fakhry, 1944; Vale, 2014).

2.2.4 Challenges in Cultural Heritage Preservation in Siwa

Several studies have documented Siwa's historical significance, cultural assets, and ecological uniqueness (Fakhry, 1944; Vale, 2014). Research highlights Siwa's Berber heritage, including its distinct language (Siwi), architectural styles, and traditional crafts (El-Shafie, 1999). Tourism in Siwa is closely tied to its cultural and natural heritage. The development of eco-lodges, such as Adrère Amellal and Albabenshal, follows sustainable construction principles using karshif and solar energy. These lodges provide economic opportunities by hiring local workers, promoting traditional architecture, and supporting ecotourism initiatives. Despite these advantages, seasonal fluctuations in tourism, lack of proper marketing, and dependency on foreign visitors pose economic vulnerabilities for Siwa's tourism sector. Addressing these challenges requires better tourism infrastructure, digital marketing strategies, and policies promoting responsible tourism (Vale, 2014).

The literature identifies several pressing challenges in preserving Siwa's cultural and environmental assets. These include:

Urban expansion and modernization leading to the deterioration of traditional karshif architecture (Dabaieh, 2011).

[3] The Archaeological area contains temple of Oracle and the temple of Amun "Ubaydah" together with the different archaeological areas and monuments.

Mass tourism impacting fragile ecosystems and historic sites.
Water resource management issues, such as rising salinity levels, affecting agriculture and local livelihoods (Moghazy & Kaluarachchi, 2021).

Sustainable development frameworks suggest that heritage conservation should be economically viable, socially inclusive, and environmentally sustainable (IPCC, 2014). Studies on community-based tourism in similar heritage-rich locations show that integrating local governance, cultural entrepreneurship, and eco-friendly tourism can mitigate environmental degradation while benefiting local communities (Radwan, 2007).

3 Methodology

3.1 Research Methods

This study employed qualitative research methods in order to examine Siwa's cultural resources, their historical development, and their role within the local community. The research methodology involved semi-structured interviews, participant observation, and direct site assessments to gain insights into the challenges and opportunities faced by Siwa's cultural landscape. A field study conducted in March 2022 utilized semi-structured interviews with six community members, including local artisans, business owners, conservationists, and tourism professionals, facilitated by local resident and one tour guide. The author engaged in field visits to cultural heritage sites, traditional markets, and community-based tourism establishments. Observations included local economic activities, sustainability practices, and the adaptation of traditional knowledge to modern challenges. Site Assessments were made based on physical documentation of environmental degradation, architectural conservation issues, and the impact of urban development on Siwa's unique landscape.

The results of this study will guide the development of the proposed Siwa Ecomuseum by integrating local priorities with sustainable conservation approaches.

4 Evaluation of Cultural Assets in Siwa Oasis

This section evaluates Siwa's cultural assets both tangible and intangible which form the foundation for the proposed ecomuseum model. These assets are essential not only for heritage preservation but also for enhancing community-based tourism and climate adaptation strategies. By categorizing them into built heritage, natural sites, social traditions, artistic expressions, and economic contributions, we provide a roadmap for how the ecomuseum can serve as a living museum rooted in community participation and adaptive reuse of heritage resources.

Siwa Oasis is a unique and culturally significant region in Egypt. Its heritage encompasses a wide range of characteristics including historical, aesthetic, social, economic, architectural, urban, cultural, religious, scientific, geographical, and natural/geological aspects (Table 1). These diverse features not only attract tourists but also play a vital role in shaping the Siwi identity. Recognizing and integrating these values into the ecomuseum will allow for a holistic, community-driven approach to conservation that aligns with both environmental and cultural sustainability.

4.1 Tangible Cultural Assets

Tangible heritage sites in Siwa including monumental ruins, traditional architecture, and sacred landscapes can become key interpretive nodes within the ecomuseum's distributed model. These places will not only be preserved but also recontextualized through storytelling, guided tours, and community-led conservation efforts.

4.1.1 Historical and Architectural Heritage

Shali Fortress Built in the twelfth century AD from karshif (mud bricks with salt), this fortress (Fig. 3) was designed to protect against Bedouin raids. It features high walls, narrow streets, and three main gates: Bab Inshal (City Gate), Bab Atrat (New Gate), and Bab Quddumah (reserved for women). The fortress remains a significant

Table 1 Assessment of tangible and intangible values (cultural significance) in the Siwa Oasis. Source: Matar (2024)

		1	2	3	4	5	6	7
	Values	Old town (Shali Fortress)	Shali Aghurmi and The Temple of Oracle	The Temple of Umm Ubayda	Gebel Mawta	Dakrur Mountain	Springs-palm and Olive Grooves	The Lakes
1	Historic	x	x	x	x	x	x	
2	Aesthetic	x	x				x	x
3	Social	x				x	x	x
4	Economic	x					x	
5	Architectural	x			x			
6	Urban	x						
7	Culture	x				x	x	x
8	Scientific	x					x	
9	Geographical	x						
10	Religious	x						
11	Natural/ Geological						x	x

Fig. 3 Shali Fortress (photo by the author, 2022)

cultural symbol, though parts have deteriorated due to erosion and modern development (Vale, 2014).

Shali Fortress offers opportunities for hands-on conservation workshops and heritage interpretation trails.

Temple of the Oracle Dating back to the sixth century BC under King Amasis of the 26th Dynasty, this temple in Aghurmi played a crucial role in Siwan religious life. Alexander the Great visited 331 BC to seek confirmation of his divine status. The temple exhibits both Egyptian and Hellenistic architectural elements (Fakhry, 1944). The temple can anchor historical narratives that connect Siwan identity with broader Mediterranean history.

Temple of Umm Ubayda This lesser-known temple near the Temple of the Oracle was built by King Nectanebo II and dedicated to Amun. It suffered significant damage from an earthquake in 1811 and was later dismantled in 1897. The massive stone blocks used in its construction exceed those of the Great Pyramid in scale (Fakhry, 1974).

Gebel Mawta (Mountain of the Dead) This necropolis, dating to the 26th Dynasty, contains rock-cut tombs decorated with intricate wall paintings and burial chambers. Notable tombs include those of Si-Amun and Mesu-Isis, which provide insights into ancient Siwan funerary customs (Vale, 2014).

Dakrur Mountain Located 3 km southeast of Siwa, Dakrur Mountain is renowned for its healing *sand baths*, where individuals bury themselves in hot sand for rheumatic pain relief. It is also home to the *Siwa Peace Festival*, which unites local tribes in reconciliation and communal feasting. Historically, the mountain was believed to conceal treasures guarded by supernatural forces, contributing to its mystical reputation (Belgrave, 1970; Allam et al., 2018).

Sites like Dakrur Mountain and Gebel Mawta reflect cultural beliefs and healing tradition's vital themes for an ecomuseum centered on community wellness and spiritual ecology.

Traditional Architecture Siwan architecture utilizes *karshif*, a sustainable mixture of salt, clay, and mud that provides insulation against extreme temperatures. Homes feature small windows, thick walls, and palm trunk beams for structural support, ensuring durability. Olive wood is used in doors and furniture due to its insect-resistant properties. Despite the modernization of building materials, efforts to preserve traditional karshif structures continue through restoration projects (El-Shafie, 1999; Dabaieh, 2011). Traditional Architecture showcases sustainable building methods that the ecomuseum can revive through public training programs.

The Springs and Palm Groves Siwa boasts over 200 freshwater and thermal springs, including *Cleopatra's Bath (Ain Juba)*, a historically significant bathing site. Springs such as Abu Shuruf provide irrigation for Siwa's vast *date palm and olive groves*, which form the backbone of the local economy (Alkalawy, 2022; FAO, 2016) and which represent opportunities for eco-agriculture exhibits, supporting the ecomuseum's role in environmental education and food heritage.

4.2 Intangible Heritage and Community-Led Conservation

The preservation of Siwa's intangible cultural heritage language, rituals, music, healing practices, and oral history is a cornerstone of the ecomuseum model. Unlike traditional museums, ecomuseums thrive on living culture that evolves with community engagement.

The Siwan language, known as Siwi, is a Berber dialect spoken by the local population. Despite Arabic being widely used in education and formal settings, Siwi remains the primary means of communication among the locals. It is an integral part of the community's cultural identity and heritage (Vale, 2014). The Siwi language can be preserved and promoted through audio-guided tours, language classes, and children's programs hosted by the ecomuseum.

As sheik Omar Rageh[4] state that new residents in Siwa form tribes or qiblas, each led by a sheikh elected by male members, with this position often passing from father to son for life. The sheikh addresses problems and disputes within the clan. Tribes have a council, Majlis al-Ajwad, aiding in conflict resolution. Siwa has 11

[4] Sheikh Omar Rageh is the sheikh of the Awlad Mousa tribe, born in1967, and a member of the Heritage Committee.

tribes, excluding Aghourmi, El Garah, and Shuhaibat, located 120 km away. With increasing tribes, an additional council, the Majlis of the Wise, was formed to assist sheikhs. Each sheikh has a council of wise men called "Alokalaa" to aid in issue resolution.

For matters outside of his control, the sheikh may advise a resident to contact the police, or the police may refer a resident to the sheikh. Sheikhs may have less power than in the past, but they serve on an advisory council that aids the mayor of Marsa Matrouh, who is the local representative of the central government appointed by the governor (Vale, 2014). Tribal governance systems can inform the ecomuseum's participatory management structure.

One of the most significant cultural events in Siwa is the Siwa Peace Festival (Eid Al-Solh) (Fig. 4), held annually at Dakrur Mountain. This festival fosters unity and reconciliation among Siwa's tribes. The festival features communal feasts, traditional dances, and spiritual gatherings. Participants prepare special traditional foods, a traditional bread cooked in sand ovens. The festival has historical significance, as it was initiated over 150 years ago by Sheikh Mohamed Al-Madani to resolve tribal conflicts (Belgrave, 1970). The Siwa Peace Festival exemplifies how the ecomuseum can support intergenerational transmission of culture through seasonal programming and public events.

Siwa has a long history of traditional healing practices based on herbal medicine. The local community relies on medicinal plants such as olive leaves, palm pollen, and desert herbs to treat ailments ranging from digestive issues to skin diseases. Traditional healers, often elderly women, prepare herbal remedies and ointments, passing down their knowledge through generations. Siwans also practice sand baths, where individuals with rheumatic conditions are buried in hot sand for therapeutic benefits (Allam et al., 2018). Traditional healing practices using desert herbs can be documented and presented in collaboration with local practitioners through interactive exhibits and herb gardens.

Fig. 4 Gebel Dakrur. Source: Mikhail (2012)

Siwan clothing is rich in cultural symbolism, particularly for women. Traditional garments include Akbir N Wasa, a tunic decorated with intricate embroidery using five distinct colors, symbolizing the growth stages of dates. Married women wear Milaya, a full-body black sheet adorned with embroidery, signifying their marital status. Siwan silver jewelry (Fig. 5), crafted with Berber motifs, holds cultural significance as an inheritance passed from mother to daughter. The designs often incorporate protective symbols such as the Hand of Fatima to ward off the evil eye (Fakhry, 1974).

Music plays an essential role in Siwan culture, often accompanying social and religious gatherings. Traditional Siwan instruments include the mizwad (a type of bagpipe) and hand drums. The dances are typically performed in circular formations, emphasizing unity and collective expression. Songs often use metaphors related to nature, such as comparing loved ones to the grace of a gazelle or the sweetness of dates. Men and women participate in separate dance performances, in line with cultural norms (Vale, 2014).

Siwan cuisine is based on local agricultural products, particularly dates and olives. A traditional dish, shakshuka, is a spicy tomato and egg dish, while tangota'at is served after childbirth as a celebratory meal. Tea drinking is a revered practice in Siwa, following the three-cup method: the first cup is bitter, the second is balanced, and the third is sweetened, symbolizing different stages of life. Tea is brewed with fresh mint and rose water, enhancing its aromatic properties. This ritual is an essential part of social gatherings and hospitality customs (الدميرى، ٢٠١٦).

The COSPE "Tangier-Siwa for a Better Heritage" Initiative has played a significant role in training local youth in heritage documentation, digital storytelling, and conservation methods. Sheikh Omar Rajeh, a key community leader involved in the

Fig. 5 Siwa girl wearing silver jewelry. Source: Maughan (1950)

initiative, supervised the documentation process and guided youth in collecting oral histories from older Siwans. The Initiative can serve as a partner program embedded within the ecomuseum.

Local festivals, such as the Siwa Peace Festival, also play a significant role in transmitting cultural knowledge across generations. These festivals encourage youth participation in music, dance, and storytelling, reinforcing the importance of maintaining Siwan traditions (Belgrave, 1970).

Several community-led initiatives have contributed to the preservation of Siwa's cultural identity, led by Sheikh Omar Rajeh and Engineer Youssef Geary, who have documented oral histories and revived traditional crafts like silverwork, textiles, and pottery. These efforts reinforce the importance of community-driven cultural sustainability. These intangible assets are not static relics. They are living expressions of identity, and the ecomuseum will provide the institutional framework to support, display, and transmit them across generations, reinforcing both climate resilience and cultural pride.

4.3 Field Work Results

4.3.1 Interviews

Fieldwork interviews were conducted with six identified community members, selected for their roles in cultural preservation, tourism, and local crafts. Initial skepticism was encountered due to Siwa's insular and conservative culture. However, collaboration with local guide Youssef Sarhan facilitated community engagement.

Additional conversations held with key community members provided valuable depth to the understanding of Siwan identity and its link to community-based tourism. These conversations revealed how deeply interwoven heritage is with daily life, rituals, crafts, and oral history.

Stories from everyday citizens like *Hajj Essa,* inventor of the salt lamp industry, revealed the intersection of innovation, resourcefulness, and environmental awareness. His craft (Fig. 6A, B, C) originated from a simple discovery while cooking in the desert and has since grown into a locally and internationally admired artisan trade, blending carpentry with natural salt formations. These personal narratives reinforce the connection between local ingenuity, eco-friendly practices, and emerging heritage industries that support community-based tourism.

Community voices also illuminated socio-economic changes such as the decline in karshif construction due to unsuitable soil from irrigation runoff, loss of traditional professions, and changing educational values influenced by external economic pressures. The interviews collectively illustrate how the proposed ecomuseum can not only preserve heritage but also empower the community to adapt sustainably to climate and societal shifts.

Fig. 6 Lampshades and hand crafts invented by Haj Ali (photo by the author, 2022)

Sheikh Omar Rajeh emphasized the importance of preserving intangible cultural heritage, recounting efforts to document customs, proverbs, crafts, and folklore through youth engagement and institutional partnerships. He highlighted his involvement in the "Tangier-Siwa for a Better Heritage" initiative, a COSPE-supported project where youth aged 18–30 gathered oral histories from elders over 60. These recordings were reviewed, corrected, and archived with the Cultnet Foundation. Sheikh Omar is also regularly consulted about architectural restorations to ensure historical accuracy. Sheikh Omar Rajeh described the tribal structure of Siwa, the integration of Amazigh and Bedouin communities, and the importance of traditional customs such as marriage celebrations, salt-gifting, and problem-solving through communal tea gatherings. His support for the ecomuseum model highlights its potential to safeguard this heritage while also stimulating economic development.

Engineer Youssef Geary, director of the Craft Industries Center, stated the center is the only institution training young women in traditional crafts such as silverwork, embroidery, and pottery. He emphasized the importance of these crafts in sustaining Siwan identity in addition to the role of crafts and documentation in maintaining cultural continuity and identity.

An *embroiderer* at the center described how the five embroidery colors used in Siwan garments represent stages of date ripeness and natural motifs. However, she expressed uncertainty about deeper meanings and noted that many workers are unmarried women who are expected to stop working once married, unless from home (Fig. 7).

Mohammad Omran, manager of Aghormi Hotel, described his efforts to revive Siwan culture through hospitality. His hotel names rooms in the Siwi language and hosts events such as a culinary festival and a poetry competition with international participation that has attracted 60 artists. He considers this work a modern form of cultural exploration. His projects demonstrated how tourism could serve as a platform for living cultural exchange and sustainable development.

Hajj Omran, a local storyteller by Fantas Lake, uses oral narratives to share Siwan traditions with visitors, participating in festivals and cultural workshops despite having limited formal education.

Interviewees noted several key challenges: youth involved in traditional crafts often lacked symbolic understanding of heritage motifs; cultural preservation efforts varied widely due to inconsistent regulations; and modern paints used on building facades raised environmental and aesthetic concerns. Hesitation was observed in sensitive topics such as education and migration, with some respondents deferring to official responses.

Fig. 7 Products made by Siwa women for Siwa creation brand (Source: https://siwacreations.yolasite.com/)

4.3.2 Observations

During the field visit, various examples of Siwa's circular economy practices were observed. Locals reuse palm fronds for fences and fuel, craft lampshades from salt, and use date palm derivatives and olive wood for household goods. Waste is composted and used as fertilizer. These methods reflect sustainable livelihoods deeply rooted in tradition.

The emergence of new industries, such as salt-based crafts, demonstrates adaptive reuse following the decline in salt's preservation role due to refrigeration. There is potential to extend this innovation to value-added products based on dates, olives, and herbal crops. The revival of the custom of gifting raw salt also suggests cultural adaptation.

Culturally, many stakeholders demonstrated pride in preserving Siwa's identity. However, hotels often fail to reflect authentic Siwan culture, offering generic decor and food. For example, hotels such as Taziri serve non-local cuisine, misrepresenting regional heritage.

Community Voice

> We solve our problems over a pot of tea. (Sheikh Omar Rajeh, Awlad Musa Tribe)

This quote encapsulates the role of informal dialogue and social cohesion in Siwan community life, emphasizing the cultural foundations that a community-based tourism model can build upon.

4.3.3 Site Assessments

Site assessments were conducted at multiple heritage and tourism locations:

Siwa House was assessed as structurally degraded, with inaccessible areas, dusty exhibits, poor craftsmanship, and missing explanatory labels. Student visits focus mostly on architectural aspects without interactive engagement. The museum lacks security systems and proper maintenance.

The Siwa Protected Area Visitor Center, established in 1998 by the Egyptian and Italian governments with IUCN support, includes panels on Siwa's natural and cultural resources. However, it suffers from inactive programming and inadequate conservation-related activities.

Shali Souq, located near Shali Fortress, houses brands like "Siwa Scent" and "Siwa Ever Organic," producing locally sourced molasses, mint, dried molokhia, and olive oil. These products use bilingual packaging targeting international markets. However, due to COVID-19, the exhibition space was reduced by 50%.

Handicraft Shops display traditional textiles, bags, and slippers. Some items, however, include non-native symbols (e.g., elephants), raising concerns about the loss of authenticity. Additionally, one prominent cosmetics vendor is run by a Moroccan business owner, hinting at economic monopolization.

Table 2 Field insights and ecomuseum relevance

Fieldwork theme	Community insight	Implication for ecomuseum
Tribal and cultural identity	Diverse heritage of Amazigh and Bedouins preserved through oral traditions	Supports community-based interpretation and governance of heritage assets
Craft revitalization	Youth and women trained in embroidery, pottery, and silverwork	Strengthens local economy and tangible heritage continuity
Innovation and sustainability	Salt lamp industry born from local materials and craftsmanship	Aligns with eco-entrepreneurship and climate-smart tourism
Socio-cultural rituals	Marriage, mourning, gifting traditions still practiced but evolving	Ecomuseum can serve as a living archive of these intangible assets
Threats from climate and modernization	Decline in karshif use, olive production, traditional roles fading	Museum can offer climate adaptation advocacy and education

These fieldwork insights demonstrate both strengths and challenges as shown in table 2 within Siwa's cultural ecosystem, reinforcing the need for an inclusive, community-centered ecomuseum model rooted in accurate heritage interpretation and sustainable development (Table 2).

Importantly, the fieldwork revealed that many community members are already informally engaging in practices aligned with community-based tourism such as local storytelling, craft-making, and heritage focused hospitality, yet these efforts remain fragmented and under-supported. Establishing a structured ecomuseum can unify and elevate these initiatives, creating meaningful economic opportunities while preserving cultural authenticity.

Additionally, several findings highlighted the impact of climate change on Siwa's natural and cultural systems. For example, the shift from salt-based food preservation to decorative uses like salt lamps, as well as the declining olive yields due to environmental stress, reflects community responses to ecological change. These challenges underscore the need to incorporate climate adaptation strategies into heritage and tourism development through the ecomuseum framework.

Note: All interviews and community conversations were conducted with verbal consent. Identities have been included with permission or anonymized where appropriate

5 Proposed Ecomuseum Model for Siwa

The ecomuseum's physical structure will reflect Siwa's traditional karshif architecture while incorporating universal design principles to ensure accessibility for all.

The museum's strategy is designed to address the multidimensional goals of heritage preservation, sustainable development, and community engagement. A key objective is preserving Siwa's unique cultural and natural heritage through exhibits, interactive displays, and documentation efforts. Reviving heritage documentation projects, including those previously halted in 2012, the ecomuseum becomes a focal point for celebrating Siwa's traditions, architecture, and environmental knowledge. It will also develop programs targeting both foreign and Egyptian immigrants, aimed at raising awareness about the importance of preserving local heritage and fostering respect for Siwan identity.

To tackle educational dropouts, the ecomuseum will collaborate with the Ministry of Education and UNESCO to create youth programs that offer financial incentives during school breaks. These initiatives will connect students with local heritage and integrate Siwan culture into school curricula to strengthen emotional and cultural ties to the oasis. As a center for sustainable tourism, the ecomuseum will facilitate responsible tourism practices in partnership with the local community, NGOs, and the Ministry of Environment, while diversifying tourism offerings to include eco-tourism, adventure tourism, meditation tourism, and more.

Site conservation will be managed through controlled access, expert monitoring, and preservation techniques for archaeological sites. The ecomuseum will support sustainable management of Siwa's natural resources and promote indigenous flora and fauna conservation. It will also serve as a research and documentation hub, interpreting cultural and natural heritage with multimedia tools and guided tours.

Community engagement is central to the museum's strategy. The ecomuseum will enable local artisans, elders, and community members to participate in exhibition design, cultural performances, craft promotion, and guided tours. This involvement not only preserves knowledge but also generates local economic opportunities.

Furthermore, the ecomuseum will function as a heritage marketing institution. Through partnerships with local entrepreneurs, tourism agencies, and government stakeholders, it will help package and promote Siwa's cultural and economic offerings. A strong digital marketing presence, including dedicated websites and social media channels, will enhance global outreach and awareness.

Siwa faces pressing water and energy challenges due to groundwater salinity and climate extremes. The ecomuseum will collaborate with stakeholders to support the closure of high-salinity wells and the drilling of deeper, cleaner sources (Moghazy & Kaluarachchi, 2021). It will encourage the transition from flood to drip irrigation systems and support infrastructure projects such as the 33.7 km drainage canal (MENA, 2023). Furthermore, the ecomuseum will promote solar-powered desalination and renewable energy systems in partnership with local and national actors.

The ecomuseum will contribute to the enhancement of Siwa's tourism sector by promoting eco-lodges that reflect authentic Siwan culture and cuisine, developing visitor management plans to protect fragile heritage sites, and offering educational tourism opportunities such as workshops on crafts and desert ecology. It will also advocate guidelines to ensure architectural conservation and the implementation of responsible tourism practices. Recommended policies include enacting strict architectural conservation laws, developing sustainable tourism frameworks tailored for local enterprises, and expanding artisan training programs through institutions like the Craft Industries Center (see Fig. 7). Additionally, efforts will focus on improving signage and site infrastructure and integrating solar technologies into construction and public utilities.

The ecomuseum will directly respond to various local challenges. To bridge knowledge gaps, it will incorporate heritage education into school curricula and organize community workshops. For community engagement, it will host storytelling festivals, artistic competitions, and other participatory events. Artisan support will be enhanced through training in sustainable craft production and improved access to marketing channels. To strengthen heritage infrastructure, the ecomuseum will provide guidelines for restoring and maintaining cultural sites. It will also address economic vulnerability by promoting year-round tourism and the sale of local crafts. In response to climate resilience needs, the museum will advocate traditional farming techniques, eco-construction methods, and water conservation practices.

If implemented effectively, the Siwa Ecomuseum could serve as a national model, inspiring other Egyptian communities to establish their own locally tailored ecomuseums. This would shift development from a top-down to a grassroots model, rooted in sustainable resource use and cultural pride. Ultimately, this strategy will instill a deep appreciation within the community for its role in Egypt's broader civilizational narrative.

Mission: To preserve and celebrate the cultural, historical, and ecological legacy of Siwa Oasis by engaging local communities and visitors in sustainable resource management and heritage appreciation. The ecomuseum will serve as a center for academic research, education, and intercultural dialogue.

Vision: To become a leading local and international institution for heritage conservation and sustainable tourism, fostering a vibrant and inclusive community rooted in pride and resilience.

5.1 Governance: Siwa Ecomuseum's Board

The board of directors will include representatives from archaeology, history, ecology, tourism, agriculture, traditional medicine, local crafts, and community leadership. This interdisciplinary composition ensures culturally sensitive, community-driven decision-making.

5.2 Ecomuseum Website

The ecomuseum's digital platform will offer interactive virtual tours of heritage sites and maintain a digital archive of photographs, oral histories, and historical documents. It will include educational resources and downloadable lesson plans to support learning. A community forum will foster knowledge-sharing among

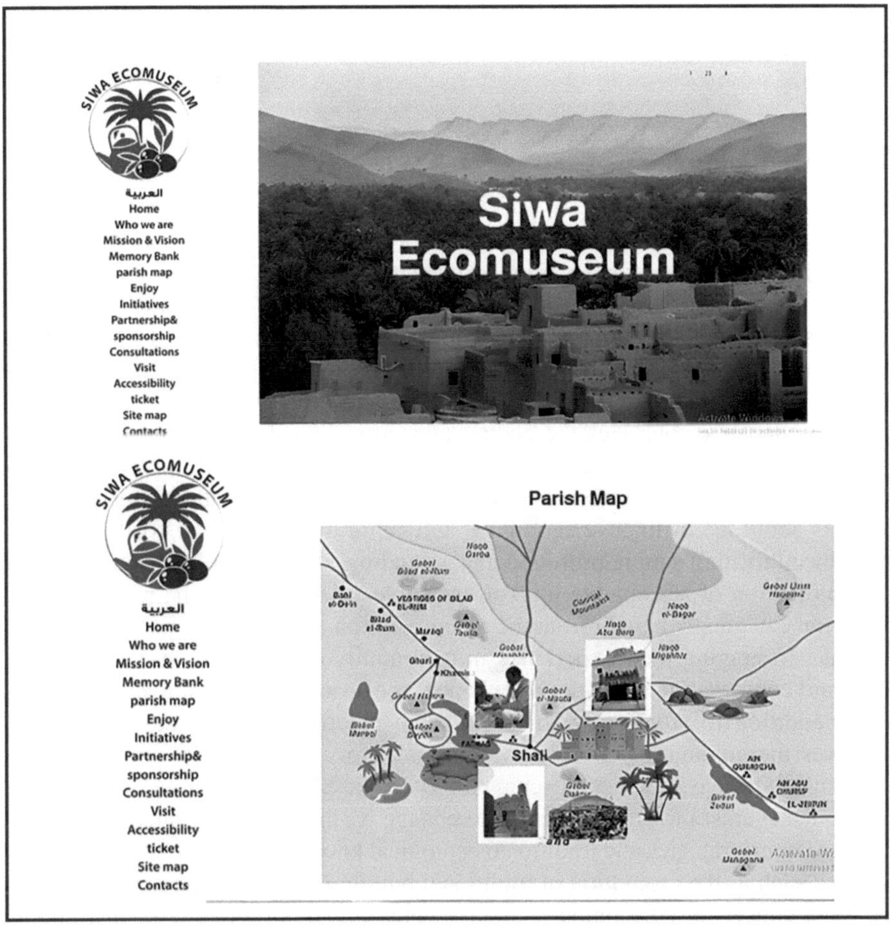

Fig. 8 Siwa Ecomuseum virtual website and parish map. Source: author

residents and researchers. To support the local economy, the website will include an e-commerce feature, allowing artisans to market and sell traditional crafts to a global audience (Fig. 8).

5.3 Parish Map as a Participatory Tool

The parish map will be developed as a dynamic and interactive representation of Siwa's cultural geography. It will be created through collaborative workshops and oral history sessions with community members, ensuring an inclusive narrative of local history and heritage. The map will feature geotagged locations of cultural, historical, and ecological significance and will integrate QR codes that provide access to multimedia narratives and archival content. It will also outline eco-friendly travel routes to encourage responsible tourism and will serve as a tool to empower local guides and enhance the visitor experience.

6 Conclusions

Creating an ecomuseum in Siwa presents a significant chance to safeguard the oasis's priceless cultural and natural heritage while enabling the local community to actively engage in its conservation and improvement. The ecomuseum has the potential to stimulate sustainable development and community-led initiatives by tackling significant issues such as educational dropouts, unsustainable tourism, and resource management.

This chapter emphasizes the ecomuseum model's substantial potential to preserve Siwa's cultural heritage and empower the local community in the context of climate change. The ecomuseum can function as a critical platform for educating visitors about the distinctive ecological and cultural characteristics of Siwa Oasis, thereby cultivating a more profound understanding of the region's heritage, through the integration of sustainable tourism principles. Additionally, the potential Siwa Museum can greatly benefit from the practices implemented in the Ecomuseum Ilha Grande, leveraging their experience in community engagement, sustainable tourism, and conservation strategies to enhance its effectiveness and impact.

In addition to advocating for sustainable practices, the ecomuseum actively involves the community in conservation endeavors as part of its climate change adaptation strategy. The ecomuseum can assist the Siwa community in developing resilience against the effects of climate change by integrating sustainable resource management strategies and traditional ecological knowledge. This encompasses initiatives such as the promotion of traditional building materials, the implementation of water conservation techniques, and the enhancement of awareness regarding the significance of preserving local biodiversity.

Furthermore, the ecomuseum framework can facilitate sustainable tourism practices that generate economic benefits for the community while simultaneously guaranteeing that tourism development does not undermine the integrity of Siwa's cultural and environmental assets. The ecomuseum can empower residents to actively participate in the tourism sector by fostering local entrepreneurship and encouraging community-led initiatives, which can facilitate equitable benefit distribution.

In summary, the successful implementation of the ecomuseum paradigm in Siwa necessitates the cooperation of the local community, stakeholders, and policymakers. In order to safeguard Siwa's cultural heritage and encourage tourism that is environmentally conscious and beneficial to the local population, it will be essential to prioritize sustainability, community engagement, and partnerships. By making these endeavors, the ecomuseum can establish itself as a model for sustainable tourism and climate change adaptation, thereby guaranteeing the future vibrancy of Siwa Oasis and its distinctive cultural identity.

It is imperative to deliberate upon the establishment of a dedicated unit within the Ministry of Antiquities and Tourism exclusively that is specifically tasked with overseeing matters pertaining to heritage instead of the general administration for preserving Cairo's heritage. It is advisable to consider renaming the ministry as the Ministry of Heritage and Tourism, as the term "Heritage" encompasses a broader and more comprehensive scope than the singular focus on antiquities.

Establishing the Siwa Ecomuseum, taking into account the study of proposals submitted during the strategic formulation process, in cooperation with the Ministry of Environment, Tourism and Antiquities, the Ministry of Planning and Finance, the Ministry of Defence, Matrouh Governorate, businessmen interested in financing heritage projects, the Arab Organization of Museums, the ICOM, ICROM, and activating consultation with the Ecomuseums Platform, the Union of Ecomuseums, and the United Nations Organization, the FAO Organization, civil society organizations concerned with preserving Siwa heritage, the Siwa Sons Association for Development, and Siwa tribal sheikhs, taking a bottom-up approach.

Based on the principles outlined in the 1972 World Heritage Protection Convention, as endorsed by UNESCO, and with the implementation of effective monitoring mechanisms to ensure compliance with these standards, I propose the nomination of the Siwa Oasis under the five designated criteria,[5] namely (i), (ii), (iii), (iv), and (v). Due to the inherent interplay between archaeological, architectural, and environmental components, as well as daily customs, it is therefore not possible to separate them when applying the laws related to heritage preservation or confronting the changes that occur to these elements of society.

The proposed approach involves resuming all previously halted initiatives through a comprehensive evaluation, analysis, and subsequent execution by a designated team responsible for project preparation. This process will prioritize the establishment of a rational and feasible budget to ensure the project's continuity, thereby mitigating the risk of encountering setbacks like those experienced in prior endeavors.

Recognizing the significance of the marketing approach is a crucial and indispensable tool for the Ecomuseum, due to its substantial influence on the preservation and transmission of heritage across generations, both locally and globally.

The integration of the evaluation approach within Egyptian museums at large would serve to answer the question: Where are we now, and what do we want to be in the future?

References

Alkalawy, N. (2022, March 15). *The legendary eye of Cleopatra...Explore the hidden beauty of Siwa Oasis in Egypt*. CNN. https://arabic.cnn.com/travel/article/2022/03/14/cleopatras-pool-siwa-oasis-egypt

[5] https://whc.unesco.org/archive/opguide83.pdf

Allam, N. M., Koura, G. M. R., Alrawaili, S. M., Hamada, H. A., & Balbaa, A. A. (2018). *The effect of Siwan therapy in management of patients with rheumatoid arthritis: A single blind randomized controlled trial*. Biomedical Research. https://www.alliedacademies.org/articles/the-effect-of-siwan-therapy-in-management-of-patients-with-rheumatoid-arthritis-a-single-blind-randomized-controlled-trial-10039.html

Belgrave, D. (1970, January 1). *Siwa: The oasis of Jupiter Ammon: Belgrave, Dalrymple C.: Free download, borrow, and streaming*. Internet Archive. https://archive.org/details/belgrave_siwa_1923

Dabaieh, M. (2011, December). *(PDF) a future for the past of desert vernacular architecture: Testing a novel conservation model and applied methodology in the town of Balat in Egypt*. Research Gate. https://www.researchgate.net/publication/247849342_A_future_for_the_past_of_desert_vernacular_architecture_testing_a_novel_conservation_model_and_applied_methodology_in_the_town_of_Balat_in_Egypt

De Almeida, G. R. (2022). *Climate actions of the Ecomuseu Ilha Grande (Brazil) for the sustainable development goals*. https://research-repository.st-andrews.ac.uk/bitstream/handle/10023/27049/Brown_2023_Community_crafts_and_cultures_Ch_CCBY.pdf?sequence=1 (pp. 189–210). Essay, Ledizioni LediPublishing.

de Varine, H. (1996). Ecomuseum or community museum? 25 years of applied research in museology and development. *Nordisk Museologi*. https://journals.uio.no/index.php/museolog/article/view/3673

EL-Shafie, M. (1999). *[PDF] the phenomenology of fit: Linking the person and environment to the subjective experience of person-environment fit. | semantic scholar*. Semantic Scholar. https://www.semanticscholar.org/paper/The-phenomenology-of-fit:-linking-the-person-and-to-Edwards-Cable/b1788721b96c01a9735eda0de1c31cffa58ed8d9

Fakhry, A. (1944). *Siwa Oasis. Its history and antiquities*. Meretseger Books. https://www.meretsegerbooks.com/pages/books/M0556b/fakhry-ahmed/siwa-oasis-its-history-and-antiquities

Fakhry, A. (1974). *Bahriyah and Farafra* (Vol. 2). American University. 2022. https://archive.org/details/bahriyahfarafra0000fakh/page/194/mode/2up

FAO. (2016, October 28). *FAO Awards siwa oasis, the globally important agricultural heritage systems (GIAHS) certificate and signs the first project within strategy to develop dates sector in Egypt*. FAO awards Siwa Oasis, the Globally Important Agricultural Heritage Systems (GIAHS) Certificate and signs the first project within Strategy to develop Dates Sector in Egypt. https://www.fao.org/neareast/news/view/fr/c/450664/

Ferretti, V., Bottero, M., & Mondini, G. (2014). Decision making and cultural heritage: An application of the multi-attribute value theory for the reuse of historical buildings. *Journal of Cultural Heritage*, 244–253. https://www.sciencedirect-com.qe2aproxy.mun.ca/science/article/pii/S1296207414000041?via%3Dihub

Hayes, W. C. (1964). *Studies in ancient Oriental Civilization - University of Chicago*. Institution of the study of ancient culture. https://isac.uchicago.edu/sites/default/files/uploads/shared/docs/saoc24.pdf

Matar, S. (2024). *Ecomuseum: The role of the socio-economic synergy in strengthening national identity in Siwa* (thesis).

Mena. (2023, April 14). Irrigation Min. follows up on development efforts in Siwa Oasis. *Egypt Independent*. https://egyptindependent.com/irrigation-min-follows-up-on-development-efforts-in-siwa-oasis/#google_vignette

Mendon, M. (2015, July 23). *Tourism research frontiers: Beyond the boundaries of knowledge: Volume 20*. Tourism research Frontiers: Beyond the boundaries of knowledge: Vol. 20 | Emerald Insight. https://doi.org/10.1108/s1571-5043201520

Moghazy, N. H., & Kaluarachchi, J. J. (2021, February 2). *Impact of climate change on agricultural development in a closed groundwater-driven basin: A case study of the siwa region, Western Desert of Egypt*. MDPI. https://www.mdpi.com/2071-1050/13/3/1578

Montalbano, C., & Petruccioli, A. (2011, January). *(PDF) siwa oasis _actions for a sustainable development*. Research Gate. https://www.researchgate.net/publication/270582220_Siwa_Oasis_Actions_for_a_sustainable_development

Ohara, K. (1998). *(PDF) the image of "ecomuseum" in Japan - Research Gate*. https://www.researchgate.net/publication/263083906_The_Image_of_'Ecomuseum'_in_Japan

Radwan, A. (2007, June 7). Women's freedom comes slowly to a sleepy oasis. *Time*. https://content.time.com/time/world/article/0,8599,1630445,00.html

Rivière, G. H. (1985). *The Ecomuseum: An evaluative definition*. Unesdoc.unesco.org. https://unesdoc.unesco.org/ark:/48223/pf0000068366

Sabrina, H. Y. (2010). *Safeguarding cultural heritage, protecting intellectual property and …* ICOM-ICME. https://icme.mini.icom.museum/wp-content/uploads/sites/16/2019/01/ICME_2005_wendland.pdf

Scheiner, T. C. (1970, January 1). Museums, communities and cultural landscapes: Towards a total relationship. *International Journal of Museum Studies: I*, 2021 [sample copy]. https://www.torrossa.com/en/resources/an/5086348

Vale, M. (2014). History and tradition. In *Siwa jewelry, costume, and life in an Egyptian oasis* (pp. 13–20). Essay, American University in Cairo Press.

Vivian, C. (2008, August). *Western Desert of the Egypt*. The Western Desert of Egypt. https://www.academia.edu/43956266/Western_Desert_of_The_Egypt

الدميري، عبد العزيز. عبد الرحمن. (2016). سيوة الماضي والحاضر. مطبعة حسن ياسو. الإسكندريه.

Case Studies and Local Approaches to Sustainable Tourism

The Impact of Home-Based Food Enterprises for Sustainable Tourism in Bali

I Ketut Sutapa, Pondang Polikarpus Nainggolan, I Nengah Subadra, and I Made Bayu Wisnawa

Abstract Home-Based Food Enterprises (HBFEs) in Bali enhance sustainable tourism by offering authentic culinary experiences that benefit tourists and local communities. This study examines their economic, cultural, environmental, and social impacts through qualitative methods: stakeholder interviews, case studies, and document analysis, processed via NVivo Pro12. Findings reveal HBFEs bolster local economies by supporting agriculture and creating jobs, preserve cultural heritage through traditional cuisine, promote environmental sustainability via eco-friendly practices, and strengthen social cohesion. Despite their contributions, the research is geographically limited to specific Balinese regions, potentially overlooking regional diversity. Future studies should expand scope and integrate quantitative metrics. The study underscores HBFEs' potential to sustain local economies during crises, enrich cultural tourism, and encourage eco-conscious practices. Policymakers are urged to provide financial aid and training to amplify these benefits. Uniquely comprehensive in scale and depth, this research fills gaps in understanding how small-scale culinary enterprises drive sustainable tourism, offering actionable insights for holistic development strategies.

1 Introduction

Tourism in Bali contributes to economic growth through increased income, job creation, economic diversification, and the addition of new products (Lin, 2024; Putri et al., 2023). In recent years, there has been a shift towards sustainable tourism methods that strive to reduce environmental effects while simultaneously increasing social and economic benefits to local communities (Rangus, 2018; Satrya et al., 2023). Home-based culinary businesses enhance tourists' experience by offering

I. K. Sutapa (✉) · P. P. Nainggolan · I. N. Subadra · I. M. B. Wisnawa
Triatma Mulya University, Bali, Indonesia
e-mail: ketut.sutapa@triatmamulya.ac.id; polikarpus.nainggolan@triatmamulya.ac.id; polikarpus.nainggolan@triatmamulya.ac.id; bayu.wisnawa@triatmamulya.ac.id

J. Marques et al. (eds.), *Tourism and Climate Change*, Tourism, Hospitality & Event Management, https://doi.org/10.1007/978-3-031-99128-8_10

traditional dishes and providing insights into the Balinese way of life, enriching their travel experience and promoting cultural exchange (Blapp & Mitas, 2018; Sukadana et al., 2022). HBFEs are small-scale operations run by local families or individuals, often from their homes. These enterprises are pivotal in showcasing Balinese cuisine's diversity and richness, deeply intertwined with the island's cultural identity. Moreover, they provide a sustainable source of income for local families and promote the use of local produce, thus supporting Bali's agricultural sector (Sedana, 2022). Home-based food enterprises are a cornerstone for local economies, not only by creating job opportunities and boosting household incomes but also by contributing to the sustainability of tourism initiatives (Marzuki et al., 2017). They enhance gastronomic tourism by offering authentic local culinary experiences and attracting tourists interested in unique and sustainable dining options (Wirawan et al., 2022). These enterprises often incorporate organic and sustainable practices, elevating the appeal of their locales and promoting environmental stewardship (Budiani et al., 2017).

In terms of community-based ecotourism, integrating local food businesses helps balance economic benefits with ecological and cultural preservation, providing a holistic approach to sustainable tourism (Christian Byczek, 2011; Wiranatha et al., 2024). They also enrich the tourist experience by offering diverse dining options that reflect local culture and traditions, increasing tourists' satisfaction and the duration of their stay (Listiani et al., 2024). By emphasizing local ingredients, these enterprises support local agriculture, enhancing food security and contributing to the sustainability of tourism (Hirakawa et al., 2024). Home-based food businesses play a crucial role in cultural preservation by maintaining traditional cooking methods and recipes (Putu Diah Sastri Pitanatri, 2016), which are vital for maintaining destination attractiveness (Sukerti & Marsiti, 2020). They are pivotal in community development, enhancing local capacity and self-reliance to ensure sustainable tourism development (Budiasa, 2014). Additionally, these enterprises adopt environmentally sustainable practices that minimize their impact, aligning with broader sustainable tourism goals (Sutama & Mudana, 2019). Lastly, the growth and sustainability of home-based food businesses rely heavily on effective policy frameworks and supportive regulations, which are essential for fostering sustainable tourism practices (Majiid et al., 2023). HBFEs are a rapidly growing sector in Bali, leveraging local wisdom and traditional culinary wealth to create a variety of food products that are not only delicious but also culturally significant. With the rise in tourism and changing consumer trends towards authentic and high-quality food products, HBFEs in Bali have become a crucial pillar in the local economy. These enterprises are typically managed by families or small groups who utilize local ingredients and traditional production methods to maintain the authenticity of taste and quality. Besides contributing positively to the economy, HBFEs also significantly preserve Balinese culinary culture and empower local communities, especially women. This discussion delves deeper into the role and contributions of HBFEs in Bali, their challenges, and the opportunities for further development.

This chapter aims to explore how HBFEs contribute to sustainable tourism in Bali. It seeks to understand these enterprises' economic, cultural, and environmental impacts within the broader context of tourism.

Table 1 displays the number of tourists visiting Bali from 2018 to 2022. In 2018, there were 6,070,473 tourists, which slightly increased to 6,275,210 in 2019. Due to the COVID-19 pandemic, the number of arrivals sharply declined to 1,069,473 in 2020, and there were only 51 arrivals in 2021 due to travel restrictions. Nevertheless, in 2022, after adjustments to visitation regulations, there was a recovery with 2,155,747 visitors. This number of tourists highlights the substantial impact of external factors and policy changes on Bali's tourism industry. Cuisine plays a central part in tourism, demanding guests seek authentic local gastronomy for both comfort and cultural experiences (Subadra, 2019). Gastronomic tourism in Slovenia plays a vital role in showcasing the country's unique culture and traditions to visitors. Focusing on sustainable practices and locally sourced ingredients supports the local economy and provides an authentic taste of Slovenian life (Janez Bogataj, 2021). This underlines the importance of culinary culture in tourism, highlighting how gastronomic experiences reflect the cultural and traditional identities of societies, thereby driving tourism development (Sormaz et al., 2016), and highlights and gastronomic tourism as a key element in diversifying tourism and supporting local economies, stressing its role in cultural tourism (Marius Boiţă, 2014). Also, gastronomy points out the comprehensive nature of gastronomic tourism, from production to post-consumption, and its importance in tourism planning and marketing (Berno & Fusté-Forné, 2022) highlights innovative gastronomic tourism as a growing trend, providing deep cultural engagement through culinary experiences.

Balinese home-based food businesses, such as those offering babi guling and fried fish, play a significant role in the local culinary scene, attracting domestic and international tourists. Despite the influx of global fast-food chains, traditional Balinese dishes like "babi guling" (roast suckling pig) remain popular, demonstrating the resilience of local culinary entrepreneurship in the face of globalization (Sukadana et al., 2022). The growth of HBFEs in nine regions of Bali increased steadily over the 5 years, from 13,909 in 2018 to 17,895 in 2021, as shown in Table 2.

Although the restaurants in Bali are not branded by subject, the particular number of HBFEs remains unrevealed. However, the Bali Statistic Agency lists 17,895 HBFEs. This Statistic recommends increasing HBFEs in Bali; traditional Balinese spices and the distinct culinary experience contribute to foreign tourists' overall satisfaction and loyalty (Hendriyani et al., 2020). The quality of traditional Balinese

Table 1 Number of tourists visiting Bali between 2018 and 2022

Year	2018	2019	2020	2021	2022
Number	6,070,473	6,275,210	1,069,473	51	2,155,747

Source: Bali Statistic Agency (2023)

Table 2 Number of HBFEs in Bali by region between 2018 and 2022

No	Regency/City	2018	2019	2020	2021	2022
1	Jembrana	1234	1345	1456	1567	1678
2	Tabanan	1567	1678	1789	1900	2011
3	Badung	2345	2456	2567	2678	2789
4	Gianyar	1789	1890	2001	2112	2223
5	Klungkung	987	1098	1209	1320	1431
6	Bangli	654	765	876	987	1098
7	Karangasem	1112	1223	1334	1445	1556
8	Buleleng	1345	1456	1567	1678	1789
9	Denpasar	2876	2987	3098	3209	3320
Total		13,909	14,898	15,897	16,896	17,895

Source: Bali Statistic Agency (2023)

Table 3 Comparison of two outstanding HBFEs in Bali

No	Criteria	Mak Beng, Sanur	Babi Guling Oka, Ubud
1	Historical significance	Operating since 1941	Operating since 1970
2	Culinary heritage	Balinese fish soup and fried fish	Traditional Babi Guling
3	Tourist attraction	Popular among tourists	Attracts local and international tourists
4	Business model	Family-run, quality maintenance	Branding and marketing success
5	Sustainable practices	Local ingredients, traditional methods	Local ingredients, traditional methods
6	Economic impact	Contribution to local economy	Significant economic contributor

Source: Processed Data Analyst (Author 2024)

cuisine, including its unique taste, presentation, and use of local ingredients, significantly impacts tourist satisfaction and purchasing decisions (Arcana et al., 2018). Focusing on these areas can enhance the tourist experience and promote Balinese culinary traditions.

In Table 3 compares two outstanding HBFEs in Bali, Warung Mak Beng Sanur and Warung Babi Guling Oka Ubud, to illustrate how they contribute to sustainable tourism. Essentially, these two restaurants (warungs) specializing in a particular local dish are successful examples of sustainable tourism. They achieve this by offering authentic culinary heritage, attracting tourists, operating with sustainable practices like using local ingredients, and contributing to the local economy. The table highlights that both Warung Mak Beng (known for its fish soup) and Warung Babi Guling Oka (famous for its suckling pig) are not just places to eat but also cultural attractions that support the local community and offer visitors a genuine taste of Bali.

2 Research Method and Theoretical Framework

2.1 *Research Method*

The research aims to investigate the impact of home-based food enterprises for sustainable tourism in Bali. The research adopts a qualitative methodology to gain deep insights into the contribution of HBFEs to sustainable tourism. The research was conducted at Warung Mak Beng Jalan Hang Tuah No. 45, Sanur Kaja, Denpasar Selatan, Bali, Indonesia and Warung Babi Guling Oka Jl. Tegal Sari No. 2, Ubud, Ubud, District Gianyar Regency, Bali, Indonesia. This research uses a qualitative approach. Primary data were gathered through face-to-face interviews with stakeholders. Conduct interviews with HBFE owners, local community members, tourists, and officials from the tourism department to collect diverse perspectives on the impacts of HBFEs. Perform detailed case studies of selected HBFEs to illustrate successful practices and challenges these enterprises face (Ahannaya et al., 2021). Review relevant documents, including policy papers, tourism reports, and previous research studies on food tourism and sustainability in Bali.

2.2 *Theoretical Framework (Fig. 1)*

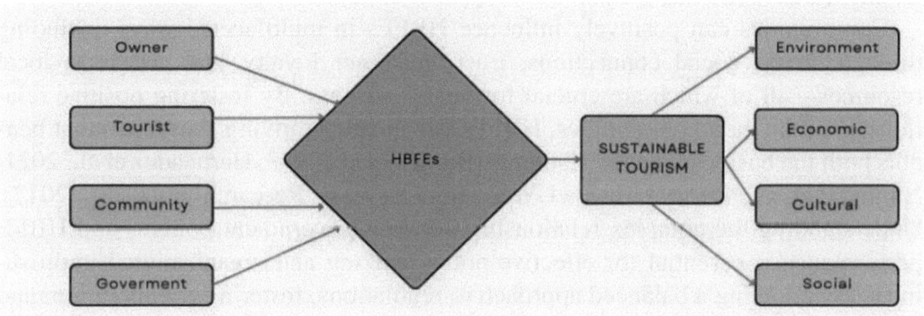

Fig. 1 Framework impact stakeholders in HBFEs to sustainability. Source: Processed Data Analyst (Author 2024)

2.3 The Impact of Stakeholder in HBFEs

HBFEs success hinges heavily on their owners' characteristics and capabilities. Several studies explore this connection. One research project investigates how owner traits like skills, experience, and education influence decision-making and ultimately impact HBFE performance (Rahadhini & Lamidi, 2023). Another study examines the challenges and opportunities the digital age presents for HBFE owners. It highlights the growing importance of digital literacy, particularly in areas like online marketing and customer engagement, for business growth (Shah & Murthi, 2021). Beyond technical skills, the research emphasizes the combined influence of culinary expertise, business management skills, and strong entrepreneurial passion on HBFE's success. Interestingly, the role of social capital (connections within the community) is explored in V et al. (2020) while not directly focused on HBFEs. This concept highlights how an owner's ability to build networks and leverage social connections can benefit their HBFE. Finally, a case study from Malaysia (Shah & Murthi, 2021) sheds light on the unique experiences and strategies employed by HBFE owners in developing countries, offering insights into their motivations and challenges. In conclusion, these studies paint a comprehensive picture of how owner characteristics play a critical role in shaping the performance and sustainability of Home-Based Food Businesses. Tourists play a significant role in the success of HBFEs. By understanding tourist preferences for authenticity and leveraging the reach of online platforms, HBFEs can thrive and contribute to a more sustainable and enriching culinary tourism experience for all (Atsız & Cifci, 2021; Graham, 2021; Krishna Kumar, 2021; Qian et al., 2023).

Communities can positively influence HBFEs in multifaceted ways, including through strong social connections, trust, customer loyalty, and access to local resources—all of which are crucial for HBFE success. By fostering positive relationships with their communities, HBFEs can create a thriving ecosystem that benefits both the business and the residents (Baruch et al., 2013; Hernández et al., 2021; Najjinda et al., 2023; Urban@UW's Homelessness Research Initiative, 2017). Understanding the complex relationship between government policies and HBFE performance is essential for effective policy-making and organizational improvement. By adopting a balanced approach to regulations, fostering collaboration, and providing support programs, governments can empower HBFEs to contribute to a thriving culinary landscape (Abd Razak et al., 2022; Bar-El et al., 2020; Janhonen & Elkjaer, 2022; Muneerali Thottoli et al., 2023).

2.4 The Impact of HBFEs to Sustainability Tourism

Based on the understanding of HBFEs and sustainable tourism, here is a breakdown of their potential environmental, economic, cultural, and social impacts (World Tourism Organization (UNWTO), 2023):

1. Cultural Impact
   (a) Preservation of Culinary Heritage: HBFEs can showcase traditional cuisines and cooking methods, promoting cultural identity and appreciation.
   (b) Tourist Experience: They offer tourists a unique and authentic culinary experience, fostering a deeper connection with the local culture.
   (c) Limited Exposure: The reach of cultural influence might be limited compared to mainstream restaurants.
2. Economic Impact
   (a) Local Economic Growth: HBFEs can stimulate local economies by supporting local food producers and creating job opportunities.
   (b) Entrepreneurship: They empower individuals to become entrepreneurs and contribute to the diversification of the tourism economy.
   (c) Limited Reach: Compared to larger businesses, HBFEs might have a smaller economic footprint.
3. Environmental Impact
   (a) Reduced Food Miles: HBFEs often source ingredients locally, reducing transportation distances and associated carbon emissions compared to large-scale food establishments.
   (b) Waste Reduction: Smaller operations may generate less food waste than restaurants.
   (c) Potential Concerns: Improper waste management from HBFEs or increased packaging for delivery services could pose environmental challenges.
4. Social Impact
   (a) Community Building: HBFEs can foster community by bringing people together through food and shared experiences.
   (b) Empowerment: They can empower women and marginalized groups by providing income-generating opportunities.
   (c) Social Equity Concerns: Unequal access to resources or training opportunities within the HBFE sector could require attention (Figs. 2 and 3).

3 Findings and Discussion

3.1 Warung Mak Beng Sanur and Babi Guling Oka Ubud

Established in 1941, Warung Mak Beng has a long history, providing a unique case study on the sustainability and resilience of family-run businesses over decades. Its ability to maintain popularity highlights successful business practices in a competitive tourism market. Specializing in traditional Balinese fish soup and fried fish, Warung Mak Beng preserves and promotes local culinary traditions, offering insights into the cultural importance of food in Bali and its role in attracting tourists.

Fig. 2 Place of research and flagship product: (1 and 2) Warung Mak Beng and Fried Fish, Sanur, Bali Indonesia, (3 and 4) Warung Bu Oka and Nasi Campur Babi Guling, Ubud, Bali, Indonesia

The Warung is a well-known destination for local and international tourists, indicating its significant contribution to the local economy and its potential role in promoting sustainable tourism. By maintaining a small, quality-focused operation, Warung Mak Beng provides a model for sustainable business practices, emphasizing the importance of quality over quantity in the food industry (Fidyanita, 2024; Wasti Atmodjo, 2005; Wayan Suadnyana, 2024) (Table 4).

Warung Babi Guling Oka is renowned for its Babi Guling (suckling pig), a traditional Balinese dish integral to local ceremonies. This information highlights the Warung's role in preserving and showcasing Balinese cultural heritage through food. The Warung exemplifies sustainable food practices using local ingredients and traditional cooking methods. The preparation of Babi Guling involves local herbs and spices, supporting local agriculture and reducing the carbon footprint. The Warung is a popular tourist destination, attracting visitors from around the world and contributing significantly to the local economy. Its success demonstrates the economic potential of home-based food enterprises in tourism-heavy regions.

Fig. 3 Conducted interviews with informants: (1) Mr. Dodit (the owner of Warung Mak Beng) and (2) Repeater costumer of Warung Mak Beng, Sanur, Bali, Indonesia, (3 and 4) Mrs. Oka, (the owner and tourist of Warung Bu Oka Ubud, Bali, Indonesia). Source: Figure by Author (2024)

Despite its humble beginnings, Warung Babi Guling Oka has gained a strong international reputation, offering valuable insights into effective branding and marketing strategies for small food businesses (Mike Aquino, 2019; Samantha Coomber, 2020; Tripadvisor, 2011). Based on the list provided below, the informants we interviewed were 14 people in this study, who are highly influential individuals who play a crucial role in the sustainability of tourism, particularly within the scope of home-based food enterprises.

The methodology employed was textual analysis, utilizing the qualitative analysis software NVivo Pro12 to classify the tweets into positive, neutral, and negative categories. This approach aimed to identify the key factors associated with each sentiment category (Reyes-Menendez et al., 2018; Saura et al., 2018; Sichewo et al., 2020) (Table 5).

Table 4 Data Informant

No	Name	Nationality	Role
1	Mr.Dodit	Indonesia	Owner Warung Mak Beng
2	Mrs.Oka	Indonesia	Owner Warung Bu Oka
3	Chef Swastika	Indonesia	Chef Community
4	Mr, Tjok Bagus Pemayun	Indonesia	Bali Tourism
5	Mr.Tito	Indonesia	Tourist Mak Beng
6	Mr.Jimmy Ong	Malaysia	Tourist Mak Beng
7	Mr.Justin	USA	Tourist Mak Beng
8	Mr.David Paterson	UK	Tourist Mak Beng
9	Mr.Antonio	China	Tourist Mak Beng
10	Mr.Bernardus Ferry	Indonesia	Tourist Bu Oka
11	Mr.Dedyis	USA	Tourist Bu Oka
12	Mr.Leonardi L	Indonesia	Tourist Bu Oka
13	Mrs.Kartika Funiculine	Indonesia	Tourist Bu Oka
14	Mr.Jimmy S	Indonesia	Tourist Bu Oka

Source: Processed Data Analyst (Author 2024)

Table 5 Nodes represent theme, concept, idea, opinions, or experience

	Description	Files	References
Cultural impact		6	14
Preservation of culinary heritage	A traditional Balinese family recipe by ensuring high-quality ingredients, being part of traditional Balinese food, distinctive sambal flavour, experiencing Balinese food culture, competition in culinary business, food enterprises (HBFEs) in preserving cultural heritage through promoting traditional and local foods.	6	6
Tourist experience	Balinese culinary heritage, family inheritance, positive tourist response, traditional recipe, Balinese spices, Balinese food culture, unique spices, comparison to Javanese cuisine, distinct Balinese food culture, unique spices, unfamiliar flavours, signature fish soup, spicy taste	6	6
Limited exposure	Competition with imitation, a specific focus on competition and differentiation. Culinary as an economic driver	2	2
Economic impact		6	14
Entrepreneurship	Unique flavour (sambal), family business, long-term employees, delicious food, cleanliness, efficient service, returning customers, family visits, marketing, social media (IG, FB), business associations, government support, HBFEs	6	6
Local economic growth	Motivations linked to village development, family success, customer satisfaction driven by reputation and quality, broader economic impact, importance of renowned eateries for tourism	6	6

(continued)

Table 5 (continued)

	Description	Files	References
Limited reach	Emphasis on product quality, importance of cleanliness, focus on service, significance of venue and location, adequate parking facilities, collaboration with financial institutions, provision of low-interest loans, mentoring programs, support in overcoming operational challenges	2	2
Environmental impact		6	18
Reduced food miles	Emphasis on product quality and cleanliness, focus on service, significance of venue and location, adequate parking facilities, and collaboration with financial institutions. Provision of low-interest loans, mentoring programs, and support in overcoming operational challenges	6	6
Waste reduction	Specifically, mention fresh fish from local sources, emphasizes using the best quality ingredients while staying local, the importance of local salt	6	6
Potential concerns	Mentions waste recycling, following regulations, observing cleanliness hygiene, reducing plastic waste	6	6
Social impact		6	14
Community building	Plastic packaging, food waste, the link between waste and employment, plastic reduction, and source-based waste management	6	6
Empowerment	Emphasizes skilled employees, customer service training, and the venue's role as a community space.	6	6
Social equity concerns	Employee treatment, on family business success, customer satisfaction factors, the broader economic impact, cultural tourism and community benefits	2	2

Source: Processed Data Analyst (Author 2024) (NVivo 12 Pro)

3.2 *Impacts of Home-Based Food Enterprises*

HBFEs are pivotal in maintaining the sustainability of tourism in Bali. This study explores the cultural, economic, environmental, and social impacts of HBFEs on sustainable tourism. Home-based food enterprises in Bali, known locally as "home industries", have become a significant economic and cultural force on the island. These enterprises range from small-scale family-run businesses producing traditional Balinese snacks to more extensive operations catering to local and tourist markets. The impact of these home-based food enterprises is multifaceted, influencing economic growth, social structures, and cultural preservation. Economically, they provide vital income streams for families and contribute to the local economy by creating job opportunities and stimulating related industries. According to The Ministry of Tourism and Creative Economy/Agency of Tourism and Creative Economic (2023), it was found that micro and small enterprises, which include many HBFEs, contributed to around 61% of Indonesia's GDP. Socially, these businesses often involve community networks and cooperation, fostering strong social ties and mutual support among residents. Culturally, home-based food enterprises are crucial in preserving Balinese culinary traditions and promoting local heritage through food. This overview explores the various dimensions of these enterprises, examining their contributions to Bali's economy, society, and culture while addressing the challenges they face in a rapidly modernizing and globalizing context.

1. Cultural Impact
   HBFEs are crucial in preserving Balinese culinary heritage by using high-quality ingredients to maintain traditional Balinese family recipes. These enterprises contribute to the cultural tapestry by offering distinctive sambal flavours and other traditional Balinese foods, allowing tourists to experience the rich culinary culture of Bali. This impact helps differentiate them from competitors in the culinary business, ensuring the survival and promotion of traditional and local foods. By highlighting these traditional recipes, HBFEs preserve cultural heritage and enhance the cultural tourism experience, making it more authentic and immersive for visitors. Tourists positively respond to Balinese culinary heritage, appreciating the unique flavours and traditional recipes passed down through generations. Balinese spices and the distinct food culture, characterized by unique spices and unfamiliar flavours, add to the tourist experience. Signature dishes like spicy fish soup showcase the distinctive taste of Balinese cuisine, which stands out compared to other regional cuisines like Javanese. This positive tourist experience is vital for the sustainability of these enterprises, as it drives repeat visits and word-of-mouth promotion. However, HBFEs face competition from imitation products and other businesses, necessitating a focus on competition and differentiation. Despite the challenges, culinary enterprises serve as significant economic drivers, contributing to the local economy by attracting tourists and promoting local food culture. Overall, HBFEs are critical in preserving Balinese culinary heritage, enhancing the cultural tourism experience, and contributing to the local economy. They face challenges but remain significant economic drivers by promoting local food culture and attracting tourists with their distinctive culinary offerings.
2. Economic Impact
   HBFEs foster entrepreneurship by offering unique flavours, such as distinctive sambal, and maintaining a family business model with long-term employees. They emphasize delicious food, cleanliness, efficient service, and a welcoming atmosphere, encouraging returning customers and family visits. Effective marketing strategies, including social media platforms like Instagram and Facebook, and collaboration with business associations and government support further bolster their success. These enterprises contribute to local economic growth by aligning their motivations with village development and family success. Customer satisfaction, driven by reputation and quality, has a broader economic impact, highlighting the importance of renowned eateries for tourism. The presence of successful HBFEs can stimulate the local economy, providing jobs and supporting ancillary businesses. While product quality, cleanliness, service, and

venue location are crucial for success, HBFEs also collaborate with financial institutions to provide low-interest loans and mentoring programs. These initiatives help them overcome operational challenges and sustain their businesses in a competitive market. Home-Based Food Enterprises (HBFEs) are pivotal in fostering entrepreneurship, supporting local economic growth, and preserving cultural culinary heritage. They succeed through unique offerings, effective marketing, customer satisfaction, and financial support. HBFEs create a welcoming environment that encourages repeat visits and positive word-of-mouth promotion by maintaining high standards in food quality, cleanliness, and service and leveraging family business models. Their contributions to job creation and village development further underscore their importance in the local economy.

3. Environmental Impact
   HBFEs contribute to environmental sustainability by sourcing fresh fish and other ingredients locally, reducing food miles, and emphasizing using high-quality, local ingredients. These HBFEs ensure the freshness and quality of the food and support local farmers and fishermen. Many HBFEs implement waste reduction practices, including recycling, following hygiene regulations, and reducing plastic waste. These practices help minimize their environmental footprint and promote sustainability within the community. Despite these efforts, concerns remain regarding plastic packaging and food waste. Addressing these issues through better waste management practices, such as reducing plastic use and implementing source-based waste management, is essential for minimizing the environmental impact of HBFEs. They significantly contribute to environmental sustainability by prioritizing the local sourcing of fresh fish and other ingredients, which reduces food miles and emphasizes the importance of high-quality, locally produced components. This approach guarantees the freshness and superior quality of their culinary offerings and supports local farmers and fishermen, thus bolstering the local economy. Many HBFEs proactively implement waste reduction practices such as recycling, adhering to stringent hygiene regulations, and minimizing plastic waste. These initiatives help to lower their environmental footprint and foster sustainability within their communities. However, challenges remain, particularly concerning the use of plastic packaging and the generation of food waste. Addressing these issues is crucial for further reducing the environmental impact of HBFEs. To achieve greater environmental sustainability, we must improve waste management practices. This includes reducing plastic use and implementing source-based waste management systems.
4. Social Impact
   HBFEs play a vital role in community building by employing skilled workers and providing customer service training. The venues often serve as community spaces, fostering social interaction and engagement. These enterprises empower their employees by treating them well and focusing on family business success and customer satisfaction. The broader economic impact of HBFEs extends to cultural tourism and community benefits, enhancing social cohesion and community well-being. HBFEs address social equity concerns by focusing on tan-

gible aspects of successful business operations, such as digital marketing and collaboration with the local community. Protecting Intellectual Property Rights (IPR) ensures that traditional recipes and culinary practices are recognized and valued, contributing to these enterprises' social equity and sustainability. Home-Based Food Enterprises (HBFEs) have a significant social impact by fostering community building and engagement. They achieve this by employing skilled workers and providing comprehensive customer service training, which enhances the service quality and empowers employees. These enterprises often serve as community hubs, encouraging social interactions and community engagements, thus strengthening social cohesion. By focusing on family business success and customer satisfaction, HBFEs empower their employees through fair treatment and supportive work environments. The broader economic impact of HBFEs extends beyond individual businesses to benefit cultural tourism and the wider community. These enterprises enhance community well-being by promoting social interaction and cultural preservation. They address social equity concerns by implementing successful business practices, such as digital marketing and local community collaboration, ensuring that the benefits of their operations are widely shared.

Moreover, protecting Intellectual Property Rights (IPR) is crucial for HBFEs, ensuring that traditional recipes and culinary practices are recognized and valued. This protection contributes to social equity and the sustainability of these enterprises by safeguarding cultural heritage and providing legal recognition for traditional knowledge. Overall, HBFEs promote social cohesion, economic growth, and cultural preservation within their communities.

4 Conclusion and Recommendations

4.1 Conclusion

This research explored how HBFEs affect sustainable tourism in Bali, considering their economic, cultural, environmental, and social impacts. The findings highlight that HBFEs significantly contribute to the sustainability of tourism in Bali by preserving cultural heritage, fostering local economic growth, promoting environmental sustainability, and enhancing social cohesion. Economically, HBFEs provide vital income streams for local families, support local agriculture, and create job opportunities, empowering individuals to become entrepreneurs and contributing to economic diversification and resilience. Their success relies on maintaining high food quality, effective marketing strategies, and support from government policies. Culturally, HBFEs play a crucial role in preserving and promoting Balinese culinary traditions, enhancing the cultural experience for tourists, and fostering a deeper connection with Balinese culture. This preservation of culinary heritage is essential for maintaining Bali's cultural identity. Environmentally, many HBFEs adopt sustainable practices, such as sourcing local ingredients and reducing food miles, contributing to environmental conservation. Waste reduction practices and adherence to hygiene regulations further minimize their ecological footprint, although challenges such as plastic waste and improper waste management must be addressed. Socially, HBFEs foster community building and empowerment by providing employment opportunities and creating spaces for social interaction, contributing to social equity by promoting fair treatment of employees and supporting local community collaboration. Protecting the intellectual property rights of traditional recipes ensures cultural preservation and recognition.

4.2 Recommendations

Several recommendations are proposed to enhance the impact of HBFEs on sustainable tourism in Bali. Government policies should continue to support HBFEs through low-interest loans, training programs, and marketing assistance to ensure their sustainability and growth. HBFEs should enhance their waste management practices, reduce plastic use, and implement comprehensive recycling programs. Effective marketing strategies, mainly through digital platforms, should be adopted to attract more tourists and promote the unique cultural heritage of Balinese cuisine. Strengthening community ties through collaboration and mutual support can enhance the overall impact of HBFEs on local development and tourism sustainability. Continued efforts to preserve traditional recipes and cooking methods are essential for maintaining Bali's cultural identity and attractiveness as a tourist destination. By addressing these recommendations, HBFEs can further enhance their contribution to sustainable tourism in Bali, ensuring long-term economic, cultural, environmental, and social benefits.

Implementing these recommendations can provide practical benefits for HBFEs and the broader tourism industry in Bali. Supporting policies and training programs can empower local entrepreneurs, helping them to scale their operations and improve their business practices. Enhanced waste management and sustainable sourcing can reduce environmental impact, making these enterprises more attractive to eco-conscious tourists. Effective digital marketing can increase visibility and attract a global audience, boosting tourist numbers and local revenues. Stronger community ties and cultural preservation efforts can enrich the tourist experience, fostering loyalty and repeat visits, thereby sustaining the tourism industry's growth. Future research should explore the long-term impacts of HBFEs on the sustainability of Bali's tourism industry, focusing on specific metrics such as economic growth, environmental conservation, and social cohesion. Studies could investigate the effectiveness of various government policies and training programs in supporting HBFEs, providing data-driven insights for policymakers.

Additionally, the research could examine consumer behaviour and preferences related to sustainable and culturally authentic food experiences, helping HBFEs tailor their offerings more effectively. Comparative studies between Bali and other regions with similar tourism dynamics could also yield valuable lessons and best practices. Finally, exploring the role of technology in enhancing the operational efficiency and market reach of HBFEs can provide strategic guidance for future development.

Acknowledgement The authors thank Triatma Mulya University Bali, Indonesia, for providing the necessary facilities and support, and extend their sincere gratitude to the following individuals for their invaluable contributions and support: Mr. Dodit, Owner of Warung Mak Beng, Mrs. Oka, Owner of Warung Bu Oka, Mr. I Made Tjok Bagus Pemayun, Head of the Bali Tourism Office, Indonesia, and Chef Swastika, representing the Bali Chef Community, Indonesia.

References

Abd Razak, S. S., Ying Tuan, L., & Teck Chai, L. (2022). Food safety legal issues in home-based food business. *International Journal of Academic Research in Business and Social Sciences, 12*(1). https://doi.org/10.6007/IJARBSS/v12-i1/12133

Ahannaya, C., Fijabi, E., Fijabi, L., Olatunde, O., & Ajani, A. (2021). Environmental Impacts and Organization Business Survival on SMEs. *GSJ, 9*, 1703–1711.

Arcana, I. N., Wiratnaya, I. N., & Ariesta Budiani, N. P. (2018). Determinant factors affect tourist satisfaction towards traditional Balinese cuisine at restaurants in Ubud tourism resorts, Gianyar, Bali. In *Proceedings of the 2nd international conference on tourism, gastronomy, and tourist destination (ICTGTD 2018)*. https://doi.org/10.2991/ictgtd-18.2018.25

Atsız, O., & Cifci, I. (2021). Can we imagine the meal-sharing economy without service providers? The impact of COVID-19. *Journal of Hospitality and Tourism Management, 49*, 172–177. https://doi.org/10.1016/j.jhtm.2021.09.011

Bar-El, R., Hatsor, L., & Tobol, Y. (2020). Home production, market substitutes, and the labor supply of mothers. *Journal of Economic Behavior & Organization, 171*, 378–390. https://doi.org/10.1016/j.jebo.2020.01.007

Baruch, G., Buchaltzev, E., Grinberg, M., Hani, G., Karavani, S., & Raz, A. (2013). *Community kitchens: From idea to recipe*.

Berno, T., & Fusté-Forné, F. (2022). Gastronomy. In *Encyclopedia of tourism management and marketing* (pp. 370–373). Edward Elgar Publishing. https://doi.org/10.4337/9781800377486.gastronomy.

Blapp, M., & Mitas, O. (2018). Creative tourism in Balinese rural communities. *Current Issues in Tourism, 21*(11), 1285–1311. https://doi.org/10.1080/13683500.2017.1358701

Budiani, N. P. A., Suastuti, N. L., & Massenga, L. M. (2017). Marketing strategies for organic restaurant in Ubud Tourist area - Bali. *Journal of Business on Hospitality and Tourism, 2*(1), 393. https://doi.org/10.22334/jbhost.v2i1.74

Budiasa, I., & A. G. (2014). Community based agro-tourism as an innovative integrated farming system development model towards sustainable agriculture and tourism in Bali. *Journal of International Society for Southeast Asian Agricultural Sciences, 20*, 29–40.

Christian Byczek. (2011). Blessings for all? Community-based ecotourism in Bali between global, national, and local interests – A case study. *Austrian Journal of South-East Asian Studies, 4*(1).

Fidyanita, S. (2024). Warung Mak Beng in Sanur, Bali: Recognize among the 100 most iconic restaurants by TasteAtlas. *Social Expat*. https://www.socialexpat.net/warung-mak-beng-in-sanur-bali-recognized-among-the-100-most-iconic-restaurants-by-tasteatlas/

Graham, S. C. (2021). Authentic culinary tourism experiences: The perspectives of locals. *Journal of Gastronomy and Tourism, 5*(2), 65–82. https://doi.org/10.3727/216929720X15968961037926

Hendriyani, I. G. A. D., Budiarsa, M., Antara, M., & Sudiarta, N. (2020). The loyalty model of foreign tourists consuming traditional Balinese food. *Global Business Finance Review, 25*(3), 34–48. https://doi.org/10.17549/gbfr.2020.25.3.34

Hernández, P. A., Galli, F., Prosperi, P., Šūmane, S., Duckett, D., & Almaas, H. E. (2021). Do small food businesses enable small farms to connect to regional food systems? Evidence from 9 European regions. *Global Food Security, 29*, 100505. https://doi.org/10.1016/j.gfs.2021.100505

Hirakawa, S., Masuyama, H., Sudiarta, I. P., Suprapta, D. N., & Shiotsu, F. (2024). Initiatives and prospects for sustainable agricultural production in Karangasem regency, Bali, Indonesia. *Sustainability, 16*(2), 517. https://doi.org/10.3390/su16020517

Janez Bogataj. (2021). *Gastronomic tourism (An opportunity to discover the diversity of local and regional cultures)* (1st ed.). Routledge.

Janhonen, K., & Elkjaer, B. (2022). Exploring sustainable food education as multi-professional collaboration between home economics and School Food Catering. *Journal of Education for Sustainable Development, 16*(1–2), 19–41. https://doi.org/10.1177/09734082221120101

Krishna Kumar, G. M. (2021). Holistic cuisine and gastronomic tourism. In *Gastronomy and food science* (pp. 205–225). Elsevier. https://doi.org/10.1016/B978-0-12-820057-5.00011-X

Lin, W. L. (2024). Tourism and economic growth: Assessing the significance of sustainable competitiveness using a dynamic panel data approach. *International Journal of Tourism Research, 26*(5). https://doi.org/10.1002/jtr.2760

Listiani, W. F. M., Ningdyah, A., & Juju Rohaeni, A. (2024). Desire to revisit: Memorable experiences drive domestic tourists to return to Bali. *Jurnal Kajian Bali (Journal of Bali Studies), 14*(1), 147. https://doi.org/10.24843/JKB.2024.v14.i01.p07

Majiid, M. A., Bagia, R. B. P., Komaladewi, A., Wijonarko, P. B., Assriakhun, G. S., Salsabila, S. N., & Reinhart, H. (2023). Land conversion analysis in Buleleng District, Bali: An outlook for sustainable tourism development. *E3S Web of Conferences, 468*, 10004. https://doi.org/10.1051/e3sconf/202346810004

Marius Boiţă, A. A. B. C. (2014). Gastronomic tourism management revitalizer factor of the tourism offer. *Annals - Economy Series (Academica Brâncuşi), 1*, 398–401.

Marzuki, M., Sunarpi, S., Amir, A., & Ridwan, M. (2017). Analysis of the potential for sustainable development based food products to support tourism fisheries and economic Community in the Province of Bali. *Environmental Management and Sustainable Development, 6*(2), 22. https://doi.org/10.5296/emsd.v6i2.8554

Mike Aquino. (2019). Warung Ibu Oka: An authentic Balinese dining experience. *TripSavvy*. https://www.tripsavvy.com/warung-ibu-oka-balinese-dining-experience-1629937

Muneerali Thottoli, M., Suhail Al Mazroui, T. S., Amur Al Hoqani, D. S., & Al-Alawi, M. M. (2023). Marketing potential of homemade food: A new entrepreneurial Endeavour. *Multidisciplinary Business Review, 16*(2), 53–65. https://doi.org/10.35692/07183992.16.2.5

Najjinda, S., Sendawula, K., Otengei, S. O., Walugembe, A., & Kimuli, S. N. L. (2023). Social capital and sustainable growth of full-service restaurants in the global south: Testing the mediating role of dynamic capabilities. *Journal of Work-Applied Management, 15*(2), 273–291. https://doi.org/10.1108/JWAM-04-2023-0031

Putri, A. R. A., Fadhila, P. G., & Furqan, A. (2023). Tourism impact on economic growth in Bali. *Archives of Business Research, 11*(6), 94–101. https://doi.org/10.14738/abr.116.14869

Putu Diah Sastri Pitanatri. (2016). No longer invincible: Women's role in promoting Ubud as culinary destination in Bali. *J. of Tourism and Hospitality Management, 4*(3). https://doi.org/10.17265/2328-2169/2016.06.002

Qian, J., Lin, P., Wei, J., Liu, T., & Nuttavuthisit, K. (2023). Cooking class travel in Thailand: An investigation of value co-creation experiences. *SAGE Open, 13*(2), 215824402311769. https://doi.org/10.1177/21582440231176994

Rahadhini, M. D., & Lamidi, L. (2023). The influence of entrepreneurship characteristics and business environment on the performance and sustainability of SMEs culinary in Surakarta. *International Journal of Social Science And Human Research, 06*(01). https://doi.org/10.47191/ijsshr/v6-i1-52

Rangus, M., B. B., & P. T. J. (2018). *Sustainable tourism development in rural areas: The role of stakeholders*. Academica Turistica - Tourism and Innovation.

Reyes-Menendez, A., Saura, J. R., & Alvarez-Alonso, C. (2018). Understanding #WorldEnvironmentDay user opinions in twitter: A topic-based sentiment analysis approach.

International Journal of Environmental Research and Public Health, 15(11), 2537. https://doi.org/10.3390/ijerph15112537

Samantha Coomber. (2020). Ibu Oka 3 (Warung Babi Guling). *Frommers*. https://www.frommers.com/destinations/ubud/restaurants/ibu-oka-3-warung-babi-guling

Satrya, I. D. G., Kaihatu, T. S., Budidharmanto, L. P., Karya, D. F., & Rusadi, N. W. P. (2023). The role of ecotourism in preserving environmental awareness, cultural and natural attractiveness for promoting local communities in Bali, Indonesia. *Journal of Eastern European and Central Asian Research (JEECAR), 10*(7), 1063–1075. https://doi.org/10.15549/jeecar.v10i7.1386

Saura, J. R., Reyes-Menendez, A., & Alvarez-Alonso, C. (2018). Do online comments affect environmental management? Identifying factors related to environmental management and sustainability of hotels. *Sustainability, 10*(9), 3016. https://doi.org/10.3390/su10093016

Sedana, G. (2022). *Rural development on the agricultural institution basis: Case of the agricultural development in Bali Province, Indonesia. Rural development-education, sustainability*. Intech Open.

Shah, D., & Murthi, B. P. S. (2021). Marketing in a data-driven digital world: Implications for the role and scope of marketing. *Journal of Business Research, 125*, 772–779. https://doi.org/10.1016/j.jbusres.2020.06.062

Sichewo, P. R., Vander Kelen, C., Thys, S., & Michel, A. L. (2020). Risk practices for bovine tuberculosis transmission to cattle and livestock farming communities living at wildlife-livestock-human interface in northern KwaZulu Natal. *South Africa. PLOS Neglected Tropical Diseases, 14*(3), e0007618. https://doi.org/10.1371/journal.pntd.0007618

Sormaz, U., Akmese, H., Gunes, E., & Aras, S. (2016). Gastronomy in tourism. *Procedia Economics and Finance, 39*, 725–730. https://doi.org/10.1016/S2212-5671(16)30286-6

Subadra, I. N., S. I. K., A. I. W. A., Y. L. H. K., & S. M. (2019). Investigating push and pull factors of tourists visiting Bali as a world tourism destination. *International Journal of Multidisciplinary Education Research, 8*, 235–269.

Sukadana, I. M., Ardhana, I. K., & Subrata, I. W. (2022). Balinese culinary entrepreneurship amid globalization. *International Journal of Health Sciences*. https://doi.org/10.53730/ijhs.v6nS2.5077

Sukerti, N. W., & Marsiti, C. I. R. (2020). Developing culinary tourism: The role of traditional food as cultural heritage in Bali. In *Proceedings of the 2nd international conference on social, applied science, and Technology in Home Economics (ICONHOMECS 2019)*. https://doi.org/10.2991/assehr.k.200218.029.

Sutama, I. K., & Mudana, I. G. (2019). Application of green home stay based on Bali local wisdom in Kuta Village of Bali. *Journal of Travel, Tourism and Recreation, 1*(1), 18–24. https://doi.org/10.22259/2642-908X.0101004

The Ministry Of Tourism And Creative Economy/Agency Of Tourism And Creative Economic. (2023, February 15). *Minister Sandiaga: 2023 Tourism & Creative Economy Target Achievement Needs A Deregulation Support*. Kemenparekraf/Baparekraf RI. https://kemenparekraf.go.id/en/articles/minister-sandiaga-2023-tourism-creative-economy-target-achievement-needs-a-deregulation-support

Tripadvisor. (2011). Babi Guling Bu Oka. *Tripadvisor*. https://www.tripadvisor.com/Restaurant_Review-g297701-d1182744-Reviews-Babi_Guling_Bu_Oka-Ubud_Gianyar_Regency_Bali.html

Urban@UW's Homelessness Research Initiative. (2017). *Home-grown success: Exploring the social embeddedness of home-based food businesses*. Evans School of Public Policy & Governance.

V, S., T, K. H., G, S., & J, E. (2020). 'Home baking': An enterprising Area in the entrepreneurial arena – A study. *Saudi Journal of Economics and Finance, 04*(03), 126–132. https://doi.org/10.36348/sjef.2020.v04i03.006

Wasti Atmodjo. (2005). *Warung Mak Beng: Sanur, Bali*. Indonesia Travel Guide. https://indonesia-logue.com/about-indonesia/warung-mak-beng-sanur-bali.html

Wayan Suadnyana. (2024). *Warung Mak Beng Sanur – Famous fish soup & spicy fried fish in Bali*. Wira Water Sport. https://www.water-sports-bali.com/warung-mak-beng-sanur/

Wiranatha, A. S., Suryawardani, I. G. A. O., Purbanto, I. G. R., Yudiastina, I. G. B. A., & Bantacut, T. (2024). A conceptual model of rural tourism in Bali to build up global competitiveness. *Systems, 12*(7), 245. https://doi.org/10.3390/systems12070245

Wirawan, P. E., Supartini, N. L., & Sudiana, I. M. (2022). Moringa leaf-based gastronomic tourism in Taro Village, Gianyar, Bali. *International Journal of Social Science and Business, 6*(4). https://doi.org/10.23887/ijssb.v6i4.52065

World Tourism Organization (UNWTO). (2023). *8th UNWTO world forum on gastronomy tourism.* 8th UNWTO world forum on gastronomy tourism on 5–7 October 2023 in Donostia-San Sebastian.

Capturing Laduma Tourism Village: A Renewable Energy-Based Tourism Destination

L. K. Herindiyah Kartika Yuni and I Made Bayu Wisnawa

Abstract Laduma tourism village in Pinge village, Bali, Indonesia, has excellent potential as a sustainable tourism destination that supports environmental conservation and local economic development. This study aims to create a business model for Laduma tourism village using the business model canvas (BMC) and renewable energy as its foundation. This research identifies the village's potential challenges and development goals through a qualitative descriptive approach. The results indicate that the BMC effectively organizes the village's business model, encompassing key partners, activities, value propositions, customer relationships, segments, resources, channels, cost structure, and revenue streams. Collaboration with the government, communities, travel agents, and educational institutions is crucial for the success of this model. The business model offers tourists a sustainable and enriching experience while promoting environmental sustainability and empowering the local economy. This research serves as a blueprint for developing sustainable tourism villages in Bali and as a guide for other village managers.

1 Introduction

In recent years, sustainable tourism has emerged as a significant concern in the global tourism industry. Recent research suggests sustainable tourism approaches benefit the environment and contribute to the local economy and community wellbeing (Risfandini et al., 2023). A study conducted in 2021 found that tourists seek environmentally friendly tourism experiences that positively impact local communities (Gupta et al., 2021).

This situation emphasizes community involvement, promoting equitable economic gains and environmental preservation. It encourages sustainable practices like renewable energy use, effective waste management, and cultural sensitivity,

L. K. H. K. Yuni (✉) · I. M. B. Wisnawa
Triatma Mulya University, Bali, Indonesia
e-mail: herindiyahkartikayuni@gmail.com; bayu.wisnawa@triatmamulya.ac.id

J. Marques et al. (eds.), *Tourism and Climate Change*, Tourism, Hospitality & Event Management, https://doi.org/10.1007/978-3-031-99128-8_11

ensuring that tourism positively impacts local communities while reducing the overall environmental footprint (Daneshwar & Revaty, 2024). Furthermore, research demonstrated that sustainable tourism can effectively preserve cultural heritage and natural ecosystems (Jayaprakash Lamoria et al., 2024). However, challenges persist in implementing sustainable tourism, including issues related to funding, policy, and stakeholder involvement (Mechinda et al., 2012). Therefore, promoting ecotourism as an up-and-coming tourism model on a broader scale is crucial.

Sustainability is the fundamental principle of ecotourism, aiming to balance economic growth with environmental conservation. Recent research highlights the importance of sustainable practices in ecotourism operations, such as minimizing waste, conserving water, and using renewable energy sources. A study emphasized the role of ecotourism operators in adopting environmentally friendly practices, stating that "sustainable ecotourism requires a commitment to reducing environmental impact through operational changes". Community involvement is another important principle of ecotourism, recognizing the importance of local communities in developing and managing ecotourism activities (Karimov et al., 2024). According to Borah (2024), community-based ecotourism initiatives have the potential to economically and socially empower local communities, fostering a sense of ownership and responsibility towards natural resources. Additionally, education is a key component of ecotourism, aiming to raise awareness and understanding of environmental issues among tourists and local communities. As Ibnou-Laaroussi et al. (2020) mentioned, ecotourism should serve as a platform for education, where visitors learn about the environment and the importance of contributing to broader conservation efforts.

Ecotourism is an up-and-coming tourism model that should be more widely promoted. The demand for environmentally friendly experiences is increasing as tourists become more concerned about the impact of tourism activities on nature and society (UNWTO, 2019). Ecotourism offers solutions to meet these needs, significantly benefiting nature conservation and local community empowerment. One of the main advantages of ecotourism is its ability to preserve the natural environment, which is the main attraction for tourists. By implementing sustainability principles, education, and community involvement, ecotourism destinations can protect biodiversity, vulnerable ecosystems, and natural resources that form the basis of their attractions (Buckley, 2009).

Competition among tourism villages in Indonesia is intensifying, with each local entity striving to attract more tourists (El Moslem Badr, 2022). The emergence of social media platforms has levelled the playing field, allowing even the smallest and most remote villages to promote their unique attractions and experiences to a global audience. As a result, tourism villages need to be more creative and innovative in their marketing strategies, offering unique packages and promotions to stand out from their competitors. Additionally, the proliferation of online travel agents and booking platforms has made it easier for tourists to compare prices and services, further increasing the competitive dynamics among tourism village entities. In this highly competitive environment, only the most adaptive and responsive tourism villages will be able to survive and thrive, while those that fail to innovate and adapt risk being left behind.

The shift towards sustainable and environmentally friendly tourism means that modern tourists seek visually beautiful destinations and expect experiences that promote conservation efforts and the use of renewable energy (Ankit Prakash & Vikas Sharma, 2023; El Moslem Badr, 2022). This shift towards sustainable and environmentally friendly tourism presents exciting opportunities for tourism villages to attract visitors by leveraging new energy sources. One increasingly popular energy source is solar energy. Laduma tourism village, located in Pinge village, Bali, Indonesia, is an example of a village that has embraced solar panel technology to generate electricity for tourism infrastructure and local residents. Tourists are attracted to the concept of a "solar village" where they can witness firsthand how solar energy is utilized to support daily activities.

As a tourism village that has embraced renewable energy, specifically solar power, Laduma tourism village in Pinge village faces challenges regarding aesthetics that may diminish its appeal to tourists. While technologically effective, installing solar panels is viewed as disrupting the natural scenery and creating an overly "technical" or less harmonious environment. Many tourists seek rural destinations embodying traditional values and local wisdom, so the presence of noticeable renewable energy infrastructure may detract from the village's inherent beauty. Furthermore, the improper placement of solar panels can also obstruct or harm the village's picturesque views, which are a significant draw for visitors.

The development of renewable energy-based tourism villages has become a top priority for enhancing the appeal of tourist destinations that have become overcrowded with famous tourist spots like Kuta Beach, Nusa Dua Beach, and Sanur Beach. However, there are significant challenges to ensuring the sustainability of tourism, primarily due to economic pressures prioritizing short-term, unsustainable profits. Therefore, it is increasingly urgent to develop innovative and sustainable business models to support the growth of Laduma tourism village in Pinge village, Bali. In this context, the business model canvas (BMC) is an effective tool for designing, developing, and analysing innovative and sustainable business models. The BMC allows us to accurately visualize the business we currently operate or plan to establish. It simplifies complex business concepts by consolidating them into a single canvas sheet that includes a business plan with nine key integrated elements encompassing the company's internal and external strategic analyses (Osterwalder & Pigneur, 2010). The BMC covers various crucial aspects of business decision-making, such as identifying target market segments, defining value propositions, determining resource structures, outlining key activities, establishing customer relationships, and managing supplier relationships. By utilizing the BMC, the objective is to establish a business model that will effectively facilitate the growth of eco-friendly tourism villages powered by renewable energy. This initiative aims to impact both the environment and the local community positively.

2 Theoretical Framework

2.1 *Definition and Importance of Sustainable Tourism*

Sustainable tourism is "tourism that contributes to economic development while preserving the environment and enhancing the well-being of local communities" (Gössling, 2002). This definition highlights the importance of tourism activities that are environmentally friendly, economically viable, and socially beneficial. According to Weaver (2008), sustainable tourism is significant because it addresses the increasing concerns regarding the negative environmental impacts of conventional tourism, including pollution, habitat destruction, and cultural erosion.

2.2 *Impact of Sustainable Tourism*

Numerous research studies have demonstrated the substantial positive effects that sustainable tourism can have on the environment and local communities. For instance, Nyaupane and Poudel (2011) conducted a study revealing that ecotourism initiatives in Nepal have resulted in heightened efforts to conserve biodiversity and have positively impacted the livelihoods of local residents. Likewise, studies conducted by Tehseen et al. (2024) have emphasized how sustainable tourism practices can effectively decrease carbon emissions and contribute to safeguarding both natural and cultural heritage sites.

2.3 *Strategies for Promoting Sustainable Tourism*

There have been several proposed strategies for enhancing tourism sustainability. These strategies include the development of eco-friendly infrastructure, promoting responsible travel behaviour among tourists, and implementing policies that support the local economy while protecting the environment (Anastasopoulos et al., 2024; B & M S, 2024). Furthermore, the paper emphasizes a comprehensive approach to innovation management in sustainable tourism, advocating for structured frameworks like ISO 56001:2024 to address barriers, including inadequate leadership and poor stakeholder communication, essential for enhancing tourism sustainability through effective strategies (Chountalas et al., 2024). Renewable energy sources, such as solar, wind, and hydroelectric power, are increasingly utilized in various aspects of the tourism industry. Hotels and resorts are installing solar panels to generate electricity, reducing reliance on fossil fuels and lowering operational costs (Gong et al., 2020; Rathore & Singh, 2017). In remote tourist destinations, wind turbines provide off-grid electricity, enhancing the sustainability of eco-lodges and adventure camps (Shezan et al., 2015). Hydroelectric power plants near waterfalls and rivers offer opportunities for hydroelectric power generation, contributing to the local economy and supporting sustainable tourism initiatives (Yüksel, 2010).

2.4 *Benefits of Renewable Energy in Tourism*

Adopting renewable energy in tourism offers several benefits, including environmental conservation, cost savings, and enhanced guest experiences. Tourism businesses can significantly reduce their carbon footprint by utilizing renewable energy, contributing to global efforts to mitigate climate change (Gössling et al., 2023). Economically, there are benefits, such as reduced energy bills for operators and new jobs in the renewable energy sector. Moreover, accommodations that prioritize sustainability attract tourists, resulting in increased bookings and revenue for businesses that invest in renewable energy technologies (Pace, 2016).

3 Research Methods

This research employs a qualitative descriptive approach to develop the business model of Laduma tourism village in Pinge village, Tabanan Regency, Bali. The BMC, as proposed by Osterwalder and Pigneur (2010), serves as the primary analytical tool, allowing for a structured exploration of key components in building a sustainable tourism business. This model has been validated in previous studies, such as by Harianto and Soenardi (2022), who applied BMC to evaluate sustainable business strategies, and by Asyuri and Setiawan (2022), who investigated the integration of renewable energy sources in tourism to improve community welfare. These studies highlight BMC's versatility in addressing environmental and economic goals within the tourism sector.

In this research, the qualitative approach involves gathering data through interviews with local stakeholders, direct field observations, and document analysis to understand Laduma's operational environment comprehensively. The data collection process allows researchers to explore the unique challenges, opportunities, and strategies needed for implementing a renewable energy-based business model. Findings from these data sources are then analysed through the BMC framework to identify sustainable practices that could enhance Laduma's competitiveness and community impact.

4 Results and Discussion

The Results and Discussion section presents the findings derived from applying the BMC framework to the development of Laduma tourism village in Pinge village, Bali. This section details each component of the BMC, providing a structured analysis of the key partners, activities, value propositions, customer segments, and other elements that contribute to the village's renewable energy-based business model. By employing a qualitative descriptive approach, this study highlights the role of sustainable practices in shaping the operational and financial viability of Laduma

Table 1 Business model canvas of Laduma tourism village

Key partner	Government, community, partnerships with travel agents, educational institutions
Key activities	Installation and maintenance of solar energy infrastructure, development of renewable energy-based tourism packages, tourism village management, development of supporting tourism infrastructure
Value preposition	Environmentally friendly tourism experience, tangible contribution to environmental sustainability, unique and memorable experience, economic empowerment of village communities, education and learning about renewable energy
Customer relationship	Community of environmentally friendly tourism enthusiasts, partnerships with educational institutions and students, community involvement
Customer segment	Environmentally conscious tourists, educational institutions, green tourism enthusiasts, families and communities, domestic and international tourists
Cost structure	Investment and maintenance of renewable energy infrastructure, operational costs, marketing and promotion, training and development for local communities
Key resources	Renewable energy technology (solar panels), natural resources, local human resources, institutional support
Channels	Online platforms and social media, travel agents and tours, eco-friendly travel packages
Revenue streams	Entrance fees and tours, renewable energy education programs, community-based homestays, eco-friendly product sales

Tourism Village. Additionally, the discussion addresses how the BMC framework can be applied to tourism villages with similar objectives, offering insights into strategies for integrating renewable energy to enhance ecotourism and community engagement. The subsequent analysis provides a comprehensive overview of the elements necessary for achieving a prosperous, environmentally conscious tourism village model (Table 1).

4.1 Customer Segments

Customer segments are individuals or groups that can be targeted and reached by businesses. Business operators must determine the specific consumers they aim to serve and may identify one or multiple customer segments (Gajdošík, 2020). In this study, the author has identified several customer segments for a tourist village in Bali that is focused on renewable energy. First, there are ecotourism travellers. These tourists are nature-oriented and outdoor enthusiasts seeking destinations that prioritize environmental conservation. Next, some academics and researchers require locations for research, testing, and experiments. Lastly, we have millennial and Gen Z travellers. This particular group consists of tourists between the ages of 18 and 40 who are both environmentally conscious and active on social media and digital platforms. The unique selling points of our tourist village include a strong focus on renewable energy, a commitment to environmental conservation, and a personalized, community-based approach to tourism.

Customer segment refers to an individual or group that a business can reach and target, and these segments share common characteristics such as interests, needs, or demographics (Weber, 2019). In the case of the tourist village of Laduma, which focuses on renewable energy, the primary customer segments are ecotourists and individuals interested in learning about renewable energy technologies (Chin et al., 2014; Hillary, 2020; Samuil & Ionică, 2021). Eco-tourists are travellers intensely interested in environmentally sustainable tourism experiences that minimize their environmental impact (Corzo-Arévalo & Cuadra, 2020). This segment is attracted to offerings that support local communities and conservation efforts. The second segment, those interested in renewable energy, includes consumers curious about renewable technologies and individuals with specialized interests, such as students, researchers, or industry professionals. By targeting these two segments, the business aligns with its unique value proposition of providing an immersive ecotourism experience centred around renewable energy, a proposition that is sure to intrigue and interest potential stakeholders (Gajdošík, 2020). To reach these customer segments, the tourist village can partner with travel agencies and online platforms specializing in sustainable tourism (Sutoni et al., 2021). These partnerships will provide access to a broader network of eco-conscious travellers and facilitate effectively promoting the village's renewable energy-focused offerings. Additionally, the village can directly market and promote its renewable energy focus through various channels, such as social media, its website, and local tourism events, which will help attract the segment interested in renewable energy technologies, including students, researchers, and industry professionals, and inform them about the unique educational and experiential opportunities available at the tourist village (Beer et al., 2018).

Aside from the customer segments, the key resources required to deliver the value proposition include renewable energy technology, the natural environment, and cultural assets of the village, as well as knowledgeable staff to manage operations and educate visitors (Anggarani Pribudi, 2020; Corzo-Arévalo, 2020; Lubis et al., 2023; Oka et al., 2021).The key activities would involve operating and maintaining the renewable energy infrastructure, providing eco-friendly accommodations and activities, and educating visitors about renewable energy. The tourist village would generate revenue from visitor fees, accommodation charges, and possibly the sale of local products and souvenirs. In order to ensure financial sustainability, the business would need to balance the costs of maintaining the renewable energy infrastructure and offering an engaging ecotourism experience with the revenue generated from visitors. This emphasis on financial sustainability is designed to make the audience feel secure and confident in the business's long-term viability (Astawa & Suamba, 2021; Chin et al., 2014; Mandryk et al., 2016; Popescu et al., 2022).

A customer segment refers to an individual or group a business can reach and target (Sotiriadis & Nduna, 2014). Based on the renewable energy focus of the tourist village of Laduma, the primary customer segments would be eco-tourists and those interested in learning about renewable energy technologies (Hillary, 2020; Samuil & Ionică, 2021). Eco-tourists are travellers intensely interested in environmentally sustainable tourism experiences that minimize their environmental impact (Corzo-Arévalo, 2020). This segment is attracted to offerings that support local communities and conservation efforts. The second segment, those interested in renewable energy, would include general consumers curious about renewable technologies and those with a more specialized interest, such as students, researchers, or industry professionals. Targeting these two segments aligns with the value proposition of providing an immersive ecotourism experience centred around renewable energy.

To reach these customer segments, the tourist village can partner with travel agencies and online platforms catering to sustainable tourism (Lubis et al., 2023). These partnerships will provide access to a broader network of eco-conscious travellers and facilitate effectively promoting the village's renewable energy-focused offerings. Additionally, the village can directly market and promote its renewable energy focus through various channels, such as social media, its website, and local tourism events, which will help attract the segment interested in renewable energy technologies, including students, researchers, and industry professionals, and inform them about the unique educational and experiential opportunities available at the tourist village (Beer et al., 2018). Apart from the customer segments, the key resources required to deliver the value proposition include renewable energy technology, the village's natural environment and cultural assets, and knowledgeable staff to manage operations and educate visitors. The key activities would involve operating and maintaining the renewable energy infrastructure, providing eco-friendly accommodations and activities, and educating visitors about renewable energy. As the village grows and establishes itself as a leading destination for sustainable tourism, there is potential for expansion and the development of new offerings, presenting long-term opportunities for investors and collaborators.

Furthermore, enhancing word-of-mouth communication can significantly increase visitor engagement, leveraging the village's unique tourism services delivered by residents. This strategy involves providing an authentic experience where every visitor is treated as part of the family, embedding them in the community's daily life and traditional practices. Such a personalized approach fosters emotional connections, creating memorable experiences that visitors will likely share and promoting the village through organic, word-of-mouth channels. This community-based, family-inclusive model not only highlights the village's commitment to sustainability through renewable energy and cultural immersion but also makes the audience feel connected and involved, adding another layer to its value proposition.

4.2 Value Propositions

Value propositions are a business's unique and superior qualities that differentiate it from others, adding value and attracting many customers (Stacchini et al., 2022). They explain why customers prefer a particular product over others. The first proposition is the eco-friendly tourism experience. The tourism village of Laduma utilizes solar power as the energy source in the tourist area, integrating activities and attractions with solar energy, providing a unique and memorable experience for tourists and contributing to environmental sustainability by reducing carbon emissions and ecological footprints through renewable energy.

Furthermore, the accommodations and tourism facilities showcase eco-friendly designs and concepts, adding to the overall appeal of the experience. Lastly, the value proposition includes economic empowerment of the village community, offering economic opportunities and jobs for residents, and the development of entrepreneurship and MSMEs (Micro, Small, and Medium Enterprises) based on local resources, leading to increased income and community welfare. In addition to these benefits, the value proposition has an educational aspect, providing inspiration and motivation for visitors to apply renewable energy in their everyday lives (Čirjevskis, 2019) (Fig. 1).

In the context of Laduma tourist village, which is based on renewable energy, the value proposition can be summarized as an eco-tourism experience that is environmentally sustainable and provides an opportunity for visitors to learn about renewable energy technologies. This support aligns with the growing consumer demand for green and sustainable tourism experiences and the increasing interest in renewable energy technologies (Sarkar & Sinha, 2015; Simanjuntak, 2021). Furthermore, recent research has emphasized the importance of developing attractive value propositions for sustainable technologies to increase market appeal and facilitate commercialization (Vernay et al., 2020). The tourist village of Laduma can leverage its renewable energy focus to create a unique value proposition that resonates with eco-conscious travellers.

To deliver this value proposition, the key activities of the business model will likely include operating and maintaining renewable energy infrastructure, providing eco-friendly accommodations and activities, and educating visitors about renewable energy. The key resources required would include renewable energy technology, the village's natural environment and cultural assets, and knowledgeable staff to manage operations and educate visitors (Nosratabadi et al., 2019; Vernay et al., 2020). The primary customer segments for this business model would be eco-tourists and those interested in learning about renewable energy (Astawa & Suamba, 2021;

Fig. 1 Solar panel installation as a renewable energy source in Laduma tourism village

Fig. 2 Tourists participating in traditional cultural and environmental activities in Laduma tourism village

Sotiriadis, 2018). These customers can be reached through partnerships with travel agencies and online platforms that cater to sustainable tourism and through direct marketing and promotion of the village's renewable energy focus (Samuil & Ionică, 2021; Astawa & Suamba, 2021). In addition to renewable energy use, the "Live-in" concept enriches the value proposition by inviting tourists to engage directly with the local way of life. Unlike typical commercial accommodations, visitors stay in residents' modest, clean homes, immersing themselves in Laduma's rural lifestyle. This experience includes participating in agricultural activities, such as farming and gardening, and engaging in cultural practices that reflect the village's heritage. This unique proposition enables tourists to connect authentically with the community, offering a profound contrast to conventional luxury hotels, thereby creating a memorable, rustic experience that prioritizes cultural immersion (Fig. 2).

Balancing the costs of maintaining renewable energy infrastructure and providing immersive ecotourism experiences with visitor-generated revenue is crucial for financial viability. These revenues will come from visitor fees, accommodation charges, and the sale of local products and souvenirs (Zvikonyaukwa et al., 2023).

4.3 Channels

Channels are how businesses can communicate with customers and ensure that the value proposition is widely conveyed to the customer segments being served. Some channels used include the official tourism website as a source of information about the tourism village (Kurniawan et al., 2021). In addition to websites, social media also plays a significant role in the growth of the tourism sector, including Instagram, WhatsApp, Facebook, and other social media platforms. Furthermore, Google Maps can facilitate access to information for new visitors by detecting Laduma tourism village in Pinge Village and providing various supporting photos and visitor reviews.

For the tourist village of Laduma, key channels include partnerships with travel agencies and online platforms that cater to sustainable tourism. These partnerships can access a wider network of eco-conscious travellers and enable the effective promotion of the village's renewable energy-focused offerings (Darmana & Koerniawan, 2019). Direct marketing and promotion: The village can promote its focus on renewable energy through various channels, including social media, its website, and local tourism events (Sari et al., 2019). This approach can attract individuals interested in renewable energy technologies, such as students, researchers, and industry professionals, while informing them about the distinct educational and experiential opportunities available. Referrals and word-of-mouth: Satisfied visitors have the potential to become brand ambassadors, sharing their experiences and encouraging others to visit the tourist village (Kurniawan et al., 2021). Presence at local and regional tourism fairs and events: By participating in tourism-focused events, the village can raise awareness about its exceptional offerings and draw in new visitors (Kurniawan et al., 2021).

4.4 Customer Relationship

Customer relationship refers to how businesses build relationships with customers who have already used their product (Maggon & Chaudhry, 2015). It is crucial for businesses to effectively manage these relationships and interactions in order to foster loyalty (Sitinjak et al., 2020). The tourism village of Laduma cultivates relationships with tourists through a community of eco-friendly village enthusiasts and partnerships with local stakeholders. For the tourist village of Laduma, establishing strong customer relationships is crucial for creating a loyal customer base and ensuring repeat visits. Some key elements of the customer relationship strategy for the tourist village could include: (1) Personalized service and attention: Providing a high level of personalized service and attention to each visitor can create a memorable and enjoyable experience, leading to positive word-of-mouth and repeat visits (Chi & Qu, 2008; Chu & Shou, 2018). (2) Educational and immersive experiences: Offering in-depth educational programs and hands-on, interactive experiences related to renewable energy technologies can foster a deeper connection between visitors and the tourist village.

By providing opportunities for visitors to engage with and learn about sustainable energy solutions actively, the village can encourage them to return and share their newfound knowledge and enthusiasm with others in their networks so these opportunities can help attract a wider audience of eco-conscious travellers, students, researchers, and industry professionals who are interested in renewable energy and seek out unique educational and experiential opportunities (Dewi et al., 2018). (3) Feedback and engagement: Actively seeking feedback from visitors and engaging with them through various channels, such as social media, can help the tourist village continuously improve its offerings and strengthen its relationship with customers. By regularly soliciting input and ideas from visitors, the village can gain valuable insights to enhance the overall experience and better cater to the needs and preferences of its target audience. Additionally, ongoing engagement through platforms like social media can foster a sense of community and encourage visitors to share their experiences, fostering deeper connections and potentially driving repeat visits and referrals (Megawati et al., 2020). (4) Loyalty programs: Implementing a loyalty program that rewards repeat visits, referrals, and engagement with the village's sustainability initiatives can incentivize customers to maintain a long-term relationship with the tourist village. This Loyalty program could offer discounts, exclusive experiences, or special recognition for customers who consistently support the village's renewable energy focus through visits, word-of-mouth promotion, and active participation in educational programs or community events (Megawati et al., 2020).

4.5 Revenue Streams

A revenue stream refers to the income or revenue generated by a business (Rust & Chung, 2006). For the tourist village of Laduma, the primary source of revenue comes from the sale of package tours. Additional revenue is generated from selling food and beverages, camping sites, tracking and environmentally related products. Offering tours around the renewable energy-based village area provides an additional income for the business. A revenue stream refers to the income or revenue that a business generates. There are several potential revenue streams for the tourist village of Laduma. Accommodation fees charged for overnight stays in the village's eco-friendly lodging options, such as traditional-style houses or tents, can generate a significant portion of the revenue (Anand et al., 2012). Other potential revenue streams for the tourist village of Laduma may include fees from educational and hands-on renewable energy workshops and programs, revenue from the sale of locally produced sustainable products, and income generated from hosting events or conferences focused on renewable energy and sustainable tourism (Duarte et al., 2023). Additionally, the village could explore revenue opportunities by offering guided tours, renting out facilities for special events, or providing transportation services to visitors. Workshop and program fees that charge for participation in educational workshops, hands-on demonstrations, and other renewable energy-focused programs can generate revenue while also providing valuable learning experiences for visitors.

Sustainable product sales can generate revenue from locally produced, sustainable products, such as agricultural goods or renewable energy-related souvenirs, contributing to the village's income and supporting local artisans and producers (Anggarani Pribudi, 2020). Event hosting can host conferences, workshops, or other events focused on renewable energy, sustainable tourism, or community development, attracting a specialized audience and providing a revenue stream for the village (Sutoni et al., 2021). Transportation services can offer transportation options, such as shuttles or guided tours, generating additional revenue while enhancing the overall visitor experience (Sarmiadi et al., 2021).

4.6 Key Resources

Key resources are the essential assets a company requires to operate its business, including human resources, technology, finance, and physical resources (Galbreath, 2002). These resources are crucial for ensuring the effective functioning of the business model. The necessary human resources for the tourism village of Laduma include tour guides, employees, and managers. It is also important to preserve natural resources by maintaining the pristine environment of the village. Technology is another important resource that needs to be acquired and managed by business operators, as it is vital for enhancing visitor arrivals through online promotional activities. For the success of Laduma tourist village, several key resources will be essential. For the tourist village of Laduma, several key resources will be essential to its success:

1. Renewable energy infrastructure: Ensuring a reliable and efficient renewable energy system, such as solar panels, wind turbines, or a microgrid, will be a critical resource to power the village. This renewable energy infrastructure will also demonstrate its sustainable energy focus to visitors (Darmana & Koerniawan, 2019).
2. Skilled personnel: Hiring and training a team of knowledgeable and passionate individuals, including renewable energy experts, tour guides, hospitality staff, and local artisans, will be crucial for delivering a high-quality experience to visitors. It will also help to manage the village's operations effectively (Ningrum et al., 2018).
3. Natural assets: Leveraging the village's natural assets, such as scenic landscapes, unique flora and fauna, or cultural heritage, will be essential for attracting visitors. This natural asset will provide them with authentic and immersive experiences.
4. Financial resources: Securing funding, whether through government grants, private investments, or community-based initiatives, will be necessary to invest in the village's infrastructure, renewable energy systems, and ongoing operations (Sari et al., 2019).

4.7 Key Activities

Key activities are essential actions that business operators must undertake and master to define their business model (Sari et al., 2019). The operators of the tourism village of Laduma exemplify these activities by focusing on providing a comfortable tourist destination while preserving the natural environment. These key activities involve strategically installing aesthetically pleasing solar panels in optimal locations that do not obstruct tourists' views without compromising their functionality. Developing tourism packages highlighting the eco-friendly energy aspects of renewable energy is also crucial. Effective management of the tourist village involves training and empowering local community members to serve as managers and service providers. Furthermore, creating environmentally friendly facilities and amenities is important in developing and supporting tourism infrastructure.

For the tourist village of Laduma, key activities may include managing the renewable energy system, curating visitor experiences, engaging with the community, marketing and promoting the village's unique offerings and sustainable focus to attract targeted visitors, building brand awareness, and encouraging repeat visits and referrals (Afala & Rahayu, 2023; Asyuri & Setiawan, 2022; Dewi et al., 2018; Sari et al., 2019).

4.8 Key Partner

Key partners are external entities collaborating with the business operator to realize the value proposition, enabling the establishment of an effective and profitable business model (Harianto & Soenardi, 2022). The first key partner is the government, which can assist in maintaining and promoting tourism in the village of Laduma. The second key partner is the community. With their involvement, the tourism village of Laduma, based on renewable energy, can become a destination for tourists seeking leisure and educational experiences. Another important partnership is with travel agencies, incorporating the tourism village of Laduma into their holiday packages. Finally, collaboration with educational institutions is essential to facilitate renewable energy education programs for students and university scholars interested in educational activities related to renewable energy implementation.

Potential key partners include renewable energy providers in the tourist village of Laduma. Collaborating with reputable renewable energy companies or institutions to design, install, and maintain the village's renewable energy infrastructure will be crucial (Chen & Zhang, 2018). Local government and tourism authorities: Partnering with local government agencies and tourism boards can provide access to resources, funding, and promotional channels, as well as ensure alignment with regional development plans and sustainability goals (Campbell et al., 2011). Educational and research institutions: Forming partnerships with universities, colleges, or research centres can facilitate knowledge sharing, access to expertise, and opportunities for collaborative projects or research initiatives related to renewable energy and sustainable tourism. Community organizations and cooperatives: Engaging with local community groups, cooperatives, or non-profit organizations can help foster community involvement, leverage local knowledge and resources, and ensure the equitable distribution of benefits (Liu et al., 2023). Value Proposition: The tourist village of Laduma will offer visitors a unique, immersive, and sustainable tourism experience by showcasing its commitment to renewable energy and providing an opportunity to engage with the local community and its traditional way of life (Murtini et al., 2021).

4.9 Cost Structure

Cost structure refers to businesses' total expenses to operate their business models (Harianto & Soenardi, 2022). The expenditure costs incurred to operate Laduma village tourism businesses include investment and infrastructure, operational and maintenance, and marketing and promotion costs. Some key cost considerations for the tourist village of Laduma may include:

1. Initial infrastructure investment: The initial infrastructure investment for the tourist village of Laduma will include the significant costs associated with the installation and set-up of the village's renewable energy systems, such as solar panels, wind turbines, or a microgrid. Additionally, the village will need to invest in developing accommodation facilities, including guest rooms, common areas, and supporting amenities, as well as the construction or renovation of other necessary infrastructure, such as transportation, utilities, and waste management systems. These initial capital expenditures will be crucial to establish the village's sustainable energy focus and provide visitors with a comfortable and immersive experience (Chen & Zhang, 2018). Ongoing operational expenses: The ongoing operational expenses for the tourist village of Laduma will include recurring costs for the maintenance, repairs, and upgrades of the renewable energy system, such as regular servicing, replacement of worn parts, and investments in technology improvements. Additionally, the village will incur expenses related to personnel, including staff salaries, training, and benefits, as well as utilities like electricity, water, and waste management, and other general operational costs necessary for the smooth running of the village and its facilities (Halim & Wahyuni, 2022).
2. Marketing and promotion: Costs for developing and implementing effective marketing strategies to attract visitors, build brand awareness, and maintain a strong online presence. These include expenses for creating and managing the village's website, social media channels, and other digital platforms to showcase the unique renewable energy features and cultural experiences. Additionally, costs may be incurred for advertising campaigns, promotional materials, participation in tourism trade shows, and collaborations with influencers or travel bloggers to reach and engage the target audience (Sutoni et al., 2021).
3. Community engagement and capacity building: Expenses related to training, empowering, and supporting the local community to actively participate in the management and operations of the tourist village. These include costs for community workshops, skill-building programs, and initiatives to involve residents in decision-making processes, as well as resources dedicated to enhancing the

community's capacity to contribute to the village's sustainable development and ongoing operation (Suprapto et al., 2019).

In the business model of Laduma Village, each element plays a crucial role and is interconnected to achieve the overarching goal of sustainable tourism. The initial infrastructure investment in renewable energy systems, such as solar panels and wind turbines, forms the foundation that supports the village's primary value proposition—offering an eco-friendly tourism experience and educating visitors on renewable energy. The operational costs for maintaining and enhancing renewable energy technologies ensure a continuous energy supply for the village's operations and provide comfort for visitors, reinforcing Laduma's commitment to sustainability. Additionally, marketing and promotional expenses enable effective customer engagement through various digital channels, social media platforms, and tourism trade shows, expanding the village's reach to environmentally conscious customer segments. Meanwhile, community engagement and capacity-building costs strengthen the village's relationship with the local community, positioning them as an integral part of its operations and enhancing service quality through training and empowerment programs. By integrating these elements, Laduma village establishes a sustainable tourism model that supports environmental preservation and delivers economic and social benefits to the local community.

5 Conclusion

Laduma tourism village in Pinge village, Bali, Indonesia, showcases the potential of sustainable tourism in promoting environmental conservation and supporting the local economy. Through the utilization of the BMC, this research has successfully formulated an innovative and sustainable business model for Laduma tourism village, which integrates various essential components of a business, including key partners, activities, value propositions, customer relationships, customer segments, resources, channels, cost structure, and revenue streams. This integrated approach is critical as it aligns all these elements to achieve the primary goal of sustainable tourism. Collaboration with government entities, local communities, travel agencies, and educational institutions enhances the tourism experience and contributes to the village's long-term sustainability. Government involvement provides regulatory support and resources, while local community engagement ensures that tourism activities benefit residents and foster a sense of ownership. Partnerships with travel agencies help attract a broader range of tourists, and educational institutions add an academic dimension, supporting tourism and renewable energy education. Together, these elements work harmoniously to create a holistic and impactful business model that promotes environmental sustainability, empowers the local economy, and meets the evolving demands of eco-conscious travellers. The study concludes that adopting renewable energy, specifically solar power, offers tourists a unique and educational experience while preserving the village's natural beauty and cultural heritage.

Despite challenges related to the aesthetics of renewable energy infrastructure, strategic placement and community involvement can address these concerns. The findings emphasize the need for continuous innovation and adaptation in marketing strategies to remain competitive in the tourism industry.

In summary, this research provides valuable insights into the development of sustainable tourism villages and offers a practical framework for village managers who wish to implement similar eco-friendly initiatives. The proposed business model enhances tourist experiences, promotes environmental sustainability, and empowers the local economy in the long run.

References

Afala, L. O. M., & Rahayu, R. K. (2023). What makes village economic development successful? Evidence in two villages, Malang Regency Indonesia. *Journal of Governance, 8*(2). https://doi.org/10.31506/jog.v8i2.18893

Anand, A., Chandan, P., & Singh, R. B. (2012). Homestays at Korzok: Supplementing rural livelihoods and supporting green tourism in the Indian Himalayas. *Mountain Research and Development, 32*(2), 126–136. https://doi.org/10.1659/MRD-JOURNAL-D-11-00109.1

Anastasopoulos, I., Georgopoulos, N., Katsanakis, I., Klada, N., Konstantopoulou, C., Kopanaki, E., Stroumpoulis, A., Tsoupros, G., & Varelas, S. (2024). *Management of sustainable tourism in tourist destination ports: The use of sustainable indicators in the case study of Piraeus port, Greece*. 23–35. https://doi.org/10.2495/ST240031

Anggarani Pribudi, A. P. (2020). Community-based approach to sustain batik Tourism Village area in the special region of Yogyakarta (The case of Giriloyo Village). *Journal of Social Science, 1*(4), 113–122. https://doi.org/10.46799/jsss.v1i4.47

Ankit Prakash, & Vikas Sharma. (2023). Ecotels – A way to green sustainable tourism. *International Journal of Asian Business and Management, 2*(2), 145–152. https://doi.org/10.55927/ijabm.v2i2.3065

Astawa, I. P. M., & Suamba, I. B. P. (2021). *Interconnection and interrelation model as a sustainable tourism village development strategy*. https://doi.org/10.2991/assehr.k.210424.085

Asyuri, R. K., & Setiawan, E. A. (2022). Optimization and integration of renewable energy sources with regional tourism potentials to improve the welfare of local communities. *IOP Conference Series: Earth and Environmental Science, 1050*(1), 012007. https://doi.org/10.1088/1755-1315/1050/1/012007

B, R., & M, S. D. (2024). Sustainable development of tourism industry: A CASE study with reference to Udupi and Dakshina Kannada District. *Al-Shodhana, 12*(2), 67–76. https://doi.org/10.70644/as.v12.i2.5

Beer, M., Rybár, R., & Kaľavský, M. (2018). Renewable energy sources as an attractive element of industrial tourism. *Current Issues in Tourism, 21*(18), 2147–2159. https://doi.org/10.1080/13683500.2017.1316971

Borah, A. (2024). Community-based ecotourism and conservation of nature. In *Environmental intimacies from India's North East* (pp. 208–223). Routledge. https://doi.org/10.4324/9781003504870-18.

Buckley, R. (2009). *Ecotourism: Principles and practices*.

Campbell, J. M., Mackay, K., & Dranzoa, C. (2011). Enhancing rural livelihoods through tourism education and strategic partnerships: A Uganda case study. *Tourism Analysis, 16*(1), 5–17. https://doi.org/10.3727/108354211X12988225899921

Chen, S., & Zhang, T. (2018). Research on the development and design of rural low-carbon tourism products based on tourist experience taking Xiaozhou Village in Guangzhou as an example. In

Proceedings of the 3rd international conference on judicial, administrative and humanitarian problems of state structures and economic subjects (JAHP 2018). https://doi.org/10.2991/jahp-18.2018.116.

Chi, C. G.-Q., & Qu, H. (2008). Examining the structural relationships of destination image, tourist satisfaction and destination loyalty: An integrated approach. *Tourism Management, 29*(4), 624–636. https://doi.org/10.1016/j.tourman.2007.06.007

Chin, C.-H., Lo, M.-C., Songan, P., & Nair, V. (2014). Rural tourism destination competitiveness: A study on Annah Rais longhouse homestay, Sarawak. *Procedia - Social and Behavioral Sciences, 144*, 35–44. https://doi.org/10.1016/j.sbspro.2014.07.271

Chountalas, P. T., Chatzifoti, N., Alexandropoulou, A., & Georgakellos, D. A. (2024). Analyzing barriers to innovation management implementation in sustainable tourism using DEMATEL method. *World, 5*(4), 1004–1022. https://doi.org/10.3390/world5040051

Chu, L. C., & Shou, K. J. (2018). A study of the impact of business strategy on customers tourism behaviors, motivation, and satisfaction -a case study of hotels at Kenting, Taiwan. *International Journal of Asian Social Science, 8*(8), 486–500. https://doi.org/10.18488/journal.1.2018.88.486.500

Čirjevskis, A. (2019). Designing organizational eco-map to develop a customer value proposition for a "slow tourism" destination. *Administrative Sciences, 9*(3), 57. https://doi.org/10.3390/admsci9030057

Corzo-Arévalo, D. (2020). A guideline proposal to formulate an environmental policy for tourism in the Topocoro reservoir in Colombia. *Revista Interamericana de Ambiente y Turismo, 16*(1), 2–14. https://doi.org/10.4067/s0718-235x2020000100002

Corzo-Arévalo, D., & Cuadra, R. (2020). La integralidad del desarrollo territorial, un proceso entendido desde los conceptos: multiescalaridad, multisectorialidad y multidimensionalidad. Grupo de Investigación LIS (Liderazgo, Innovación y Sociedad), Dirección de Educación e Investigación, Centro Latinoamericano del Propósito.

Daneshwar, D., & Revaty, E. V. (2024). A paradigm shift towards environmental responsibility in sustainable green tourism and hospitality. *International Journal of Social Science and Human Research, 7*(05). https://doi.org/10.47191/ijsshr/v7-i05-33

Darmana, T., & Koerniawan, T. (2019). The development of energy tourism village in Indonesia: A case study in Nusa Penida. *E3S Web of Conferences, 125*, 11001. https://doi.org/10.1051/e3sconf/201912511001

Dewi, N. I. K., Astawa, I. P., Siwantara, I. W., & Mataram, I. G. A. B. (2018). Exploring the potential of cultural villages as a model of community-based tourism. *Journal of Physics: Conference Series, 953*, 012072. https://doi.org/10.1088/1742-6596/953/1/012072

Duarte, M., Dias, Á., Sousa, B., & Pereira, L. (2023). Lifestyle entrepreneurship as a vehicle for leisure and sustainable tourism. *International Journal of Environmental Research and Public Health, 20*(4), 3241. https://doi.org/10.3390/ijerph20043241

El Moslem Badr, M. (2022). Challenges and future of the development of sustainable ecotourism. *International Journal of Modern Agriculture and Environment, 2*(2), 54–72. https://doi.org/10.21608/ijmae.2023.214937.1006

Gajdošík, T. (2020). Smart tourists as a profiling market segment: Implications for DMOs. *Tourism Economics, 26*(6), 1042–1062. https://doi.org/10.1177/1354816619844368

Galbreath, J. (2002). Success in the relationship age: Building quality relationship assets for market value creation. *The TQM Magazine, 14*(1), 8–24. https://doi.org/10.1108/09544780210413219

Gong, G., Zhang, Z., Zhang, X., Mahato, N. K., Liu, L., Su, C., & Yang, H. (2020). Electric power system operation mechanism with energy routers based on QoS index under blockchain architecture. *Energies, 13*(2), 418. https://doi.org/10.3390/en13020418

Gössling, S. (2002). Global environmental consequences of tourism. *Global Environmental Change, 12*(4), 283–302. https://doi.org/10.1016/S0959-3780(02)00044-4

Gössling, S., Balas, M., Mayer, M., & Sun, Y.-Y. (2023). A review of tourism and climate change mitigation: The scales, scopes, stakeholders and strategies of carbon management. *Tourism Management, 95*, 104681. https://doi.org/10.1016/j.tourman.2022.104681

Gupta, N., Patel, H., Afzal, S., Panwar, N., Mittal, R. S., Guttula, S., Jain, A., Nagalapatti, L., Mehta, S., Hans, S., Lohia, P., Aggarwal, A., & Saha, D. (2021). *Data quality toolkit: Automatic assessment of data quality and remediation for machine learning datasets*.

Halim, D., & Wahyuni, A. (2022). Feasibility of rooftop solar PV program for 9 tourism villages towards green village development in Bali. *IOP Conference Series: Earth and Environmental Science, 1027*(1), 012030. https://doi.org/10.1088/1755-1315/1027/1/012030

Harianto, E., & Soenardi, E. (2022). Creating new strategy for business sustainability through business model canvas (BMC) evaluation in paper manufacturing. *International Journal of Economics, Business and Accounting Research (IJEBAR), 6*(2), 978. https://doi.org/10.29040/ijebar.v6i2.3025

Hillary, J. (2020). Factors affecting tourists satisfaction in Candirejo Tourism Village, Indonesia. *Journal of Business on Hospitality and Tourism, 6*(2), 315. https://doi.org/10.22334/jbhost.v6i2.221

Ibnou-Laaroussi, S., Rjoub, H., & Wong, W.-K. (2020). Sustainability of green tourism among international tourists and its influence on the achievement of green environment: Evidence from North Cyprus. *Sustainability, 12*(14), 5698. https://doi.org/10.3390/su12145698

Jayaprakash Lamoria, Mishra, A., Kumar, N., Colaco, P., Bhatnagar, T., Kenchappa, R. M., & Tusha. (2024). The role of sustainable tourism in enhancing cultural heritage preservation: In-depth analysis. *Evolutionary Studies in Imaginative Culture*, 870–879. https://doi.org/10.70082/esiculture.vi.1085

Karimov, N., Kulmetov, M., Safarova, N., Jumaev, K., Fayzullaev, M., Sultanov, S., Gapporov, A., & Yakhshieva, Z. (2024). The ecotourism industry's role in environmental stewardship. *Natural and Engineering Sciences, 9*(2), 293–308. https://doi.org/10.28978/nesciences.1574450

Kurniawan, A., Komara, B. D., Setiawan, H. C. B., & Setiawan, N. B. (2021). *Tourism Village Development Strategy to Increasing Tourist Visits*. https://doi.org/10.2991/assehr.k.211020.050

Liu, Y.-L., Chiang, J.-T., & Ko, P.-F. (2023). The benefits of tourism for rural community development. *Humanities and Social Sciences Communications, 10*(1), 137. https://doi.org/10.1057/s41599-023-01610-4

Lubis, I. H., Batubara, M., & Arif, M. (2023). Development strategy of village tourism based on green economy in Denai Lama Village, deli Serdang. *Asian Journal of Applied Business and Management, 2*(3), 317–326. https://doi.org/10.55927/ajabm.v2i3.5583

Maggon, M., & Chaudhry, H. (2015). Revisiting relationship marketing and customer relationship management in leading tourism and hospitality journals: Research trends from 2001 to 2013. *Journal of Relationship Marketing, 14*(1), 53–77. https://doi.org/10.1080/15332667.2015.1022110

Mandryk, O. M., Arkhypova, L. M., Pobigun, O. V., & Maniuk, O. R. (2016). Renewable energy sources for sustainable tourism in the Carpathian region. *IOP Conference Series: Materials Science and Engineering, 144*, 012007. https://doi.org/10.1088/1757-899X/144/1/012007

Mechinda, P., Serirat, S., Serirat, S., Popichit, N., & Popichit, N. (2012). The participation of community and stakeholders in environmental issue management for green tourism in Thailand. *International Business & Economics Research Journal (IBER), 11*(5), 589. https://doi.org/10.19030/iber.v11i5.6976

Megawati, V., Hananto, H., Benarkah, N., & Juniati, N. (2020). The role of technology in developing local wisdom based tourist destination in Bejijong Village. In *Proceedings of the 17th international symposium on management (INSYMA 2020)*. https://doi.org/10.2991/aebmr.k.200127.089

Murtini, S., Sutedjo, A., Sabitah, S. A., & Putri, R. D. K. (2021). *Quality of environmental resources for Eduwisata Lontar Sewu as a Sustainable Tourism Village*. https://doi.org/10.2991/assehr.k.211223.153

Ningrum, L., Hendradewi, S., & Nurbaeti, N. (2018). Healthy rural tourism with good management in Penglipuran Bali – Indonesia. *IOP Conference Series: Earth and Environmental Science, 145*, 012068. https://doi.org/10.1088/1755-1315/145/1/012068

Nosratabadi, S., Mosavi, A., Shamshirband, S., Kazimieras Zavadskas, E., Rakotonirainy, A., & Chau, K. W. (2019). Sustainable business models: A review. *Sustainability, 11*(6), 1663. https://doi.org/10.3390/su11061663

Nyaupane, G. P., & Poudel, S. (2011). Linkages among biodiversity, livelihood, and tourism. *Annals of Tourism Research, 38*(4), 1344–1366. https://doi.org/10.1016/j.annals.2011.03.006

Oka, I. M. D., Astuti, N. W. W., & Suparta, I. K. (2021). *Undagi Farmers Village*. https://doi.org/10.2991/assehr.k.210424.021

Osterwalder, A., & Pigneur, Y. (2010). *Business model generation: A handbook for visionaries, game changers, and challengers*. OSF Alexander Osterwalder and Yves Pigneur. https://books.google.co.id/books/about/Business_Model_Generation.html?hl=id&id=Bjj8G3ttLWUC&redir_esc=y

Pace, L. A. (2016). How do tourism firms innovate for sustainable energy consumption? A capabilities perspective on the adoption of energy efficiency in tourism accommodation establishments. *Journal of Cleaner Production, 111*, 409–420. https://doi.org/10.1016/j.jclepro.2015.01.095

Popescu, G., Popescu, C. A., Iancu, T., Brad, I., Peţ, E., Adamov, T., & Ciolac, R. (2022). Sustainability through rural tourism in Moieciu area-development analysis and future proposals. *Sustainability, 14*(7), 4221. https://doi.org/10.3390/su14074221

Rathore, U. C., & Singh, S. (2017). Control topologies for 3-phase SEIG used in small hydro power plant feeding isolated domestic load in remote mountainous region. *Indian Journal of Science and Technology, 10*(31), 1–7. https://doi.org/10.17485/ijst/2017/v10i31/113857

Risfandini, A., Yulianto, I., & Wan-Zainal-Shukri, W.-H. (2023). Local community empowerment for sustainable tourism development: A case study of Edelweiss Park Wonokitri Village. *International Journal of Sustainable Development and Planning, 18*(11). https://doi.org/10.18280/ijsdp.181127

Rust, R. T., & Chung, T. S. (2006). Marketing models of service and relationships. *Marketing Science, 25*(6), 560–580. https://doi.org/10.1287/mksc.1050.0139

Samuil, I., & Ionică, A. C. (2021). Business model based on community for a sustainable tourism development. *MATEC Web of Conferences, 342*, 09004. https://doi.org/10.1051/matecconf/202134209004

Sari, N., Springfield, D. R., & Sari, K. E. (2019). The sustainability factors of tourist village (case study: Ngadas Village, Poncokusumo District, Malang Regency). *IOP Conference Series: Earth and Environmental Science, 361*(1), 012018. https://doi.org/10.1088/1755-1315/361/1/012018

Sarkar, R., & Sinha, A. (2015). The village as a social entrepreneur: Balancing conservation and livelihoods. *Tourism Management Perspectives, 16*, 100–106. https://doi.org/10.1016/j.tmp.2015.07.006

Sarmiadi, Suryani, Y., & Sumiarti, E. (2021). *Community participation in Tourism Village development in Tungkal Selatan Village, Pariaman Utara District, Pariaman City*. https://doi.org/10.2991/assehr.k.210424.107

Shezan, S. K. A., Saidur, R., Ullah, K. R., Hossain, A., Chong, W. T., & Julai, S. (2015). Feasibility analysis of a hybrid off-grid wind–DG-battery energy system for the eco-tourism remote areas. *Clean Technologies and Environmental Policy, 17*(8), 2417–2430. https://doi.org/10.1007/s10098-015-0983-0

Simanjuntak, M. (2021). Designing of service dominant logic and business model canvas: Narrative study of village tourism. *Golden Ratio of Marketing and Applied Psychology of Business, 1*(2), 73–80. https://doi.org/10.52970/grmapb.v1i2.60

Sitinjak, M. F., Triagung, H., Sarjono, H., Putra, O. B. W., & Ariani, V. (2020). Implementing of customer-driven approach to develop a web-based e-CRM in rural tourism. In *2020 International Conference on Information Management and Technology (ICIMTech)* (pp. 654–659). https://doi.org/10.1109/ICIMTech50083.2020.9211292

Sotiriadis, M. (2018). Collaborative forms and strategies for business venturing in tourism industries. In *The emerald handbook of entrepreneurship in tourism, travel and hospitality* (pp. 131–150). Emerald Publishing Limited. https://doi.org/10.1108/978-1-78743-529-220181005.

Sotiriadis, M., & Nduna, L. (2014). Market segmentation of nature-based attractions: A framework for experience and activity-oriented segmentation. *Journal of Human Ecology, 46*(1), 63–71. https://doi.org/10.1080/09709274.2014.11906707

Stacchini, A., Guizzardi, A., & Costa, M. (2022). The value of sustainable tourism destinations in the eyes of visitors. *Highlights of Sustainability, 1*(3), 202–223. https://doi.org/10.54175/hsustain1030015

Suprapto, P. A., Sarja, N. L. A. K. Y., Suastika, I. N., & Wongbuangam, B. (2019). Developing rural community economies through village-owned enterprises in the Province of Bali. In *Proceedings of the International Conference on Applied Science and Technology 2019 - Social Sciences Track (ICASTSS 2019)*. https://doi.org/10.2991/icastss-19.2019.85

Sutoni, A., Legiawan, M. K., & Renaldi, A. (2021). *Tourism industry development strategy: A concept and case study of a tourism village*. https://doi.org/10.2991/assehr.k.210304.089

Tehseen, S., Hossain, S. M., Ong, K. Y., & Andrews, E. (2024). Sustainable tourism in a changing climate: Balancing growth and environmental responsibility. In *The need for sustainable tourism in an era of global climate change: Pathway to a greener future* (pp. 69–94). Emerald Publishing Limited. https://doi.org/10.1108/978-1-83608-668-020241012

UNWTO. (2019). International tourism highlights, 2019 edition. *World Tourism Organization (UNWTO)*. https://doi.org/10.18111/9789284421152

Vernay, A.-L., Sohns, M., Schleich, J., & Haggège, M. (2020). Commercializing sustainable technologies by developing attractive value propositions: The case of photovoltaic panels. *Organization & Environment, 33*(2), 220–244. https://doi.org/10.1177/1086026619853797

Weaver-Hightower, M. B. (2008). An ecology metaphor for educational policy analysis: A call to complexity. *Educational Researcher, 37*(3), 153–167.

Weber, F. (2019). *Correction to: Demand for sustainable tourism* (pp. C1–C1). https://doi.org/10.1007/978-3-030-15624-4_23.

Yüksel, I. (2010). Hydropower for sustainable water and energy development. *Renewable and Sustainable Energy Reviews, 14*(1), 462–469. https://doi.org/10.1016/j.rser.2009.07.025

Zvikonyaukwa, J., Musengi, K., & Mudzengi, C. P. (2023). The contributions of wildlife to people's livelihoods and economic development in Sub-Saharan Africa.

Climate Change Resilience in Tourism Destinations: An Exploratory Study of Tan Hoa Weather-Adaptive Tourism Village in Vietnam

Bui Thi Nga and Ho Ngoc Ninh

Abstract This chapter aims to analyze the climate change resilience in a new thriving tourism destination of Tan Hoa weather-adaptive tourism village in Vietnam, evaluate its benefits, and suggest some recommendations for other tourism destinations. The primary information came from interviews with 25 locals, 05 tourists, and 01 local government official. The finding showed that Tan Hoa—a disadvantaged village in the rural and remote areas of Minh Hoa district, Quang Binh Province turned the negative impacts of climate change into tourism development opportunities by developing climate change adaptive tourism products through providing diverse services for tourists. This model helped them survive in extreme weather conditions and created new jobs and relatively stable incomes for locals. It has also helped them understand the significance and practice of better forest protection and conservation, promoting rural and community-based values, products, and lifestyles, and promoting sustainable innovation.

1 Introduction

In Vietnam, tourism is one of the industries heavily affected by climate change. Extreme and severe weather patterns, such as floods, landslides, droughts, hurricanes, etc., directly impact tourism activities and destinations. The tourism industry is directly affected by climate change in many aspects, from damaging tourism resources and destinations to cultural heritages, infrastructure, and tourism environment.

Vietnam has a coastline of 3260 km, with 125 beaches, including many beautiful, recognized beaches in the region and the world, such as Lang Co, Hoi An, Nha Trang, … (MONRE, 2021). These beaches have attracted many tourists recently. In addition, the Red River Delta, Mekong River, bays, lagoons, estuaries, and coastal

B. T. Nga (✉) · H. N. Ninh
Hanoi, Vietnam
e-mail: btnga@vnua.edu.vn; hnninh@vnua.edu.vn

J. Marques et al. (eds.), *Tourism and Climate Change*, Tourism, Hospitality & Event Management, https://doi.org/10.1007/978-3-031-99128-8_12

sandbanks are ecosystems that are exploited and contribute to tourism development. Moreover, Vietnam has many other natural resources, such as mangrove forests in estuary areas of the Mekong Delta, Red River Delta, and coastal and island landscapes (Ha Long Bay, Phu Quoc, Con Dao…). These are all tourist resources and destinations that attract many tourists. Climate change with unusual weather phenomena, including hurricanes, floods, and sea level rise… caused many adverse effects on these tourism resources. It damages, destroys, or reduces the value of these tourism resources and destinations.

Moreover, many tourist destinations are on riverbanks, lakes, seas, islands, mountains, waterfalls, etc. Climate change causes many types of extreme natural disasters that frequently occur, last long, on a large scale, and have sudden changes that have had a profound impact, both directly and indirectly, on all aspects of the tourism industry, activities, and destinations.

Besides, this lowlands ecosystem is home to long-standing residential communities, and the natural habitats of many native species, such as nature reserves and biosphere…, are very attractive for ecological tourism. Lowland areas will be lost when sea levels rise, affecting many aspects of local communities and habitats. Biodiversity degradation is becoming more and more profound; 21% of mammal species, 6.5% of bird species, 19% of reptile species, 24% of amphibian species, 38% of fish species, and 2.5% of vascular plant species are already threatened (MONRE, 2021).

Prolonged storms and floods damage many cultural and historical relics and tourism resources in Vietnam (DMA, 2024). Tourism service facilities are damaged or degraded due to heavy storms, floods, and extreme temperature and humidity conditions. Material and intangible cultural heritages are also affected by climate change. Annually, floods cause damage to architectural relic complexes, ancient towns, temples, and tower systems. Many architectural works are affected by termites and mold under humid conditions. Many historical and cultural relics, especially architectural and archaeological relics, were degraded and damaged due to extreme weather phenomena. Landslides and saltwater intrusion are increasingly complicated and affect a large area. In 2018, natural disasters damaged 65 cultural heritages, 127 cultural and historical relics, and many cultural equipment worth Vietnam dong (VND[1]) 1.9 billion (DMA, 2024).

Not only does climate change affect marine and lowland tourism destinations, but it also directly impacts mountainous tourism destinations. Many famous waterfalls are constantly dry due to weather changes. Forest fires, landslides, coastal erosion, floods, and rising temperatures cause changes in tourism resources, and cultural and historical relics are quickly collapsed, degraded, and damaged. The lifespan and value of these resources will be reduced (DMA, 2024).

[1]VND is Vietnamese currency. 1 USD = 24,288 VND according to exchange rate by State Bank of Vietnam. https://www.sbv.gov.vn/TyGia/faces/TyGia.jspx?_afrLoop=7033564868093755&_afrWindowMode=0&_adf.ctrl-state=mmzmyic1g_4, consulted on 13/11/2024.

Many infrastructures serving tourism activities, such as roads, hotels, resorts, cable cars, outdoor entertainment areas, etc., are destroyed and degraded due to climate change (DMA, 2024).

In addition, tourism activities are relatively stable, travel trips normally predetermined, booked, programmed from sightseeing areas, restaurants, and stay… The impact of the increasing unpredictability of climate leads to stagnation and cancelation of programs, plans, and tours, especially tours to lowland areas, beaches, islands, etc. The greater the volatility, the more significant the negative impacts are because tourism products cannot be stored or moved and are easily damaged. Canceling, suspending, or postponing tourists' travel plans will affect a series of stages and parts of the tourism industry. These caused fluctuations in the number of tourists, unstable revenue, and costs incurred in operating. This, in turn, reduced tourism profits. Climate change damages crops, houses, and technical facilities, stagnating production and business (DMA, 2024), thereby reducing income, increasing difficulties in people's lives, and increasing risks, especially for people in mountainous areas, farmers, and fishermen. The difficulties and risks will reduce their chance of going on tours. In addition, even if they can go on a tour, climate change can damage their tours due to heavy storms, floods, landslides, etc.

Tan Hoa village is a rural and remote area in Minh Hoa district, Quang Binh Province, in the central region of Vietnam. Situated in a large, flat valley surrounded by limestone mountains, the village is a flooding center, and floods occur every year in October and November. In recent years, like many other destinations in Vietnam, Tan Hoa has faced unusual patterns of climate change more often with extreme effects, such as in 2007, 2010, 2011, 2017, 2020, 2021, and 2024. Floods occurred with high water levels up to 12 meters deep in 2010 (Tien, 2021), negatively impacting the local people's social lives.

In order to adapt to climate change, Tan Hoa has implemented many solutions. It has achieved specific results in improving natural and social systems' resilience and adaptive capacity. It was recognized as one of the best tourism villages in 2023 by the United Nations World Tourism Organization (Tan Hoa, 2024).

This chapter aims to analyze the way climate change resilience in Tan Hoa weather adaptive tourism village in Vietnam, evaluate its benefits, and suggest some recommendations for other tourism destinations. After the introduction, the chapter came with the literature reviews and methodology. Next, the results presented the impact of climate change in Vietnam, analyzed the climate change resilience in Tan Hoa weather adaptive tourism village in Vietnam, and evaluated the benefits of climate change resilience in Tan Hoa weather adaptive tourism village in Vietnam. Finally, the chapter provides some conclusions and recommendations for other tourism destinations.

2 Literature Review

Scientists and organizations did much research and recognized that climate change influences the tourism industry worldwide in general and tourism destinations in particular (Abegg & Elsasser, 1996; Amelung et al., 2007; Berrittella et al., 2006; Moen & Fredman, 2007; Scott & McBoyle, 2007; Scott et al., 2012; Simpson et al., 2008; Rixen et al., 2011; Gossling et al., 2012; Dogru et al., 2019; Barbhuiya & Chatterjee, 2020; Adekuajoa et al., 2023).

Climate change can have both short-term and long-term impacts on tourist destinations. For example, in the short term, climate change directly affects housing prices in coastal areas (Hamilton, 2007), while in the long term, it can impact the tourism labor market and local unemployment rates (Mishev & Mochurova, 2010).

While the landscape is a significant driver of tourist arrivals (Saarinen et al., 2012), climate change can create both opportunities and challenges for a destination (Ehmer & Heymann, 2008) because of the changing tourism patterns (Hamilton et al., 2005).

Destinations developed based solely on nature-based activities may be more severely affected by climate change than others (Tervo-Kankare, 2011). Developing countries are more vulnerable to climate change than developed countries. They usually develop a tourism industry based on natural landscapes, which are quickly impacted by climate (Saarinen et al., 2012).

Coghlan and Prideaux (2009) supposed that climate change would directly affect tourist satisfaction and perceptions of destinations. Tourists can adjust their behavior due to climate changes in one destination by using alternative tourism activities, changing areas, and changing times (Dawson & Scott, 2010).

In order to cope with climate change, local people, authorities, and tourism businesses in many tourist destinations have adaptive strategies to minimize risks and seek opportunities for tourism development.

A study by Narain et al. (2011) in the Swiss Alps analyzed tourism stakeholders' attitudes and adaptation strategies and found that the local people and authorities tried to promote technical and marketing strategies for the climate change resilience of winter tourism. Another study by Wyss et al. (2014) examined existing business models to understand how perceptions of climate change influence future adaptation and explore the role of business in implementing climate change resilience at local and regional levels.

Tervo-Kankare (2011) studied a Finnish winter destination through interviews with tourism stakeholders and found that respondents expressed their views supporting climate change adaptation efforts. However, destination climate change-related activities were still in their infancy, reactive, and highly dependent on individual motivation.

Research by Scott et al. (2012) on the vulnerability of Caribbean Community countries revealed that around 29% of resorts would be partially or fully inundated by a one-meter sea level rise; approximately 49% and 60% of those face the risk of beach erosion damage in these sea level rise scenarios.

In a study on climate change concerns and adaptation in coastal provinces of Spain, Bujosa et al. (2015) found that the popularity of these areas was reduced, and they recommended that adaptation solutions should be implemented in affected areas.

Buzinde et al. (2010) argued that if adaptation solutions were not considered, destinations may engage in misleading marketing to attract tourists, leading to dissatisfaction. Therefore, local governments, citizens, and tourism businesses play an essential role in proposing appropriate strategies to adapt to climate change and need to implement effective marketing (Gossling et al., 2012; Scott & McBoyle, 2007).

Some climate change resilience models for tourist destinations have been researched and implemented in Vietnam. A project to build embankments, dredge riverbeds, and create ecological landscapes to protect the ancient town of Hoi An (Quang Nam province) is part of the supporting program for adapting to climate change. Thua Thien Hue province invested over 2.6 billion VND to install lightning protection systems at Hue Royal's intangible cultural heritage (VNAT, 2024).

In some destinations, the locals turned climate change's negative impacts into new tourism development opportunities. Climate change tours are organized. Tourists will be taken to places that have been and are being heavily affected by climate change. Some typical tours are tours to landslides and loss of protective forests in Thanh Phu district; tours to saltwater intrusion in Lach market, Ba Tri and Binh Dai districts, Ben Tre province (VTV, 2017); rain experience tours in Hue, flood rowing tour in Hoi An (VNAT, 2024). Many tourist destinations exploited the unusual and exciting weather phenomena that occurred in recent years, such as freezing, snowfall, and cold, in Sapa, Mau Son, and the Northern provinces—which is a precious tourism resource in tropical regions like Vietnam, to provide unique tours to attract curious, explorative, and adventurous tourists. These tours have attracted many visitors to these areas each time they occur (Mai, 2021).

Climate change adaptation tourism destination models were created, such as integrated farming with ecotourism model, climate-smart village model, and natural adaptive villages (DMA, 2024).

Mountainous, ecological, and sea tourism destinations have created new tours to attract residents from the extreme temperatures and weather regions. During the unusual heat waves in recent years, millions of tourists have come to the beaches in the central region, the cities in the highlands, or the mountainous areas of Da Lat, Sapa, and Ba Na to avoid the sun and relax (DMA, 2024).

3 Research Methodology

Tan Hoa weather adaptive tourism village was chosen as an exploratory study of climate change resilience in tourism destinations in Vietnam because it is flood-prone in a rural and remote area in Minh Hoa district, Quang Binh Province, in the central region of Vietnam, which has often faced unusual patterns of climate change with extreme effects and has transformed climate vulnerabilities into strengths, improved the resilience and adaptive capacity of natural and social systems, developed community-based tourism, ensured environmental conservation, and upgraded innovative infrastructure, and was recognized as one of the best tourism villages in 2023 by the United Nations World Tourism Organization.

Primary data were collected through in-depth interviews to provide qualitative information on the different points of view about climate change resilience in Tan Hoa. The interviewees included 15 locals working in tourism-related jobs, such as homestay owners, tour guides, porters, cooks, and agricultural experience service staff; 10 other locals; 05 tourists; and 01 local government official.

Additionally, informal conversational interviews were exploited to explore a broad field related to climate change resilience and gain insight into problems in Tan Hoa. Moreover, field research notes and observations were used to cross-check the information obtained from other sources, collecting related information about infrastructure, hygienic conditions, and the attitudes of those interviewed.

Secondary information was collected from reports from Tan Hoa commune authorities, research articles, and related reports from the Ministry of Natural Resources and Environment of Vietnam.

After cleaning, the descriptive analysis method was applied to present the study site's primary conditions and the characteristics of the significant actors, using analysis tools in the next section.

4 Results

4.1 Climate Change and Its Impacts in Vietnam

As one of the world's top five most vulnerable countries to climate change (AUSAID, 2022), annually, Vietnam faces many hazardous weather conditions such as typhoons, floods, droughts, landslides, etc. The frequently threatening types of climate change affect various aspects of Vietnam: economy, society, culture, environmental lives, and tourism destinations.

The average temperature in Vietnam increased by 0.89 °C over 60 years from 1958 to 2018 (MONRE, 2021). That means it increased around 0.15 °C per decade during this period, at the low average global level of 0.15–0.2 °C per decade in the recent period (IPCC, 2019). The growth rate in the first 27 years (1958–1985) was very low, averaging 0.056 °C/decade, while the next 33 years (1986–2018) increased by an average of 0.22 °C/decade (MONRE, 2021). There were fluctuations in temperature between regions, following different trends in Vietnam. The highest difference temperature measured at the monitoring station of Vung Tau and Lam Dong was 9–10 °C. The extreme average temperature measured in Vung Tau at 27.9 °C, while the lowest average temperature in Lam Dong was recorded at 18.4 °C (Fig. 1).

Unseasonal rain and unusually heavy rain occur more frequently. The number of heavy storms (above level 12) tends to increase slightly. The storm season ends later, and more storms cause landfall in the southern region (MONRE, 2021). The

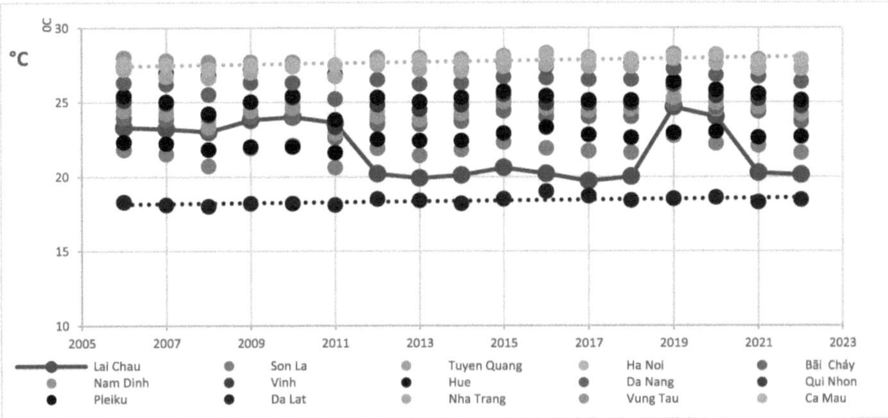

Fig. 1 Average temperature at the monitoring stations in Vietnam (GSO, 2024)

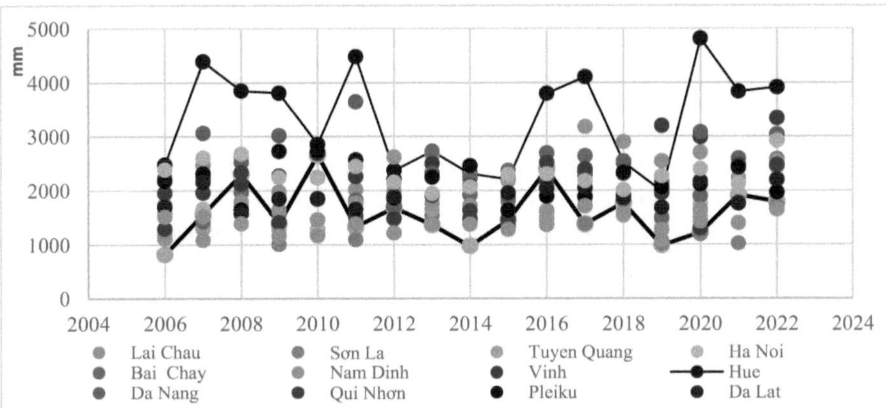

Fig. 2 Rainfall at the monitored stations in Vietnam (GSO, 2024)

rainfall measured at the monitoring stations varies significantly over the years and tends to increase. The most enormous variation was recorded in Hue, with the highest level of 4812.8 millimeters (mm) in 2020, while the lowest was 1984.6 mm in 2019 (Fig. 2).

Heavy storms tend to increase slightly and land more in the Southern region. The activities of hurricanes affecting Vietnam in recent years have been very unusual (MONRE, 2021).

The water levels measured at the monitoring stations on the largest significant rivers of Vietnam showed a decreasing trend. The Mekong River was recorded as the lowest level in Vietnam, with the deepest level of −68 cm in 2007, 2011, and 2012 (GSO, 2024). The water level of the largest rivers in Vietnam tends to decrease due to the construction of lakes and hydropower dams. Meanwhile, the sea level tends to increase slightly over the years. On average, water levels at coastal oceanographic stations in Vietnam tend to increase by about 2.7 mm/year (Fig. 3).

Climate change is causing many negative impacts on this region, such as droughts, landslides, erosion, storms, floods, etc. These negative influences affect every aspect of the social life of local people, in which households and ethnic minorities who depend mainly on agriculture are the most vulnerable people.

The statistics on the economic losses caused by natural disasters due to climate change in Vietnam in recent years showed a clear increasing trend. The losses greatly and negatively affected the development of the country's economic sectors in general and the tourism industry in particular.

Climate change also causes many social losses in Vietnam, including human death, missing, and injuries; impacts on education and health; transportation damage; and affects other aspects of life. On average, the number of deaths due to natural disasters is around 200 people per year. The most severe number was recorded in 1996, with 1128 deaths. Besides that, an average of 30 people went missing each year, and nearly 400 people were injured by natural disasters. In 2017, natural disasters caused 235 deaths, 61 missing people, and 664 injuries (Fig. 4).

Natural disasters also negatively affect and cause significant damage to culture, communication systems, transportation systems, and other fields. For instance, in 2018, natural disasters destroyed more than 1101 km of embankments, quai, canal, and irrigation systems; 718 km of roads in the transportation system; 4 communication stations collapsed and broke 390 antennae and 1639 electricity poles, broke off 31,970 m of electric wire (Fig. 5).

In addition, there are also infectious diseases caused by contaminated drinking water and mental health problems due to psychological trauma, anxiety, and stress.

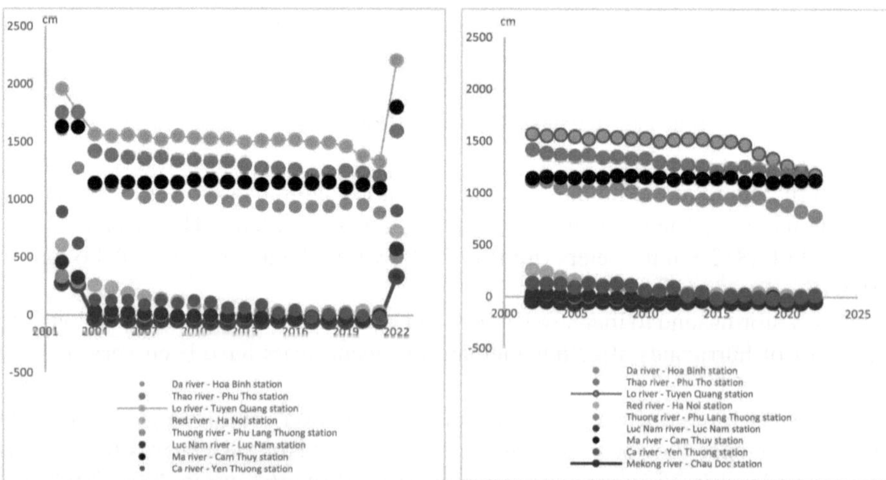

The deepest (mm) The most shallow (mm)

Fig. 3 The water level of major rivers (GSO, 2024)

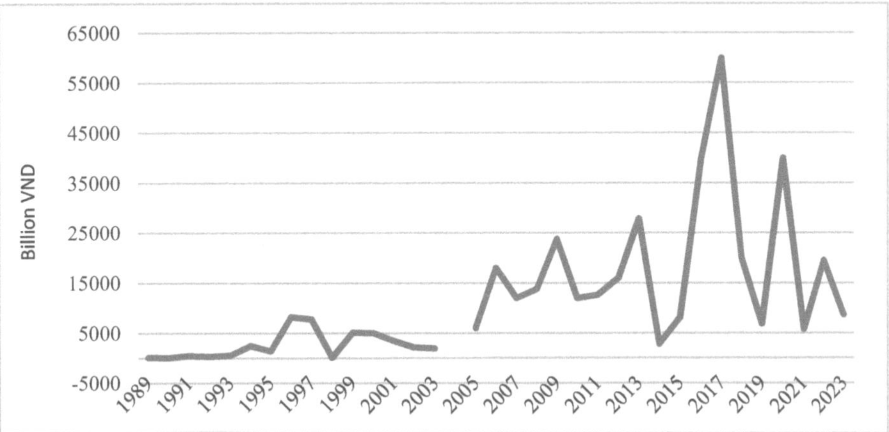

Fig. 4 Economic losses caused by disasters in Vietnam (DMA, 2024)

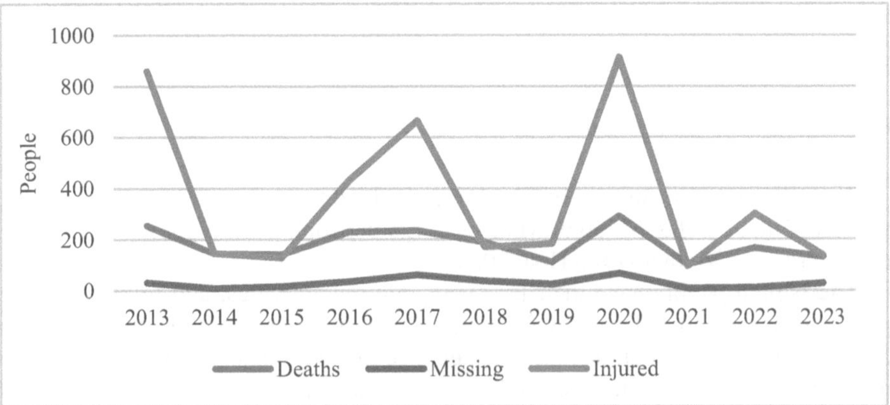

Fig. 5 Humans affected by disasters in Vietnam (DMA, 2024)

Extreme weather in 2012, 2018, 2020–2021 destroyed 204,485 hectares of forest. From 2011 to 2016, the area of mangrove forests in the Mekong Delta decreased by nearly 10%, from 194.7 thousand hectares in 2011 to 179.3 thousand hectares in 2016 (a decrease of about 15.3 thousand hectares). The Mekong Delta region has 24 frequently eroded areas with a total length of about 147 km, erosion speed from 5–45 m/year, and an average loss of about 500 hectares of land/year (DMA, 2024).

4.2 *Climate Change in Tan Hoa*

Tan Hoa village is in a rural, remote, and deprived region in central Vietnam, Minh Hoa district, Quang Binh Province (Fig. 6).

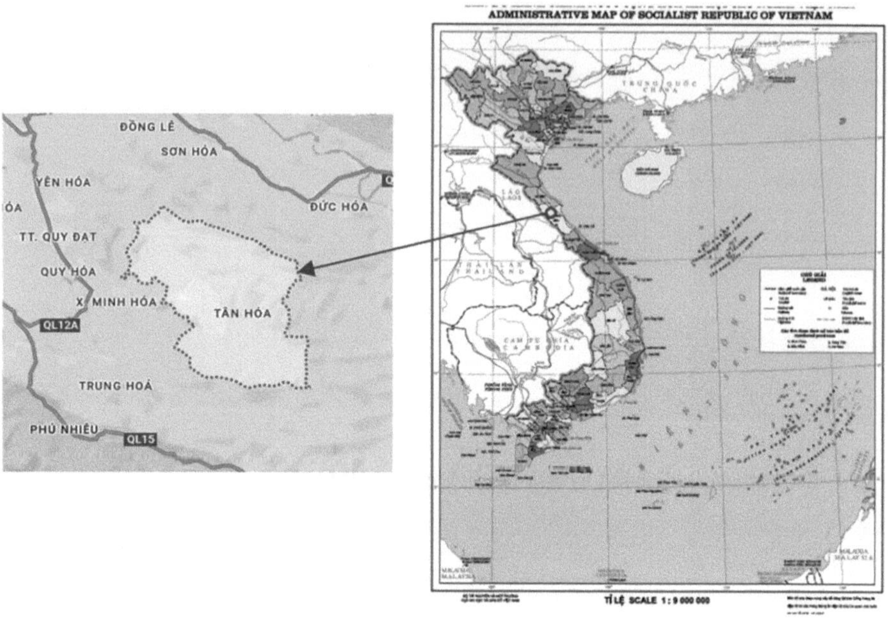

Fig. 6 Map of Tan Hoa village (Google Earth, 2024; CSMD, 2024)

The village covers an area of 72.4 km^2, of which more than 2/3 is mountainous, with a population of 3355, all of whom are Nguon people, who belong to the Muong ethnic group of Vietnam (Tan Hoa, 2024).

The Nguon people have their language, cultural activities, and folklore. In the past, the locals made a living by farming, fishing, hunting, and collecting honey. In addition to these occupations, they also engage in natural harvesting activities such as catching snails, picking "cà Lào" (Laotian fruit), picking wild vegetables, and digging local potato-like to supply themselves daily.

Tan Hoa residents mainly settle along the sides of the Rao Nan River. This river originates on the Vietnam-Laos border, runs underground for almost three kilometers in Ruc Cave, flows along the limestone mountain to Rao Nam River, and flows the Gianh River. Due to its geographical location and specific terrain structure, Tan Hoa is in the west of the Hoang Son range and is dominated by limestone mountains, so it has a subtropical climate. It is hot, humid, rainy in the summer, and cold in the winter. The rainy season usually comes from July, ending in November of the lunar calendar. For many years, Tan Hoa also suffered from prolonged cold weather, causing significant damage to the locals' livestock. The dry season usually occurs from March to June of the lunar calendar. In these months, the sun is extremely hot; the average number of sunshine hours per day is 6.5 h, and the highest day reaches 9.9 h. This season also has southwest winds that blow from Laos and are influenced by the Phoenix effect (also known as the Lao wind or southwest Phoenix wind). Tan Hoa is known for its natural beauty, cultural richness, and the Tu Lan Cave System, which was chosen as the filming location for several domestic and international movies (Tan Hoa, 2024).

Surrounded by limestone mountains on all four sides, Tan Hoa is a large and flat valley. Every time it rains heavily, water flows from all directions to this village, often causing floods here, and Tan Hoa becomes a flood-prone area. Every year, Tan Hoa village often faces 3–4 floods, generally from October to November.

> Because this is a flood-prone area, after a few days of heavy rain, this area floods. In less rainy years, it floods two or three times per year; in more rainy years, it floods more often, sometimes up to 5–6 times. On average, there are 3–4 floods each year. Floods usually occur from the end of June to September of the Lunar calendar (around August to November in the solar calendar). There were times when two floods occurred within ten days, and we could not react (In-depth interview with a female resident).

Limestone mountains in Tan Hoa have many caves, such as Chuot (Rat) Cave, Bat Cave, Tung Cave, and Tu Lan Cave… These are excellent sources of benefits for ecotourism and cave tourism. However, with many caves, the water drains slowly when floods occur, causing difficulties in economic development, production, and people's daily lives. Due to climate change's impact, floods have become heavy and unusual.

> Floods have become more unusual in recent years, especially in the last 15 to 20 years. The number of deep floods has increased, and there have been many times the entire village was submerged. The years with the most floods were 2007, 2010, 2011, 2020, 2021, and 2024. In the past, the flood water rose rather slowly, but now, the flood water often rises very quickly, and people only have time to escape to the mountains. Buffaloes can swim, which is still okay, but countless pigs, cows, and chickens have drowned. Sometimes, houses were swept away by floodwaters and hung on treetops. (In-depth interview with a resident, male).

In 2007, the village experienced a recorded flood at 9.57 m high. In 2010, Tan Hoa witnessed a historic flood with high water levels up to 12 m, submerging 90% of houses in the village (Tien, 2021). In October 2011, the flood with the highest water level was 4 m, and many houses were submerged (Ha et al., 2011). The floods typically drowned out everything the locals owned: their houses, fields, crops, animals, etc. In October 2020, the flood was up to 4 m in depth and, in some places, up to 10 m, causing more than 600 houses in Tan Hoa commune to be submerged in water. During the flood in September 2024 (Fig. 7), nearly 400 houses (56%) in Tan Hoa were flooded, from 0.5 to 2 m deep (Hung et al., 2024).

> In the past, whenever we recognized signs of flooding, we rushed to climb the mountain, stayed in a cave, and waited for the water to recede before coming home. There were times when the water rose to the roof of the house, and it took some days, sometimes even a week, for the water to recede. Many times, the water rose so quickly that we only had time to move people, not enough time to bring food, and everyone was isolated. We had to go to the crevices around the cave to find suitable leaves to eat to stave off hunger. (In-depth interview with a male resident).

> People here used to be very poor, often not having enough cereals to live for the whole year because they could only grow crops for one season; the rest were short-term crops like corn and cassava. After the flood, the mud killed all the grass, and there was nothing to feed buffaloes and cows; local people sometimes had to go to Laos (25–30 km away) to cut grass for ruminant animals. Some people could not survive, so they had to leave the village to make a living elsewhere. (In-depth interview with a resident, female).

Fig. 7 Flood in Tan Hoa on 20/9/2024

4.3 Climate Change Resilience in Tan Hoa

Living in flooded conditions, local people have devised ways to adapt to climate change. "In the past, every time it rained heavily, the village would call each other to climb the mountain. If the flood water rose quickly, we would only have time to move people. If the flood water rose slowly, we would take animals with us." (In-depth interview with a male resident).

> If it rains a lot when the water rises, the elderly, women, and children quickly go up the mountain to avoid the flood first. Strong men will cut banana tree trunks to make rafts to put food on for the living. (In-depth interview with a female resident).

For ruminants, Tan Hoa people have a different way. They built barns in a concentrated area for ruminants (cows, buffalo) in higher areas, generally at the foot of the mountain. "We do not build a barn behind our house like in other regions in Vietnam, but the whole village sets aside a large area of land near the mountain, and each family builds a barn for the buffaloes and cows to gather in. They can bring the buffaloes and cows up the mountain faster when it floods." (In-depth interview with a female resident).

Some people have different ways of doing things. "Some families tie one end of the rope to the buffalo's nose, and the other is hung on a high tree. Wherever the water rises, the buffalo will swim up there. If the rope is tied so that the buffalo's nose is always facing the sky, it can still breathe, it will not die or be swept away." (In-depth interview with a female resident).

> We have learned how to prepare for floods. Food (rice, potatoes, corn, cassava, etc.) is stored in a warehouse, with an empty plastic barrel attached underneath and tied tightly to a high pole so that wherever the water rises, the warehouse will float along the pole. (In-depth interview with a male resident).

After the historic flood in October 2010, many proposals were made to help people reduce the impact of flooding. Two notable options were: (1) explosives to expand the Chuot Cave to help water drain faster when there was a flood or (2) relocating the village to another place. However, the option of relocating the villagers was challenging to implement because the villagers had lived there for hundreds of years and did not want to move elsewhere, while the solution of breaking the mountain to create a flood drainage channel required a considerable amount of funding and could affect the natural ecosystem. Therefore, neither the provincial leaders nor the local people approved the option. The local government built two large communal houses on the mountain for people to stay in case of a flood.

After many discussions, Tan Hoa was oriented to develop following the model of a climate change resilience tourist village that adapts to the weather, providing diverse experiences for tourism from homestay, experience farming, dining at local people's houses, souvenir stalls, and other diverse services for tourists.

In 2011, based on their flooding food warehouse, local people in the village took the initiative to make a raft with some sealed empty barrels attached underneath and tied tightly to two high poles to allow the raft to move up and down the poles wherever the water rises. After that, they improved the raft to small floating house models next to their traditional house to live in case of floods in 2012. When the water rises, the house will float, and people and animals can live. In the dry seasons, a floating house stores their food and equipment. The locals also built simple flooding rafts for their animals, such as chickens or pigs.

> After the historic flood in October 2010, we came up with the idea of making floating houses based on the foundation of food warehouses. The houses are made of light materials placed on about 20-30 sealed empty barrels. These barrels act as buoys so that when the water rises, the houses rise as well. To prevent the houses from drifting away with the water, we place two poles 6 to 9 meters high and tied iron straps to keep the floating houses from moving up and down along the poles. (In-depth interview with a male resident).

Now, all Tan Hoa locals have built floating houses next to their ordinary houses to protect themselves and their belongings in case of a harsh flood (Fig. 8). They use Inox for the frame of the houses, heat insulation tôle for covers and roofs, and empty barrels attached underneath to allow the house to float.

> Currently, all family in the village has a floating house. When the rainy season comes, we store food and clean water in the floating houses that are enough for 7 to 10 days. Whenever it rains heavily and there is a flood, we move to the floating houses, so we do not have to run into the mountains like before. Now, Tan Hoa residents no longer feel panic when floods come, but instead, they feel that floods are normal and even happily welcome them. (In-depth interview with a female resident).

In 2011, the Quang Binh provincial government granted a license to Oxalis Holiday Company Limited to survey and pilot adventure tours of the Tu Lan cave system. By 2014, the Tu Lan Discovery Tour officially operated with nine tours at different levels.

> In 2014, the Quang Binh provincial government coordinated with a tourism company to build a sustainable development strategy for Tan Hoa to encourage local people to participate in tourism activities to enhance them to take control of tourism services gradually;

A floating house

An ordinary house together with a floating house

Fig. 8 Houses in Tan Hoa

> develop tourism products to explore the Tu Lan cave system; create off-road motorbike tours to explore Lim forest and other products to diversify services to serve the tourists gradually. (In-depth interview with a local government leader, male).

Soon after that, this model attracted tourists to experience the flood season. Recognizing the chance to develop, the village has provided diverse services for tourists.

> This area has several caves in the mountains where we used to take shelter during floods, such as Tu Lan cave, Tien cave, Chuot cave… In the past, people only knew the cave gate, no one dared to go deep into it. The caves are quite beautiful, now they have become attractive caves for tourists to experience and explore. (In-depth interview with a local tour guide, male).

Now, nine tours are developed within this region, of which four main tourism products and services exist. The first is a tourism program to explore more than 20 limestone caves, including the gentle Rao Nan River, with different tours ranging from 1 day to 4 days.

The second product and service are a tour to experience adventure and discover the palm forest, the rare and preserved ironwood forest, and the Myrtle Flowers forest.

The third product is a tour to experience the culture and lifestyle of a local village and to have dinner at a local home with authentic Vietnamese cuisine prepared by the host family of a local ethnic group with the unique culture of the Nguon people, and learn more about local history, traditions, customs, crafts and join them in their daily activities, such as farming, fishing, cooking, and weaving. Tourists can experience the unique lodging model that adapts to harsh weather conditions in Nguon's rural homestay, equipped with modern facilities and suitable for 1–3 people.

The fourth tourism product and service are a cycling tour around the peaceful village to explore the natural beauty of the rice fields, the palm forest, the ironwood forest, and the myrtle flowers forest or to visit the caves, the tunnel home of Kong, and the waterfalls.

> Currently, we have four major tourism products including exploring more limestone caves, experiencing adventure to discover the palm forest, experience the culture and lifestyle of a local village, and cycling tour around the peaceful village to explore the natural beauty. (In-depth interview with a local tour guide, female).

As a result, within 10 years, from 2013 to 2023, the total number of visitors reached 63,115 (QBPEIP, 2023). Because this is an isolated place, even in the peak of the COVID-19 epidemic in 2021, 3508 tourists were coming to experience Tan Hoa. In 2022 and 2023, the number of visitors increased rapidly to 9304 and 9437, respectively (Tho et al., 2023). Preliminary statistics for the first 9 months of 2024, more than 11,000 tourists were coming to Tan Hoa (Nam, 2024).

In November 2022, the local authorities of the village, in cooperation with the Department of Tourism of Quang Binh Province, developed a project to turn Tan Hoa into a community tourism village operated by local people themselves. This includes homestays, dining with local dishes, and tours to visit the cultural beauty of the Nguon community. The project aims to expand community involvement in tourism, where locals own homestays, provide dining and farming experiences, and live at floating houses and souvenir stalls.

The village has cooperated with Oxalis Adventure to create high-quality tourism products and services that benefit visitors and the locals. They exploit the natural beauty, cultural richness, Tu Lan Cave System, and Hang Tien Cave... to make Tan Hoa Village the ultimate destination for adventure seekers and nature lovers (Tan Hoa, 2024).

Recently, the village has nearly 700 floating houses with a new design and decoration that can accommodate a family/tourist group of at least six people and their necessities to survive in floods. This has made Tan Hoa a climate change adaptive tourism village. Tan Hoa has become a popular destination for both domestic and foreign tourists. On October 19th, 2023, the United Nations World Tourism Organization recognized it as one of the best tourism villages (Fig. 9).

4.4 Benefits of Climate Change Resilient Tourism in Tan Hoa

The change from the negative impact of vulnerability to seasonal flooding into a tourism development opportunity provides economic, cultural, social, and environmental benefits to tourists and locals.

Not only can the tourists experience the living of the floating village and interact with the local people, but they also support their livelihoods. This model has helped locals survive in extreme weather conditions and created new jobs and stable income for many people in the village, such as tour guides, porters, cooks, drivers, and homestay owners.

Fig. 9 Welcome gate of Tan Hoa

In addition, tourists can enjoy unique dishes made from honey, green beans, black beans, peanuts, wild bamboo shoots, blood grass, and Gastrodia. These local specialties are a must-try for anyone visiting Tan Hoa. Tourists can also purchase these delicacies as gifts for their friends and family. This is an excellent way to support the local community and share the delectable flavors of Tan Hoa with others.

Tan Hoa village encourage local engagement. The people actively participate in tourism planning and execution. They provide products and services to serve tourists and share cultural knowledge. This model allows the community to continue hosting tourists even during periods of high water. This contributes to developing the local economy to save residents and reduce urban migration.

> There are hundreds of local people participating in tourism services nowadays: 10 are homestay owners, some serve as porters for tourists, and some help them experience our lives. This gives us regular jobs and stable income, ensuring our lives all year round. (In-depth interview with a local homestay owner, male).

Currently, there are nearly 100 local people engaged in tourism activities in Tan Hoa village who become chefs, safety supervisors, and tourist service staff on Tu Lan cave exploration tours and off-road motorbike driving tours, with salaries of 6–8 million VND/person/month. In addition, there are ten homestay owners with an average of 15–20 nights of hosting guests each month, bringing in an average income of nearly 7.2 million VND/month, ten households providing home-cooked meals with an average income of more than 12.3 million VND/month. The model also creates indirect jobs for some food suppliers. Tan Hoa people have recently earned more than 9.6 billion VND annually (Nam, 2024).

Since the transformation into a weather-adaptive tourism village, the capacity of the local people has constantly improved. Community involvement is vital to maintaining resilience and enhancing adaptive capacity. Community members are trained in first aid, weather monitoring, and sustainable tourism practices, equipping them with the skills to manage tourism under varying conditions. Locals also receive training on sustainable tourism practices, helping to maintain the area's natural resources and cultural heritage.

Many villagers have participated in training courses to become chefs and porters serving tourists on Tu Lan discovery tours, learning everything from cleaning rooms to what to do and what not to do when meeting tourists. The Quang Binh Provincial Department of Tourism also opened training courses on basic tourism skills for employees of tourism businesses and local people, focusing on basic English communication skills and product development, marketing, and operation of tourist destinations. "They teach detail, with proper testing, only those who pass can work as homestays." (In-depth interview with a local homestay owner, male). "I was thoroughly taught about food hygiene and safety, and how to renovate the kitchen to meet standards." (In-depth interview with a local chef, female).

Some families in the village have sent their children to study tourism-related training programs in some big cities. Every year, in October, teachers from big cities also come to the village to teach local people. This will create a high-value labor force to boost the village's development.

The Tan Hoa model's development also supports preserving indigenous cultural values. People are more aware of preserving local products and developing traditional techniques and their cultural values, beliefs, and lifestyles to promote sustainable tourism development. They provide traditional dishes for the tourists, but do not forget to preserve biodiversity in the area. "With snail dishes, Tan Hoa people only catch male snails, rarely catching female snails to eat to help them continue to reproduce." (In-depth interview with a chef, male).

The climate change resilience model has also helped local people understand the significance and practice of better forest protection, conservation, promotion of rural and community-based values, products, and lifestyles, and promoting sustainable innovation. It directly benefits the community, creating incentives for conservation and disaster preparedness. "The model not only changes our socio-economic situation and likelihood but also changes our awareness. From exploiting forests and living off forests, we now have the awareness to protect forests, considering forest development the most sustainable livelihood." (In-depth interview with a porter, male). "In this village, even during the flooding, homestay guests are not too worried about environmental sanitation like in many other places because buffaloes and cows are raised in the mountains." (In-depth interview with a male tourist).

5 Conclusion and Recommendations

This chapter aims to analyze climate change resilience in Tan Hoa weather adaptive tourism village in Vietnam, evaluate its benefits, and suggest recommendations for other tourism destinations. The primary information came from in-depth interviews with 15 locals currently working in tourism-related jobs, such as homestay owners, tour guides, porters, cooks, and agricultural experience service staff; 10 other local people; 05 tourists; and 01 local government official. The finding showed that Vietnam is among the countries most affected by climate change, facing rising sea levels, intense rainfall, and increased flooding. For tourism destinations, these impacts translate to disrupted operations, economic losses, and reduced tourist interest during extreme weather seasons.

Tan Hoa—a village in a rural and remote area in Minh Hoa district, Quang Binh Province in the central region of Vietnam, is in a large, flat valley surrounded by limestone mountains and experiences floods yearly in October and November. In recent years, Tan Hoa faced unusual patterns of climate change more often with extreme effects and posed negative impacts on the social lives of the local people. In order to adapt to climate change, Tan Hoa orients to develop the model of a climate change resilience tourist village that adapts to the weather, providing diverse experiences for tourism from homestay, experience farming, dining at local people's houses, souvenir stalls, and other diverse services for tourists. Thanks to this adjustment, Tan Hoa has transformed climate vulnerabilities into strengths, improved the resilience and adaptive capacity of natural and social systems, developed community-based tourism, ensured environmental conservation, and upgraded innovative infrastructure, and was recognized as one of the best tourism villages in 2023 by the United Nations World Tourism Organization.

Tan Hoa's weather-adaptive tourism village provides valuable insights into a new approach to building climate resilience in tourism destinations and how climate change continues to affect tourism. From this model, some recommendations for other tourism destinations are provided as follows:

First, local people should learn and explore how to adapt to climate change conditions to survive. Rural and ecotourism can be boosted sustainably by resilience to environmental challenges. Solutions to promote community involvement, sustainable infrastructure, and responsible resource management are necessary to reduce climate risks and enhance sustainable tourism. The construction of floating houses by the Tan Hoa people is a valuable lesson for other tourism destinations affected by climate change. Even in harsh conditions, people can survive with a low-cost floating house made from local and available materials (barrels, tôle).

Second, it is necessary to change production models to be compatible with climate change. Through resilience, tourism destinations can adjust their orientation to develop their traditional lives while opening a new opportunity to continue to attract travelers who value environmental sustainability and community engagement. Tan Hoa people have learned to shift from traditional agricultural production, such as growing crops, raising livestock, and exploiting natural resources from forests to a new model combining agricultural production and tourism development to create sustainable livelihoods.

The third one is to renew thinking and lifestyles from passive to proactive in adapting to climate change. In the past, the locals were passive whenever there was a flood and had to wait for social support. With the new model, they are proactive in living well and can still generate income by attracting tourists to experience living in a floating house during the flood season.

The fourth one is that locals should preserve indigenous culture and exploit it to develop tourism products. In addition to attracting tourists, it is necessary to create unique products from indigenous culture to keep and satisfy tourists' demand, thereby increasing income and, at the same time, having conditions to preserve their indigenous culture better.

Finally, local people had better learn new occupations to adapt to climate change. In the case of Tan Hoa, many new jobs have arisen thanks to this model, such as porters, tour guides, cooks, serving tourists, and the accommodation business.

6 Limitations and Further Studies

Due to the study's limited sample size, more comprehensive and quantitative studies are needed to make more accurate conclusions.

Future studies could investigate how to maintain visitor safety and comfort in extreme weather conditions and ensure sustained resilience and adaptation to changing climate risks.

References

Abegg, B., & Elsasser, H. (1996). Climate, weather, and tourism in the Swiss Alps. *Geographische Rundschau, 48*, 737–742.

Adekuajoa, I. O., Udehb, C. A., Okonkwoc, F. C., Nnabugwud, O. C., & Daraojimbae, C. (2023). Climate resilience in tourism: a synthesis of global strategies and implications for U.S. destinations. *Ecofeminism and Climate Change (EFCC), 4*(2), 128–137. https://doi.org/10.26480/efcc.02.2023.128.137

Amelung, B., Nicholls, S., & Viner, D. (2007). Implications of global climate change for tourism flows and seasonality. *Journal of Travel Research, 45*, 285–296.

AUSAID. (2022). *Vietnam climate change country profile Vietnam Climate Change Country Profile* | U.S. Agency for International Development (usaid.gov) consulted on 29/04/2024.

Barbhuiya, M. R., & Chatterjee, D. (2020). Vulnerability and resilience of the tourism sector in India: Effects of natural disasters and internal conflict. *Tourism Management Perspectives, 33*(1), 100616. https://doi.org/10.1016/j.tmp.2019.100616

Berrittella, M., Bigano, A., Roson, R., & Tol, R. S. (2006). A general equilibrium analysis of climate change impacts on tourism. *Tourism Management, 27*(5), 913–924. https://doi.org/10.1016/j.tourman.2005.05.002

Bujosa, A., Riera, A., & Torres, C. M. (2015). Valuing tourism demand attributes to guide climate change adaptation measures efficiently: The case of the Spanish domestic travel market. *Tourism Management, 47*(4), 233–239. https://doi.org/10.1016/j.tourman.2014.09.023

Buzinde, C. N., Manuel-Navarrete, D., Kerstetter, D., & Redclift, M. (2010). Representations and adaptation to climate change. *Annals of Tourism Research, 37*(3), 581–603. https://doi.org/10.1016/j.annals.2009.10.018

Coghlan, A., & Prideaux, B. (2009). Welcome to the wet tropics: The importance of weather in reef tourism resilience. *Current Issues in Tourism, 12*, 89–104.

CSMD (Center of Survey and Mapping Data, Vietnam Department of survey, mapping and geographic information). (2024). *Administrative Map of the Socialist Republic of Vietnam*. https://bandovn.vn/vi/Document/Download/1

Dawson, J., & Scott, D. (2010). Systems analysis of climate change vulnerability for the US Northeast Ski sector. *Tourism and Hospitality Planning and Development, 7*(3), 219–235. https://doi.org/10.1080/1479053X.2010.502383

DMA (Vietnam Disaster and Dyke Management Authority). (2024). *Announcing data on damage caused by natural disasters nationwide*. http://phongchongthientai.mard.gov.vn/FileUpload/2024-02/qiha4zUpzEO6OYrhThi%e1%bb%87t%20h%e1%ba%a1i%202023%20%c4%91%c4%83ng%20web.pdf, downloaded on 10/7/2024 (Vietnamese).

Dogru, T., Marchio, E. A., Bulut, U., & Suess, C. (2019). Climate change: Vulnerability and resilience of tourism and the entire economy. *Tourism Management, 72*(6), 292–305. https://doi.org/10.1016/j.tourman.2018.12.010

Ehmer, P., & Heymann, E. (2008). *Climate change and tourism: Where will the journey lead?* Deutsche Bank Research, Current Issues, Frankfurt.

Gossling, S., Scott, D., Hall, C. M., Ceron, J.-P., & Dubois, G. (2012). Consumer behaviour and demand response of tourists to climate change. *Annals of Tourism Research, 39*(1), 36–58.

GSO (General Statistics Office of Vietnam). (2024). *Administrative units and climate*. Statistical data – General Statistics Office of Vietnam (gso.gov.vn).

Ha, T., Giang, L., & Linh, N. (2011). *Quang Binh: running to the cave to escape from the flood* (tuoitre.vn), download on 14/07/2024 (Vietnamese).

Hamilton, J. M. (2007). Coastal landscape and the hedonic price of accommodation. *Ecological Economics, 62*(3–4), 594–602. https://doi.org/10.1016/j.ecolecon.2006.08.001

Hamilton, J. M., Maddison, D. J., & Tol, R. S. (2005). Climate change and international tourism: A simulation study. *Global Environmental Change., 15*, 253.

Hung, N., Quang, L, & Nguyen, D. (2024). *Quang Binh: Tanhoa flood center is deeply submerged*. https://quochoitv.vn/cum-tin-quang-binh-ron-lu-tan-hoa-ngap-sau-236756.htm (Vietnamese).

IPCC (Intergovernmental Panel on Climate Change). (2019). *Special report on Ocean and Cryosphere in a Changing Climate (SROCC)*. https://www.ipcc.ch/srocc/

Mai, N. (2021). *Crowds flock to travel on extremely cold days, 'sold out' snow-watching tours in the northern mountains of Vietnam*. https://tienphong.vn/do-xo-du-lich-ngay-ret-dam-chay-ve-tour-ngam-tuyet-o-mien-nui-phia-bac-post1305603.tpo (Vietnamese).

Mishev, P. & Mochurova, M. (2010). Impacts of climate change on tourism. In V. Alexandrov, M. Gajdusek, C. Knight, A. Yotova, (Eds.), *Global environmental change: Challenges to science and society in Southeastern Europe*. Springer, Dordrecht. https://doi.org/10.1007/978-90-481-8695-2_15

Moen, J., & Fredman, P. (2007). Effects of climate change on alpine skiing in Sweden. *Journal of Sustainable Tourism, 15*(4), 418–437. https://doi.org/10.2167/jost624.0

MONRE (Ministry of Natural Resources and Environment). (2021). *National climate assessment report*. https://imh.ac.vn /files/doc/2022/s%C3%A1ch/Bao%20cao%20dGKHQG_Web.rar, downloaded on 13/07/2024

Nam, H. (2024). *Think differently to succeed*. https://tienphong.vn/tu-duy-khac-biet-de-thanh-cong-post1680400.tpo (Vietnamese).

Narain, U., Margulis, S., & Essam, T. (2011). Estimating costs of adaptation to climate change. *Climate Policy, 11*(3), 1001–1019. https://doi.org/10.1080/14693062.2011.582387

QBPEIP (Quang Binh Provincial Electronic Information Portal). (2023). *Tanhoa Village was voted as the World's Best Tourist Village in 2023*. https://quangbinh.gov.vn/chi-tiet-tin-1/-/view-article/1/13848241113627/1697775764860 (Vietnamese).

Rixen, C., Teich, M., Lardelli, C., Gallati, D., Pohl, M., Ptz, M., & Bebi, P. (2011). Winter tourism and climate change in the Alps: An assessment of resource consumption, snow reliability, and future snowmaking potential. *Mountain Research and Development, 31*, 229–236.

Saarinen, J., Hambira, W. L., Atlhopheng, J., & Manwa, H. (2012). Perceived impacts and adaptation strategies of the tourism industry to climate change in Kgalagadi South District, Botswana. *Development Southern Africa, 29*, 273–285.

Scott, D., & McBoyle, G. (2007). Climate change adaptation in the ski industry. *Mitigation and Adaptation Strategies for Global Change, 12*(8), 1411–1431. https://doi.org/10.1007/s11027-006-9071-4

Scott, D., Hall, C. M., & Gossling, S. (2012). Tourism and climate change: Impacts, adaptation, and mitigation. In *Contemporary geographies of leisure, tourism and mobility* (Vol. 10). Routledge.

Simpson, M., Gossling, S., Scott, D., & Hall, C. (2008). *Climate change adaptation and mitigation in the tourism sector: Frameworks, tools and practices.*

Tanhoa. (2024). *About the village*. Tanhoa Village - Weather Adaptive Tourism Village (tanhoavillage.com), consulted on 28/04/2024.

Tervo-Kankare, K. (2011). The consideration of climate change at the tourism destination level in Finland: Coordinated collaboration or talk about weather? *Tourism Planning and Development, 8*(4), 399–414. https://doi.org/10.1080/21568316.2011.598180

Tho, H., Trung, H., & Thanh, N. (2023). *Tanhoa: From flood center to the best tourist village.* https://tuoitre.vn/tan-hoa-tu-ron-lu-den-ngoi-lang-du-lich-tot-nhat-20231024102434798.htm (Vietnamese).

Tien Pham. (2021). *Floating house at Tanhoa flood village)*. Ethnic and Development Newspaper (baodantoc.vn), consulted on 28/04/2024 (Vietnamese).

VNAT (Viet Nam National Authority of Tourism) (2024). *Weather-adaptive tours in Central Vietnam.* https://moitruongdulich.vn/tour-du-lich-thich-ung-thoi-tiet-tai-mien-trung/ (Vietnamese).

VTV (Vietnam Television online) (2017). *Additional climate change tours in Ben Tre*. https://vtv.vn/trong-nuoc/them-tour-du-lich-bien-doi-khi-hau-o-ben-tre-20171016135629982.htm (Vietnamese).

Wyss, R., Abegg, B., & Luthe, T. (2014). Perceptions of climate change in a tourism governance context. *Tourism Management Perspectives, 11*, 69–76. https://doi.org/10.1016/j.tmp.2014.04.004

Tourism and Agro-Silvopastoral Systems as Alternatives to Counteract Climate Change Case: Guasaganda and Pucayacu Province of Cotopaxi–Ecuador

Marco Antonio Toledo, José Luis Núñez, Katherine Joselyn Villarreal, and Nancy Carolina Montenegro

Abstract This study presents a proposal that combines tourism and agro-silvopastoral systems applied in the parishes of Guasaganda and Pucayacu, in the province of Cotopaxi (Ecuador), with the aim of meeting the needs of the population and contributing to the development of sustainable strategies that not only serve to counteract climate change but also to adequately manage natural resources, agriculture, and rural tourism. A review of literature related to climate change, sustainable actions, agro-silvopastoral systems, and agrotourism was carried out, and successful experiences in countries such as Italy, Colombia, and Costa Rica were also cited. By applying participatory methodologies, such as community mapping and interviews, it was possible to identify the main needs and perceptions of the inhabitants of the two parishes, which are summarized in two proposals that are complementary; the design of an agrotourism route and the application of agro-silvopastoral systems along the route. These two proposals constitute actions whose objective is to fight against climate change and contribute to sustainable development, being one of the best examples in the province of Cotopaxi and that could be applied in other sectors of Ecuador.

1 Introduction

Climate change represents one of the greatest challenges of the twenty-first century, impacting not only ecosystems but also the economy and well-being of rural communities. In this context, tourism and agro-silvopastoral systems emerge as viable

M. A. Toledo (✉) · J. L. Núñez · K. J. Villarreal · N. C. Montenegro
Universidad Técnica de Cotopaxi–Ext. La Maná, Ciudad La Maná, Ecuador
e-mail: marco.toledo6833@utc.edu.ec; jose.nunez@utc.edu.ec;
katherine.villarreal0992@utc.edu.ec; nancy.montenegro5374@utc.edu.ec

J. Marques et al. (eds.), *Tourism and Climate Change*, Tourism, Hospitality & Event Management, https://doi.org/10.1007/978-3-031-99128-8_13

alternatives to counteract the negative effects of this phenomenon. According to the World Tourism Organization (WTO, 2013), sustainable tourism is defined as tourism that meets the needs of current tourists and host communities while protecting and enhancing opportunities for the future. This perspective is essential for areas such as the parishes of Guasaganda and Pucayacu, province of Cotopaxi, Ecuador, where the interrelationship between agriculture and tourism can offer comprehensive solutions to the climate crisis. This study focuses on the analysis of tourism practices and agro-silvopastoral systems, aiming to understand how these practices can contribute to the development of sustainable strategies. The integration of participatory methodologies, such as community mapping and interviews, allows for the capture of local knowledge and perceptions, aspects that are fundamental for formulating effective policies (Berkes, 2017). Furthermore, a review of existing literature on sustainable agricultural practices, rural tourism, and biodiversity conservation will establish a theoretical framework to support the research. Through the valuation of regional resources, two concrete strategies are proposed: the design of a tourism route and the implementation of agroecological actions, such as agro-silvopastoral systems (SAP). These actions not only aim to mitigate the impact of climate change but also promote social cohesion and local economic development (González et al., 2020). The results of this study highlight the importance of implementing sustainable agricultural practices and promoting responsible tourism in the parishes of Guasaganda and Pucayacu. These findings could have significant implications for the design and development of local and regional policies and strategies, with the potential to be applied in other geographical areas aiming to contribute to conservation and implement practices to minimize the effects of climate change. In this context, the urgency of sustainable strategies becomes evident, especially as climate change continues to exacerbate environmental and socioeconomic vulnerabilities in rural areas. The impacts of climate change, including extreme weather events, altered rainfall patterns, and rising temperatures, threaten agricultural productivity and the livelihoods of communities that rely on these ecosystems. Therefore, adopting sustainable practices is not merely an option but a necessity for ensuring food security and resilience in the face of ongoing environmental changes. The convergence of sustainable tourism and agriculture provides a unique opportunity to address these challenges holistically. By integrating agroecological practices with responsible tourism initiatives, it is possible to reduce the environmental footprint of human activities while generating resilient livelihoods for local communities. This dual approach fosters the creation of synergistic solutions where agriculture and tourism coexist as mutually reinforcing pillars for sustainability. For instance, promoting farm-to-table tourism can enhance local food systems while providing tourists with authentic experiences that celebrate the region's agricultural heritage. Moreover, the incorporation of local communities in the planning and execution of these strategies ensures that their voices are heard, fostering a sense of ownership and stewardship over their resources. This participatory approach not only enhances social cohesion but also empowers communities to be active agents in mitigating climate change impacts. As such, this study underscores the need to prioritize actions that align with global climate goals while tailoring them to the specific needs of places such as Guasaganda and Pucayacu parishes, which serve as a model for other territories facing similar challenges. By investing in sustainable tourism and agriculture, we can pave the way for a more resilient, equitable, and environmentally conscious future.

2 Literature Review

2.1 Climate Change and the Urgency of Sustainable Strategies

Climate change has established itself as one of the greatest contemporary global challenges, negatively affecting ecological, agricultural, and social systems. Climate change directly affects tourism (WTO, 2013); and highlights that climate change affects both nature and tourism demand. (Buckley, 2009), according to the IPCC (2022), climate phenomena such as rising temperatures and extreme weather events demand integrated strategies to mitigate their impacts. In this context, Agro-Silvopastoral Systems (SAPs) emerge as a key solution, integrating agriculture, forestry, and livestock to capture carbon, regenerate soils, and conserve biodiversity (Jose, 2009). On the other hand, sustainable tourism has proven to be an effective mechanism for promoting local economies and preserving natural and cultural resources. The convergence of rural tourism and sustainable agriculture offers opportunities to address climate change by diversifying income sources and promoting environmental education. In diverse regions like Ecuador, implementing tourist routes that integrate agroecological elements not only attracts visitors interested in sustainable experiences but also generates direct economic benefits and fosters environmental sustainability (Sharpley & Telfer, 2014; Barkin, 2017). Together, the integration of SAPs with sustainable tourism represents a promising approach to strengthen climate resilience in rural communities while promoting ecosystem conservation and sustainable development. Scheyvens (2002) acknowledges that environmental threats, including climate change, must be integrated into the development of local and community-based tourism.

2.2 Agro-Silvopastoral Systems (SAPs) and Agrotourism: Benefits and Applications

Agro-silvopastoral systems, which combine trees, crops, and animals in the same space, have proven to be an effective strategy for addressing climate change challenges. The key benefits include improving soil fertility through nutrient recycling, mitigating greenhouse gas emissions by capturing atmospheric carbon, and diversifying income and food sources, strengthening climate resilience (Nair, 2011).

In Ecuador, SAPs have been implemented in vulnerable areas like Cotopaxi province, promoting regenerative agricultural practices and combining them with tourism activities that diversify local incomes.

Despite their potential, SAPs face challenges such as lack of funding, insufficient public policies, and local resistance to new productive systems. Collaboration between communities, local governments, and international organizations is essential to overcome these barriers.

Regarding agrotourism, it contributes to connecting visitors with agricultural life and rural traditions, generating additional income, conserving natural and cultural resources, and promoting environmental education (Barbieri & Mshenga, 2008). The key benefits include: the diversification of rural income, reducing dependence on vulnerable agricultural activities; the conservation of cultural and natural resources, through activities like farm visits and organic farming workshops; the environmental education, fostering behaviour changes towards more sustainable practices; and strengthening community resilience, attracting tourists interested in authentic and ecological experiences.

The combination of agrotourism with agro-silvopastoral systems (SAPs) represents an innovative strategy for promoting sustainable rural development and addressing climate change and it's been implemented already in some countries. Costa Rica, for example, has shown that integrated systems of tourism and agroecology generate positive impacts on both environmental conservation and local economic growth (Pérez-Méndez et al., 2021). In Italy, the "Agriturismo" model combines rural tourism with sustainable agriculture, offering activities like grape harvesting and culinary workshops, preserving traditional agricultural practices (Sonnino, 2004). In Colombia, agrotourism routes have been implemented in the Colombian Andes, integrating visits to agroecological farms, reforestation activities, and workshops with local producers. These initiatives have strengthened local economics and promoted environmental education (García-Cortés et al., 2020). These experiences highlight how agrotourism can be a transformative tool when combined with sustainable agricultural systems, creating replicable models that benefit both communities and the environment.

Climate change is one of the greatest contemporary global challenges that affect negatively ecology, agriculture, and social systems. According to the Intergovernmental Panel on Climate Change (IPCC), climatic alterations, including rising temperatures and extreme weather events, demand integrated approaches to mitigate their impacts (IPCC, 2022). In this context, agro-silvopastoral systems (SAPs) emerge as a key tool for climate resilience. These systems, which integrate agriculture, forestry, and livestock, are recognized for their ability to improve carbon capture, conserve biodiversity, and regenerate agricultural soils (Jose, 2009). Guasaganda (2020) highlights that they contribute to forest conservation, enhancing the resilience of rural communities. Furthermore, their potential to be combined with rural tourism activities offers a unique opportunity to promote sustainable development.

3 Methodology

The methodology used in this research begins with a literature review related to the topic of study. According to Kitchenham and Charters (2007), literature review is useful for identifying gaps in knowledge and justifying the relevance of the research. Also, the use of participatory mapping was a key tool in the research, using methodologies such as Participatory Rural Appraisal (PRA), proposed by Chambers (1994). This approach allowed the integration of local communities in planning and decision making, promoting their role in the development of solutions. PRA includes techniques such as community mapping, diagramming, and participatory analysis, which are essential for identifying strategic points that have the potential to become tourist attractions. This method not only strengthened collaboration, but also ensured that local perspectives and knowledge were at the heart of the research.

The interviews conducted with about 30 key stakeholders in the sector, including tourism entrepreneurs, farmers, leaders, and representatives, were unstructured and applied at the discretion and convenience of the researcher. The days allocated for these interviews were carried out over the course of a month, specifically on weekends when the interviewees were commonly present in their places of residence. The information provided by the interviewees was reinforced using qualitative methodologies such as content analysis (Krippendorff, 2004). This allowed for systematic analysis of the data, identifying patterns, categories, and interpretations that provided a multidimensional view of the challenges and opportunities related to agrotourism resources.

To choose and determine the key tourist resources in the design of the tourist route, the methodology established by the Ministry of Tourism of Ecuador was used. This process consists of assigning a weighting and hierarchy to the resource, complemented by techniques such as Multi-Criteria Analysis (MCA). This approach has been widely used in tourism planning to classify and prioritize resources based on criteria such as accessibility, attractiveness, sustainability, and the cultural or natural relevance of the location.

For the implementation of agro-silvopastoral systems, an agricultural evaluation called Systems Analysis was conducted. This methodology will allow for the assessment of interactions among the system components (trees, pastures, and crops) and their impact on production and environmental health. Additionally, environmental sustainability indicators, such as those proposed by Bell and Morse (2008), will be used to measure positive impacts in the short and medium term.

The use of Excel facilitated the systematization of the information obtained. In addition, the use of tools such as QGIS Desktop 3.28.1 software allowed the entry of the geographic points obtained from the GPS and visualized on a map of the tourist route in a more understandable way for both the community and the scientific public.

The study's focus on climate change mitigation was aligned with the framework of the United Nations Sustainable Development Goals (SDGs), specifically SDG 13

(Climate Action) and SDG 15 (Life on Earth). The incorporation of these goals as a strategic guide in the methodology reinforced the relevance and applicability of the project, linking it to global sustainability efforts.

3.1 Study Area

The community of Guasaganda is home to approximately 3866 inhabitants, while Pucayacu has a population of 1960, bringing the total to around 5826 residents (Instituto Nacional de Estadística y Censos [INEC], n.d.). These rural communities primarily rely on agricultural and livestock production as their main livelihood, with products being traded in local, provincial, and national markets. Both communities are located in the canton of La Maná, in the province of Cotopaxi, Ecuador, with approximate geographic coordinates of −0.9416° S, −79.2216° W (see Fig. 1).

This region is situated in a privileged environment, characterized by its biological diversity and tropical landscapes, including waterfalls, rivers, and a wide variety of endemic flora and fauna. Particularly, the birds of the region, such as toucans and hummingbirds, are an essential component of these ecosystems and hold significant value for conservation efforts (BirdLife International, 2023). From a cultural perspective, the communities of Guasaganda and Pucayacu possess significant heritage wealth, especially due to the archaeological ruins of Malqui Machay, which, because of their importance, have the potential to become a key reference for both historical research and sustainable cultural tourism in Ecuador. These parishes feature exceptional natural attractions, such as rivers, waterfalls, endemic species, and more, positioning the area as a potential destination for activities like hiking, birdwatching, and biodiversity conservation (Krippendorff, 2004; United Nations, 2015). However, the current tourism impact remains limited due to the lack of adequate infrastructure and the early stage of organized project development. Despite this, local initiatives have demonstrated that tourism can complement the agricultural economy, diversify income sources, and improve the quality of life for its residents.

4 Results

Key findings from the interview conducted on the topic: tourism, agro-silvopastoral systems, and climate change. As for profile of respondents, 80% are farmers, 15% are tourism entrepreneurs, and 5% are community leaders and local representatives. Average age is 55 years, with experience in their respective roles. Regarding education level, 40% have an intermediate level of knowledge about climate change, tourism, and agro-silvopastoral systems.

Regarding perception of climate change and its impacts, 85% of farmers reported significant changes in the rainy seasons (December–April), affecting the production of local crops. However, only 30% are implementing adaptation practices such as

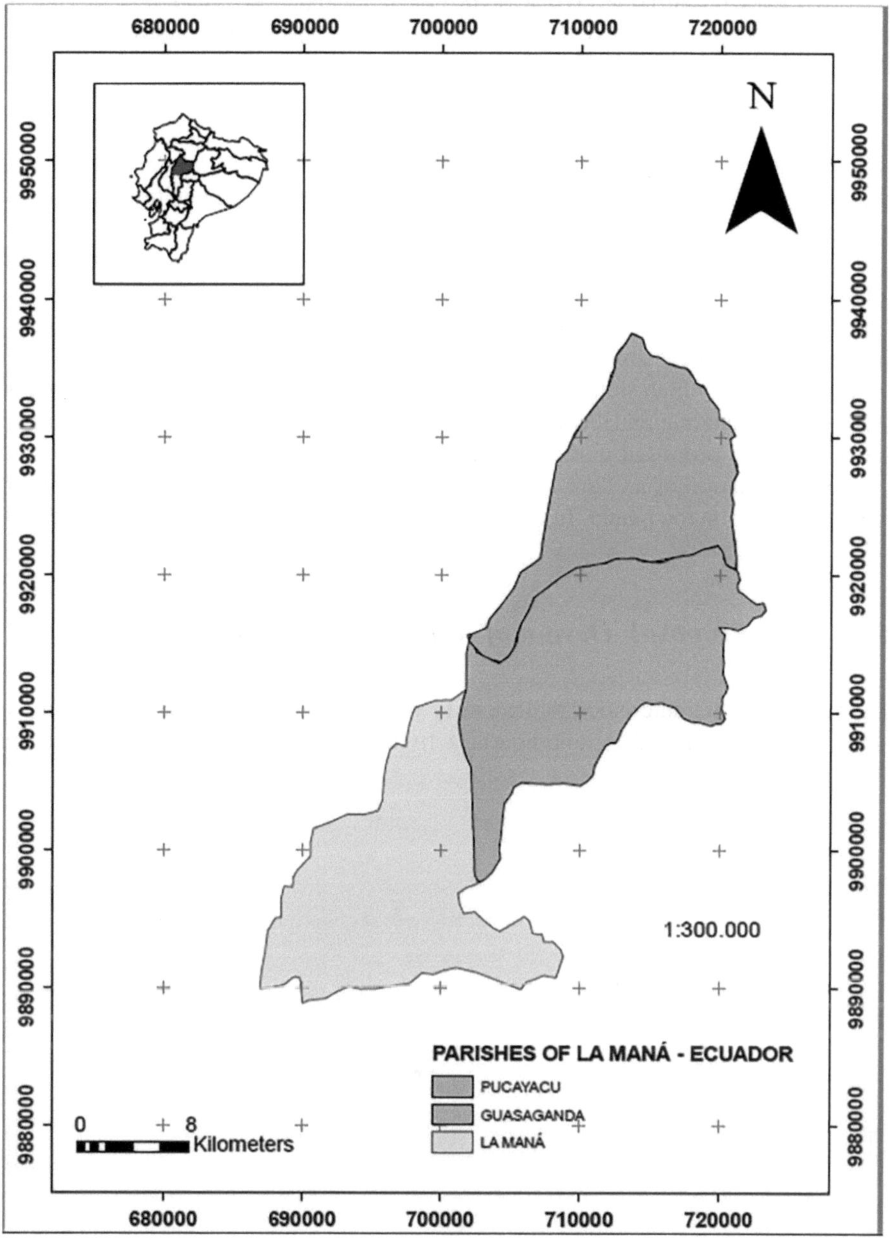

Fig. 1 A geographic location of Cotopaxi Province with respect to South America and Ecuador. B Study area of rural community with respect to the province of Cotopaxi

agro-silvopastoral systems. The rest do not adopt these practices due to lack of knowledge and resources. As for role of tourism as a sustainable alternative, it contributes to environmental conservation and supports the socio-economic development of local communities.

The 60% of tourism entrepreneurs perceive that the tourism potential in the parishes of Guasaganda and Pucayacu is underutilized due to a lack of promotion and resources. Additionally, 85% agree that the creation and implementation of an agro-tourism route is a good development alternative.

The 90% of community leaders identified the lack of specific public policies and funding as the main barriers to implementing sustainable strategies. They also expressed that strengthening sustainable tourism and providing training in agro-silvopastoral practices are essential to mitigate the effects of climate change.

These results, which stem from the investigation of a real need of the local population of Guasaganda and Pucayacu, are the starting point and correspond to two specific actions proposed in this project: the design of the agro-touristic route and the implementation of an agro-silvopastoral system that minimizes the problem of climate change on the planet. In this context, the following proposals are suggested.

4.1 *First Proposal: Design of Agrotourism Route (Fig. 2)*

The route has characteristics that identify it as agro-touristic. Due to its components, the attractions and entrepreneurial activities of the parishes of Guasaganda and

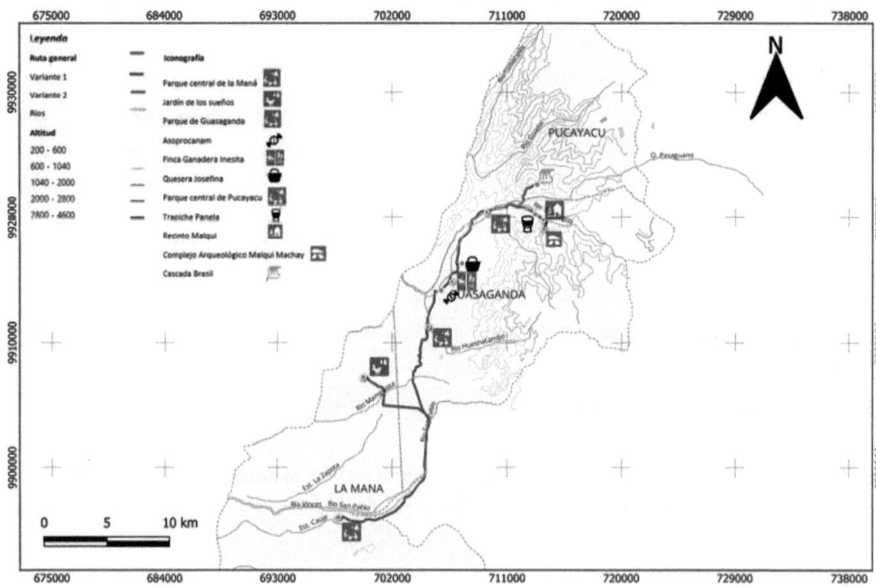

Fig. 2 Agro-touristic route map

Table 1 Coordinates of the strategical touristic points

Strategic points	Coordinates X	Coordinates Y	Altitude m.o.s.l.
1-1 La Maná central park	696,931	103,856	212
1-2 "Dream Garden" Lodge	699,763	9,907,315	573
1-3 Guasaganda park	704,684	9,911,230	517
1-4 Agricultural association of National Cocoa Producers of La Mana. (ASOPROCANAM)	706,213	9,914,601	562
1-5 Inesita Cattle Ranch	706,436	9,914,861	515
1-6 Josefina Cheesse farm	707,208	9,916,324	546
1-7 Pucayacu park	709,241	9,920,516	645
1-8 Malqui-Machay Archaeological ruins.	715,947	929,918,438	1010

Table 2 Agro-touristic route

Hour	Itinerary
07 h30–08 h20	Meeting point at the "La Maná central park"
08 h20–10 h00	Tour by the "Dream Garden" protected forest
10 h35–10 h50	Visit to the al park and the Guasaganda church
11 h00–11 h30	Visit and degustation at the "ASOPROCANAM"
11 h35–12 h00	Tour by the "Inesita" cattle ranch
12 h05–12 h30	Visit and degustation at the cheese farm "La Josefina"
12 h35–13 h30	Lunch at the "AUCAS" dinner
13 h45–14 h00	Visit to the "San Pablo" Pucayacu's Church
14 h30–16 h00	Tour by Malqui-Machay Archaeological ruins
16 h30–17 h30	Return to the La Maná town and end of the Tour

Pucayacu have been ranked and georeferenced to identify the key sites for designing the tour, as shown in the following data. (see Table 1).

The tourist who arrives to this sector can perform different activities, therefore the following itinerary is proposed: (See Table 2).

Photographs of the Main Attractions Are Shown Below:

The Central Park of La Maná is an emblematic space where the community gathers to enjoy moments of recreation and connection. Surrounded by trees and gardens, the park features areas for relaxation, a central fountain, and spaces for cultural and social events. Its tranquil and welcoming atmosphere makes it an ideal place for both locals and visitors, offering an authentic experience of daily life in La Maná (Fig. 3).

Fig. 3 La Maná Central Park

Fig. 4 Dream Garden

The Garden of Dreams is a natural and tourist space that stands out for its scenic beauty and biodiversity. For nature lovers, it offers areas for rest and perfect spaces for the contemplation and study of various species of flora and fauna (Fig. 4).

Like most parks in the urban and rural parishes of the La Maná canton, the central park of Guasaganda occupies a central location and offers its inhabitants a space for relaxation surrounded by green areas and resting spots; there are also playgrounds for children. Additionally, it serves as a meeting point where various civic, cultural, or sports events are held (Fig. 5).

The central park of the Pucayacu parish, in the La Maná canton, is a strategic space characterized by its location in front of a Catholic temple, with a religious monument featuring the figure of Saint Paul, the patron of the parish, at the other end. It is important to mention that this park is a very busy place, especially on Thursdays when commerce takes place in the parish, attracting hundreds of people to buy and sell local products (Fig. 6).

The Malqui Machay Archaeological Complex has great historical and cultural value, considered one of the most enigmatic places in the province as it is linked to the history of the Incas. It is believed that the remains of the Inca emperor Atahualpa rest here, which gives it a special significance as a ceremonial and sacred site (Estupiñán, 2011). Surrounded by lush Andean vegetation, Malqui Machay stands out for its stone structures and architectural design that reflects Inca influence (Fig. 7).

4.2 *Second Proposal: Impact of Agro-Silvopastoral Systems on Climate Change Along the Agrotourism Route*

As part of the actions carried out in this proposal, agro-silvopastoral systems (SAP) are presented as a sustainable alternative to traditional agricultural production, offering benefits both for climate change mitigation and ecosystem improvement in

Fig. 5 Guasaganda central park

Fig. 6 Pucayacu central park

Fig. 7 Archaeological ruins of Malqui-Machay

general. These systems consisted of planting trees, grasses, and agricultural crops in spaces within the agrotourism route, focused on three factors:

- Carbon capture: The trees and grasses of the SAP capture CO_2 from the atmosphere and store it in their biomass, contributing to reduce the greenhouse effect.

- Reduction of methane emissions: Ruminants such as cows generate methane, a potent greenhouse gas, which is mitigated by the SAP by improving animal digestion and forage quality, thereby reducing methane emissions.
- Reduced fossil energy use: The SAP have been successful in promoting production diversification, reducing dependence on external inputs such as synthetic fertilizers and pesticides, the manufacture of which requires large amounts of fossil energy (see Tables 3 and 4).

Table 3 Socioeconomic and environmental benefits of the project

Impact	Description	Rates
Income increasing	Income increasing for the crop's diversification and touristic activities	30% of increasing in annual incomes for the SAP and agro-touristic implementation
Tourism increasing	Visitors number increasing in the agro-touristic route	50% rural tourism increasing in the first year of implementation
Carbon capture	CO_2 capture of the atmosphere and biomass storage	Even 20 tons of carbon for hectare by year
Methane emission reduction	Digestion and forage quality improvement, reducing the methane emission	Significant reduction of traditional agricultural systems
Decrease in fossil energy use	Diversification in the products, and decreasing of external supplies dependence	Less fertilizers and synthetic pesticides use that required a high fossil energy consuming
Biodiversity	Biodiversity increasing of species in the ecosystem	Implementation of 255 Fernan Sanchez and Guayacan trees, 150 orange and tangerine plants and over 600 Morera and Mata Raton legumes
Total economic benefit	Benefit economical approach for the community	$200,000 annual generated by the tourism combination and the sustainable agriculture production

Table 4 Species planted on the "Bolón" and "Malqui-Machay" farms as part of the agro-silvopastoral practices on the agrotourism route

Properties of the seeded species	Quantities	Scientific and common names
Study area of the forestry species	225	Fernán Sánchez (*Triplaris cumingiana*)
Study area of the forestry species	225	Guayacán (*Tabebuia chrysantha*)
Agricultural plants of interest	150	Orange (*Citrus sinensis*)
Agricultural plants of interest	150	Tangerine (*Citrus reticulata*)
Shrubby legumes	600	Morera (*Morus alba, L.*)
Shrubby legumes	600	Mata ratón (*Gliricidia sepium*)

It is expected that by the end of the year the socioeconomic contribution to the community and the environment will be reflected in the following results (Figs. 8 and 9).

5 Discussion of Results

People's need of escape from routine has led them to look for different alternatives to relax and have fun. Thus, it is essential to value the different resources that a territory possesses to develop diverse tourism activities (Castillo Vizuete et al., 2021).

The agrotourism route in the rural parishes of Guasaganda and Pucayacu has a positive approach to promote sustainable tourism and local economic development. Considering that community participation and the offer of authentic services in the sector are key to the success of the route. In this sense, the feasibility of creating an agrotourism route in two specific rural sectors has been discussed, considering factors such as infrastructure, service offerings, and community participation.

Agrotourism as a complement to human needs must be appropriate in terms of sustainability, suitability, embellishment, and comfort (Lara Vásconez et al., 2020). In order for an agrotourism route to be successful, it is essential to have adequate and accessible infrastructure. According to Manuel Martinez mentions that "tourism accessibility has become a specific factor of tourism quality" (Martinez, 2018), you can no longer have a quality experience when you do not have adequate access towards the tourist destination; then it is necessary to ensure that the route is well

Fig. 8 Forestry species seeding (Quinatoa, 2024)

Fig. 9 Capacitation of the agro-silvopastoral systems SAP (Quinatoa, 2024)

connected through roads and public transportation, and that points of interest are easily accessible to visitors.

The supply of services is also crucial to the success of an agrotourism route. According to the study of the tourist offer: theoretical precisions for its analysis by the University of Alicante San Vicente del Raspeig, the tourist offer is a combination of services and products that are presented for consideration by a tourist client who wishes to make a trip (Naranjo, 2022). For which, it is important to offer a variety of services, such as accommodations, restaurants, and recreational activities, that meet the needs and preferences of visitors.

Historically, indigenous and campesino organizations and communities have managed to implement several development projects (Lara Vásconez et al., 2020). Therefore, community participation is fundamental for the success of an agrotourism route. According to the Magazine Universidad y Sociedad en el desarrollo del turismo comunitario en Ecuador: reflexiones necesarias, community tourism is described as an activity that generates economic benefits for the communities involved and that in turn contributes to the conservation and strengthening of their ethnic and cultural diversity as it allows local families to diversify their economy and improve their quality of life (Mullo Romero et al., 2019). Therefore, it was found that local residents are essential for participation in the planning and management process of the route, which has contributed to its success.

There are examples of successful agrotourism routes in rural sectors. For example, the agrotourism route of the concha prieta (Pilataxi Guamán et al., 2022), which has managed to attract several visitors each year, thanks to its range of services and adequate infrastructure.

These processes of change in the localities generate a development alternative leveraged on agrotourism and actions such as in this specific case of SAP agro-silvopastoral systems, promote respect for conservation, and offer benefits both for climate change mitigation and for the improvement of the ecosystem in general. The implementation of agroforestry systems not only contributes to environmental sustainability, but also has a significant impact on the economic growth of local communities. The figures reflect how these systems can improve incomes, reduce costs, and generate employment, resulting in more robust and sustainable economic development.

6 Conclusions

The Guasaganda Pucayacu agrotourism route has been designed to offer a unique and enriching experience for both nature lovers and those interested in agriculture and cultural sites. One of the main objectives of this route is to promote rural tourism and the local economy by providing farmers and producers with a platform to showcase their hard work and dedication. Direct interactions between tourists and local inhabitants not only enrich visitors' knowledge of the agricultural sector, but also strengthen the community by creating bonds of support and collaboration.

Through visits to selected points, tourists are able to pick fruit and taste products. This helps the development of socio-productive activities and economy of the locality and also provides a tourist offer different from what usually has the La Maná city. The SAP agro-silvopastoral systems offer benefits both for climate change mitigation and for the improvement of the ecosystem in general and this constitutes a real contribution of the project to the populations of the sector and their quality of life in environmental terms.

The relationship between the community, its heritage, and the surrounding environment is a web of interactions that shape its cultural identity. Agricultural practices, deeply rooted in daily life, not only guarantee subsistence, but are also intertwined with animal and plant elements that are fundamental in rituals and celebrations, reflecting the community's cosmovision. For example, traditional crops and the raising of native species are linked to festivities that honour the land and its cycles, strengthening collective identity. However, climate change represents an imminent threat to the biological diversity that sustains these practices, putting the very existence of these cultural elements at risk. It is therefore crucial to address these challenges to ensure the continuity of these legacies. Developing and preserving a community with a unique character requires a deep understanding of its authentic connections to the environment, fostering an approach that not only values its heritage, but also promotes its resilience in the face of global changes.

The study of tourism and agro-silvopastoral systems practices in the parishes of Guasaganda and Pucayacu has revealed significant findings that underscore the importance of integrating local knowledge and participatory methodologies in sustainable development. It has been shown that the implementation of agro-silvopastoral systems not only contributes to biodiversity conservation and climate change mitigation, but also strengthens the cultural identity of the community by linking agricultural practices with local rituals and traditions. In addition, the development of an agrotourism route has shown considerable potential to improve income and generate employment, which in turn encourages the active participation of the inhabitants in the management of their resources.

The Pucayacu and Guasaganda, with their natural and cultural wealth, are ideal for integrating agrotourism as a complementary strategy to rural tourism and agro-silvopastoral systems. Local communities can develop activities such as:

- Workshops on agroecological practices based on SAP.
- New agritourism routes that include visits to local farms and educational experiences on regenerative agriculture.

Cultural events that highlight the agricultural and culinary traditions of the region.

These activities would not only diversify local income but would also position these parishes as a sustainable tourism destination in the province of Cotopaxi and replicable in other provinces of Ecuador.

References

Barbieri, C., & Mshenga, P. M. (2008). The role of the Agritourism farm in Di versifying rural households and reducing poverty. *Journal of Rural Studies, 24*(2), 247–257.

Barkin, D. (2017). Ecotourism and sustainability: The Latin American experience. *Tourism Review, 72*(1), 1–12.

Bell, S., & Morse, S. (2008). *Sustainability indicators: Measuring the immeasurable?* Earthscan.

Berkes, F. (2017). Understanding uncertainty and reducing vulnerability: Lessons from resilience thinking. *Natural Hazards, 82*(2), 1–13.

BirdLife International. (2023). *Global bird conservation status report*. Recuperado de https://www.birdlife.org

Buckley, R. (2009). *Environmental impacts of ecotourism*. CABI.

Castillo Vizuete, D., Gabilanes Montoya, A., Chávez Valásquez, C., Benalcázar Bergara, P., & Mestanza Ramón, C. (2021, May 16). *Design of a nature tourism route in the Chimborazo Wildlife Reserve, Ecuador*. Retrieved from MDPI: https://www.mdpi.com/1660-4601/18/10/5293#B79-ijerph-18-05293

Chambers, R. (1994). Participatory rural appraisal (PRA): Challenges. *Potentials and Paradigm. World Development, 22*(10), 1437–1454.

Estupiñán, T. (2011). Los Sigchos, el último refugio de los incas quiteños. Una propuesta preliminar. *Bulletin de l'Institut Français d'Études Andines, 40*(1), 191–204; Lima.

García-Cortés, S., Jiménez, M., & Restrepo, J. (2020). Agro-tourism routes in the Colombian Andes: A sustainable model for rural development. *Sustainability, 12*(10), 4123.

González, A., et al. (2020). Sustainable tourism: A comprehensive approach. *Journal of Sustainable Tourism, 28*(6), 843–860.

Guasaganda, G. A. (2020, November 11). *PDOT-GUASAGANDA-FINAL.pdf*. Retrieved from Untitled https://guasaganda.gob.ec/cotopaxi/wp-content/uploads/2021/08/PDOT-GUASAGANDA-FINAL.pdf

IPCC. (2022). *Climate change 2022: Impacts, Adaptation and Vulnerability*. Cambridge University Press.

Jose, S. (2009). Agroforestry for ecosystem services and environmental benefits: An overview. *Agroforestry Systems, 76*(1), 1–10.

Kitchenham, B., & Charters, S. (2007). *Guidelines for performing systematic literature reviews in software engineering*. EBSE Technical Report.

Krippendorff, K. (2004). Content analysis: *An introduction to its methodology (2nd ed.)*. Thousand Oaks: Sage Publications.

Lara Vásconez, R., López Pumalema, J., & Cunalata García, A. (2020, November 3). *Agrotourism in the northern Amazon: Point of*. Retrieved from Green World Journal https://www.greenworldjournal.com/_files/ugd/dac1d8_a8a86f5584454bf8822de4f069c1726b.pdf?index=true

Martinez, M. J. (2018, February). *Accessible tourism, tourism for all. Analysis of accessibility in hotel establishments in Marrakech*. Retrieved from EUMED: https://www.eumed.net/rev/cccss/2018/02/turismo-accesible-marrkech.html

Mullo Romero, E., Vera Peña, M., & Herrera, G. (2019, June 2). *Community-based tourism development in Ecuador: Necessary reflections*. Retrieved from Scielo: http://scielo.sld.cu/scielo.php?script=sci_arttext&pid=S2218-36202019000200178

Nair, P. K. R. (2011). Agroforestry systems and environmental quality: Interconnections and advances. *Journal of Environmental Quality, 40*(3), 789–801.

Naranjo, M. R. (2022, August 5). *Tourism supply: Theoretical precisions for its analysis*. Retrieved from Encuentros.

Pérez-Méndez, A., Villalobos, C., & Alvarado, J. (2021). Agrotourism and agroforestry in Costa Rica: A sustainable synergy. *Journal of Ecotourism, 20*(3), 245–267.

Pilataxi Guamán, A. L., Prado-Carpio, E., Carvajal Romero, H., & Quezada Campoverde, J. M. (2022). *Design and feasibility of the agrotourism route of the concha prieta as an alternative for economic diversification in Bajo Alto, province of El Oro canton El Guabo, Ecuador*. Retrieved from South Florida Journal of Development, file:///C:/Users/Dell/Downloads/028+SFJD.pdf

Scheyvens, R. (2002). *Tourism for development: Empowering communities*. Pearson Education.

Sharpley, R., & Telfer, D. J. (2014). *Tourism and development: Concepts and issues*. Channel View Publications.

Sonnino, R. (2004). For a 'piece of bread'? Interpreting sustainable development through agritourism in Southern Tuscany. *Sociologia Ruralis, 44*(3), 285–300.

WTO. (2013). *Sustainable tourism: An introduction*. World Tourism Organization.

Sustainable Practices in Rural Tourism: Promoting Environmental Responsibility and Community Engagement in India

Madhu Kumari, Sandeep Guleria, Suneel Kumar, and Nisha Devi

Abstract The study provides factual support for the sustainable practices that stakeholders in rural tourism have embraced. It examines the idea and definition in this area, the factors that encourage stakeholders in rural tourism to implement them, and suggestions for improving efforts in India. This study outlines an exploratory qualitative research project examining rural tourism stakeholders' sustainable tourism practices. A methodical strategy was utilized to tackle the study's goals. This study examines how individuals interpret and engage with activities in this area, exploring their motivations, areas for enhancement, and the perceived significance of these practices. The research methodology involved conducting semi-structured interviews to gather diverse perspectives from participants. The results indicate that sustainable tourism practices significantly impact various aspects: future growth, awareness, knowledge enhancement, conservation efforts, economic benefits, cooperation, environmental protection, and more. It's noteworthy that the concept of sustainable tourism practices aligns closely with competitive offerings, high-quality attractions, and meeting traveler expectations. This suggests that the definition of sustainable tourism practices encompasses quality tourism and visitor demands in economic, environmental, and sociocultural effects. Government assistance and financial, social, and ecological factors are important forces behind sus-

M. Kumari
Department of Tourism and Hospitality Management, Chandigarh University, Ajitgarh, India
e-mail: kumarimadhu070@gmail.com

S. Guleria
Department of Tourism, Chandigarh University, Ajitgarh, India
e-mail: hod.uithm.tm@cumail.in

S. Kumar (✉)
Department of Commerce, Shaheed Bhagat Singh College, University of Delhi, New Delhi, India
e-mail: suneel.kumar@sbs.du.ac.in

N. Devi
Department of Commerce, Shri Ram College of Commerce, University of Delhi, New Delhi, India
e-mail: nisha.devi@srcc.du.ac.in

J. Marques et al. (eds.), *Tourism and Climate Change*, Tourism, Hospitality & Event Management, https://doi.org/10.1007/978-3-031-99128-8_14

tainability initiatives. Enhanced collaboration among rural tourism operators, focused efforts on conservation and preservation, highlighting destination uniqueness, promoting long-term growth, and embracing the 3Rs principles (reduce, reuse, recycle) are crucial strategies for advancing sustainable practices in rural tourism. The conclusions point to the need for players in rural tourism to embrace and improve in this area of rural locations and destinations, as well as for sustainable rural tourist destination management. Additionally, it broadens the focus of previously published research on the significance, inspiration, and sustainable tourism practices of rural areas. In India, the relevance in this area is the subject of this study. There hasn't been much empirical research in this field. From the viewpoints of the rural tourist operators, the research offers a thorough knowledge of the definition and principles behind sustainable operations. It also suggests sensible tactics for sustained rural tourist operations. The advancement and acceptance of these methods owe to critical infrastructure, governmental backing, and individuals' grasp of sustainable development principles. Lastly, this research offers valuable suggestions for fostering the sustainable expansion of rural tourism.

1 Introduction

According to Borto (2002), rural tourism refers to travel to isolated or rural locations. Refocusing efforts to revitalize the local community and environment enhances living standards, fostering improvements in economic, ecological, and sociocultural dimensions of life. In this way, rural tourism supports regional food and recreational pursuits amid stunning scenery, enabling rural communities to improve and enrich their natural and cultural resources and promoting development and value-generating opportunities. Furthermore, it offers job and income prospects to rural communities. It is crucial to the resuscitation of cultural and natural resources. As a result, rural tourism contributes significantly to the local economy. In addition to protecting the environment and advancing rural development in line with sustainable development principles, sustainable tourism also helps preserve the unique character of the place. For a location to develop economically, socially, and environmentally, it must be sustainable. By addressing the detrimental effects of tourism on the area, sustainable practices enhance rural residents' quality of life and increase the area's competitiveness. Activities that support the healthy social and economic development of rural communities while respecting the environment and society are considered sustainable forms of rural tourism (Jovanović & Manic, 2012).

However, managing rural tourism and rural destinations is a complicated matter that calls for the active participation of tourism groups and rural tourism stakeholders. The most important thing is that they value and accept sustainable practices. According to Podovac and Tončev (2016), stakeholders must be informed on how rural tourism should behave regarding natural resources. In a nutshell, sustainability is the capacity to sustain and conserve one's behaviors and resources throughout time. The notion and tenets of sustainable tourism development have been extensively employed to comprehend the sustainability of tourist locations. Sustainable tourism involves harmonizing environmental, economic, and sociocultural considerations using the triple bottom-line framework (Elkington, 1998). This approach ensures that tourism activities and developments prioritize the preservation of social, economic, and environmental integrity, safeguarding natural and cultural resources. It also aims to enhance visitor satisfaction through meaningful experiences while promoting advancements in sustainable tourism practices, as articulated by Niedziółka (2012). It is often known that the main areas of concern are environmental, social, and economic factors, as well as long-term growth without depletion or deterioration of its resources.

Adopting sustainable practices is generally done for several reasons, such as protecting the environment, cutting costs by using fewer resources, being competitive and differentiating, competing in destination branding, adhering to the law and ethics, and practicing corporate social responsibility. Adopting sustainable practices requires all of them. In addition to their distinctive beauty, India's rural locations and destinations are endowed with a wealth of natural and cultural elements. Sustainable methods are necessary to deliver authentic experiences at distinctive rural tourist attractions. In addition, the number of people traveling to rural areas to get away from the city and their jobs is increasing. (Schmidt et al., 2016) Travelers seeking to visit remote locations and those interested in rural tourism have altered their behavior during the COVID-19 outbreak (Aydin, 2022). The needs of travelers who wish to unwind and recover from the pandemic or engage in physically and mentally soothing activities in a natural setting can be well satisfied by rural tourism (Ahmad et al., 2021; Rosalina et al., 2021). As a result, rural tourism gives local people the chance to rehabilitate and diversify their rural areas' economies. Sustainability emerges as a paramount concern in rural tourism development, aiming to harmonize the needs of tourists, the local environment, and the community's economic capacity, according to Ivona in the year (2021).

Furthermore, by preserving the harmony between social fairness, economic rewards, environmental preservation, and cultural integrity, sustainability can raise living conditions (Liu et al., 2013). Furthermore, it was demonstrated by Kantar and Svržnjak, (2017) that there are four components to the integrated sustainability of rural tourism: ecological, political, social, and biological sustainability. To promote sustainable planning and development in rural tourism, the Sabah government founded the District Tourism Association. Additionally, it guarantees that partners in rural tourism adopt and use sustainable practices throughout the locations. The development of rural tourism has universally accepted sustainability as a key issue. Moreover, sustainability and the Sustainable Development Goals (SDGs), unveiled by the UN General Assembly in 2015 to encourage worldwide action for a more sustainable future, are closely related concepts. It is believed that natural and cultural resources are priceless tourism assets that must be preserved and developed using sustainable methods. Nonetheless, overseeing eco-friendly rural tourism locations is a challenging undertaking. Sustainability is multifaceted, circumstantial, subjective, and difficult (Ivona, 2021). Therefore, these locations necessitate a thorough comprehension of both the principles of sustainability and the goals of rural tourist providers. Both directly and indirectly, tourism supports the local community's way of life and the development of rural areas.

However, the progress needs to be long-term. Those engaged in rural tourism must embrace sustainable practices to achieve this. According to Durana et al. (2015), sustainable rural development strategies are often the economic, ecological, and human facets of existence. Four components make up the integrated sustainability of rural tourism: ecological, sociological, political, and biological (Kantar & Svržnjak, 2017). Consequently, it has been affirmed that human, ecological, and economic considerations are pivotal in advancing sustainable rural tourism. By embracing sustainable tourism practices, cultural heritage can be preserved, authentic visitor experiences can be cultivated, and local populations in tourist areas can benefit socioeconomically. Similarly to this, evaluating destination competitiveness from a sustainable development perspective is crucial in addition to economic, socioeconomic, and community development aspects (Luštický & Bednářov, 2018). Similar to how sustainable business practices have emerged as the new standard in response to societal and governmental demands, they have also been linked to social sustainability and economic performance (Duong, 2022), and consumers' awareness of current business practices has grown (Chabowski et al., 2010; Kotler, 2011).

1.1 Problem Statements

Rural tourism has enormous potential to restore local communities, conservation of cultural heritage and promotion of environmental sustainability (Borto, 2002). However, the sustainable development of rural tourism in India faces important challenges. Despite being more accepted for sustainability, there is a limited understanding of how stakeholders in rural areas perceive and implement sustainable guidelines (Ivona, 2021). This gap consists of the 19th rural tourism demand due to travelers seeking natural environments for relaxation and rehabilitation (Chunid, 2016; Ahmad and the group 2021), planning and insufficient awareness lead to a waste of resources the imbalance of the ecosystem and culture and awareness (Podovac & Tončev, 2016). Go to use (Kantar & Svržnjak, 2017). This is a special matter because KRG comes with orientation in the conservation area. The purpose of the search is to inform you how to listen to the local farmers of Logan (Jovanović & Manić, 2012).

Evaluation of the perception of stakeholders: inspection of the sustainable tourism business, and their role in promoting environmental-human responsibility and community participation in the monitoring of tourists (Ivona, 2021; Niedziółka, 2012). Motivation and limitations: Preliminary motivation, sedimentation survey. The stakeholders' perception of sustainable practices and obstacles to their implementation (Elkington, 1998; Chabowski et al., 2010), cultural and economic dimensions (Liu et al., 2013; Durana et al., 2015) offer suggestions: Providing useful suggestions to improve sustainable guidelines for rural areas in line with Sustainable Development Goals (SDGs) and promote long-term benefits for local communities

(United Nations General Assembly, 2015; Duong, 2022) makes a difference. This finding aims to provide in-depth information that can be carried out for policymakers of tourism and local communities to expand rural tourism operations into a sustainable practice.

1.2 Objective of the Study

This study seeks to explore sustainable practices within rural tourism and hospitality in India, emphasizing their role in fostering community engagement and promoting environmental responsibility. The goal of the study is to learn more about how Indian rural tourism stakeholders view and understand sustainable tourism initiatives. Its goal is to pinpoint the main inspirations behind stakeholders' adoption of sustainable practices in rural tourist environments. In addition, the study intends to evaluate the sustainability strategies now in use in rural tourist sites and investigate areas for improvement. The aim of this synthesis of studies is to offer practical proposals and recommendations for the sustainable growth of rural tourism in India.

2 Review of Literature

Moreover, Demirović et al. (2020) explored the sentiments of rural residents regarding the expansion of sustainable tourism, while Rosalina et al. (2021) analyzed the economic impacts of rural tourism. In addition, Getz and Carlson (2000) examined the traits of rural tourist businesses, while Kastenholz et al. (2018) investigated the features of rural tourism encounters. While Podovac and Tončev (2016) looked at the significance of sustainable rural tourism development in Serbia, Kantar and Svržnjak (2017) examined rural tourist development from a sociological viewpoint. Similarly, a great deal of research has been done on how rural tourism is conceptualized (Lane, 1994), why people travel to rural areas (Park & Yoon, 2009), and what challenges arise when trying to diversify rural tourism (Sharpley, 2002). Several facets of the literature have expressed interest in sustainable practices. There has been a wide range of prior studies on rural tourism. Although a great deal of research has been done on sustainable rural tourism and its aspects from the viewpoints of customers, the environment, and the economy as a whole (An & Alarcon, 2020), tour operators' perspectives have been conspicuously absent (Hamid & Isa, 2020). The importance of engaging in sustainable tourism practices is underscored by the United Nations Environment Program and the World Tourism Organization, as highlighted in 2005. The extent to which rural tourist operators implement sustainable practices is unknown, as is the significance of the driving forces behind sustainable practices in Sabah's rural areas. Such data is arguably essential for state tourist offices and relevant policymakers to execute sustainable development in rural areas using sensible plans. Long-term sustainable growth depends on making rural tourism sites more competitive and offering high-quality tourism experiences.

2.1 Rural Tourism

The local community has a chance to develop several rural locations abundant in natural and cultural resources into appealing rural tourism destinations. By leveraging tourism to foster local development and diversify rural economies, these communities are revitalizing their unique districts. Over time, Sabah's rural tourism industry has grown significantly and quickly. It offers distinctive rural tourism experiences and benefits the local community, both of which are essential for the industry to remain competitive and experience sustainable growth. Maintaining the abundance of resources and distinctive experiences at rural locations requires sustainable rural tourism. To what extent the rural tourism community and operators have implemented the sustainability idea in the rural environment is unknown. Rural communities play a crucial role as stakeholders in tourism destinations as they actively contribute to the creation of the tourism product and have the power to greatly influence the perceptions of tourists. To put it simply, they are the proprietors and custodians of the local resources land, structures, and goods that serve as the foundation for development. Their involvement is essential because their abilities, customs, knowledge, and enthusiasm are the main resources for rural development. These are the residents of the specific rural area that serves as the foundation for sustainable development. It is their expertise on what, how, and why to implement sustainable practices. This report investigates the sustainable behaviors of stakeholders in rural tourism to fill this research gap. This research specifically examines the driving elements and underlying meaning of adopting sustainable practices in rural tourism in India from the perspective of rural communities.

2.2 Sustainability in Rural Tourism

To be competitive, vacation rentals must be environmentally friendly. According to recent research using Cucculelli and Goffi (2016), expanding sustainable tourism places importance on improving the profitability and maintaining the biological stability of the region. This requires effective tourism management (Iqbal & Ahmed 2022). In comparison with similar destinations, a competitive tourism vacation spot no longer only preserves its ecological and cultural assets but also enhances the long-term nicely being of its citizens by using presenting enriching stories (Bahar & Kozak, in the 12 months 2007; Ritchie & Crouch, within the yr 2003). Achieving sustainable growth in tourism requires long-term strategic plans and substantial investment from the public and private sectors (Coros et al., 2017). The 1987 Bruntland Report first addressed sustainable improvements that it achieves two objectives without compromising the ability of future generations to meet human needs. As time passes the concept emphasizes a balanced strategy covering financial, social, and environmental dimensions. In the first phase, sustainable tourism growth generally requires efforts in environmental protection and conservation.

2.3 Sustainability and Consumer Behavior in Rural Tourism

Sustainable consumer behavior in rural areas is to decide consciously, which reduces the environmental impact while being beneficial to the local community. Visitors are responsible for the environment, like accommodation that uses environmentally friendly guidelines such as renewable energy, water conservation, and waste management. They support real experiences, local crafts, food, and traditions. In preserving culture and economic development (2020), the selection of local products reduces the amount of carbon dioxide emitted from transport, which contributes to environmental sustainability. In addition, the ecosystem in rural areas benefits from tourists who are responsible for the conservation of resources and participation in community activities to ensure the long-term viability of rural tourism.

2.4 Sustainability Strategies and Practices in Rural Tourism

Rural tourism has developed into an important avenue for sustainable development. It integrates environmental protection, environmental and cultural enrichment, and increased investment around the world. Practices related to green living spaces, renewable energy, and waste management systems are common. For example, agritourism in Italy combines environmental protection with tourism, helping to develop nearby biological areas and reduce greenhouse gas emissions. Similarly, international locations like New Zealand emphasize low-impact tourism via strict environmental policies and community-driven conservation tasks. Tourism is gaining momentum, with sustainability at its center. Initiatives together with the "Rural Tourism Scheme" through the Ministry of Tourism promote cultural immersion while supporting neighborhood artisans and history conservation. States like Rajasthan and Kerala give priority to domestic residues. Eco hotel and city tours to increase your acquaintance with companies and reduce the gap between the city and the countryside. Develop social innovation projects for communities and partnerships with NGOs. It conceals financial flexibility in these areas. Technological innovations such as biogas construction, solar energy, and a curling farm play an important role in reducing the environmental footprint of rural tourism. Moreover, the integration of ancient knowledge into current tourism strategies is not particularly outstanding, but by enhancing cultural history, it also promotes more calculated tourism, and helps to strengthen the relationship between host and visitor.

2.5 Sustainability Practices in the Hospitality Industry and Their Role in Rural Tourism

The hospitality enterprise is central to rural tourism, with numerous activities having a significant effect on environmental, social, and economic outcomes. Sustainability in hospitality encompasses strategies to lessen resource use, lessen waste, and promote neighborhood network participation. Many worldwide hospitality vendors have followed eco-certifications, which include LEED and Green Key, which emphasize energy performance, water conservation, and reducing waste in the context of rural tourism. These practices have been adapted to preserve the natural ecosystem and cultural heritage. For example, eco-lodges and home stays in villages in India incorporate renewable energy sources such as solar power and use traditional, sustainable building materials. Responsible Tourism Mission in Kerala demonstrates this integration by promoting community-owned tourism models that empower local people while preserving environmental integrity. Hotels are also introducing zero waste initiatives, such as using bio-waste facilities and composting food scraps. For instance, Marriott switched to reusable amenities, reducing the usage of single-use plastics in the environment. Sourcing products and materials locally is another critical motion. This stimulates rural economies and reduces transportation-associated greenhouse gas emissions. In addition, sustainable hospitality regularly consists of visitor schooling programs that sell environmentally friendly behaviors, including water conservation and environmental protection. Various pollution techniques are not the only cause of harm to the environment, but they also create a deeper connection among visitors and locals. Improving awareness of rural tourism and ensuring its long-term sustainability are essential.

2.6 Challenges and Opportunities for Sustainable Rural Tourism

Sustainable rural tourism faces significant challenges that may hinder its development. The main problem is the infrastructure gap. Many rural areas lack roads, transportation, and essential amenities. This limits investment in tourism projects and discourages investment. Another challenge is reducing rural population. This is driven by seasonal employment and limited economic opportunities. This makes it difficult to retain skilled workers. Environmental degradation is also a threat. This is because poorly managed tourism can deplete natural resources and harm biodiversity. There is still a lack of knowledge and access to digital technology. Stakeholders regularly face difficulties in using technology correctly for marketing or control functions. Despite these demanding situations rural tourism gives vital opportunities to enhance sustainability. It can power monetary increase and job advancement. This is because income from tourism enables diversifying the agricultural financial system and reduces reliance on agriculture. Rural tourism can also make contributions to cultural and environmental conservation by way of promoting the protection of historical sites and landscapes. The growth in revolutionary tourism merchandise, along with agritourism, ecotourism, and digital advertising platforms, offers a new way to draw vacationers while promoting sustainability. Aligning rural tourism with the Sustainable Development Goals (SDGs) additionally promotes gender equality poverty reduction and sustainable community improvement.

3 Research Methodology

This study centers on examining rural tourism stakeholders' viewpoints and comprehension of sustainable tourism practices in India, exploring their significance, motivations, and potential areas for enhancement. The researcher used the qualitative technique in order to get more in-depth information from the participants on the connection between sustainable practices in rural tourism and its function in the area's social change. Qualitative approaches are useful for examining the contextual component of complicated issues in real-life situations using a variety of data sources and when the target group is easily accessible in their home. A set of interview questions has been created as a result of the researcher's several visits to the study area, rounds of conversations, and meetings with the local tourism stakeholders in the Garhwal regions of Uttarakhand. Purposive and snowball sampling were used to choose research participants. The researcher first spoke with one villager, after which he or she suggested the next, and so on. To capture a variety of perspectives on the growth of Rural Tourism in the area, an attempt has been made to engage participants of all ages, genders, professions, and economic levels. Based on the intended goal and information gathered from in-depth interviews, the researcher is able to identify the different variables in the Garhwal regions of Uttarakhand. Personal interviews were chosen as the suitable method for data collection. The research describes an exploratory qualitative project aimed at investigating stakeholders' sustainable tourism practices in rural tourism in India. An inductive approach was utilized to meet the research objectives, involving interviews with 29 respondents conducted from May to June 2024. The majority of participants had over 6 years of experience in rural tourism and possessed formal education. Data collection took place on-site in two districts, Chamoli, Pauri Garhwal, facilitated by rural tourism operators' facilities. The study adopted a cross-sectional design, and data collection continued until saturation was reached, as proposed by Walker (1985), indicating no new information was emerging from interviews. Permission for in-person interviews was obtained from the respondents, facilitated with the assistance of rural tourism associations in each district. The process of data analysis included identifying keywords from interview responses, categorizing and organizing emerging themes and aligning them with the study's objectives. The eight themes of sustainable practices in tour operations proposed by Hamid and Isa in

2020, as well as the pull and push elements served as the incentive factors for implementing these practices.

4 Results and Discussions

The primary objective of sustainable development in rural areas and rural tourism is to achieve a harmonious balance among ecological, social, and economic sustainability. Sustainability cannot be claimed for a region or activity without understanding and integrating these three interconnected elements. Sustainable rural tourism not only highlights the natural landscapes and features of rural areas but also preserves and promotes the cultural traditions and values of local communities. The answers provided by the people interviewed on the definition of sustainable tourism practices revealed nine main topics: conservation, growth and low environmental effect, competitive products and high-quality attractions, cooperation and protection, future generations, equity, and satisfying visitor demands. These themes are a close reflection of the descriptions of sustainable tourism activities found in the previously listed literature. A sizable portion of the respondents equated resource protection and preservation with sustainable tourism activities. For instance, respondents defined sustainable tourism as "focusing on the conservation of ecology and natural environment" and "means to conserve and protect the environment and nature," "without damaging the natural conditions and culture." Similarly, a few participants linked sustainable tourism to financial gains, expansion, and the creation of new employment opportunities: "The arrival of tourists will boost and energize the local economy," "generating revenue." The following factors were also linked to sustainable rural tourism practices: cooperation and protection; competitive goods and high-quality tourist attractions; justice; knowledge, awareness, and improvement; and attending to the requirements of visitors. As stated by the respondents, these interpretations include: "long-term growth and for a future generation[s], [it] means products have high attractiveness," "are competitive and attractive," and "collaboration between members of the association." Curiously, the rural tourist organization president understood the word to indicate "willingness to improve and indicates that sustainable practices are closely related to the rural destination's social, economic, and environmental characteristics as well as the sustainability pillars." Research indicates that sustainable practices are closely related to the rural destination's social, economic, and environmental characteristics as well as the sustainability pillars. Similarly, long-term development and supporting future generations, offering top-notch services maintaining destinations' competitiveness, attending to visitor demands, and advancing knowledge and education were all considered aspects of sustainable practices. Beyond the pillars of sustainability, sustainable practices are essential to the rural tourism destination's viability. Seven topics on the reasons for adopting sustainable rural tourism were identified by the empirical findings. The topics surrounding economic gains include growth and economic activity, managing government support and governance, protecting and conserving natural resources, local and social development, networking, collaboration, and cooperation; improving knowledge and training to develop resilience; and marketing and competition. The sustainability pillars the economy, the environment, and society as well as pull and push variables can be used to group these themes. Overall, the majority of participants demonstrated a keen interest in obtaining essential information necessary for the advancement of rural tourism and promoting sustainable growth. Similarly, some of the participants in rural tourism have been spurred to adopt sustainable practices by social and local development, including government support: "Sustainable practices enable us to collaborate better." Some said, "We are really motivated to move toward sustainable practices by the government's support." Furthermore, the participants said that information, up skilling initiatives, and awareness of sustainable practices were the main drivers of their motivations: "[We] wanted to know more; the knowledge and training about sustainable practices are our motivation." As can be seen, a large number of respondents claimed that government programs and assistance were what drove them. Sustainable practices are considered as draw factors or destination attractions that stimulate the adoption of local and social development, government assistance, and laws. This shows how important it is to provide incentives for those involved in rural tourism to adopt sustainable practices. The adoption of sustainable practices was driven mostly by the desire to protect the environment, promote social and local development via cooperation, and reap the financial rewards. Beyond this, it has been shown that adopting sustainable practices also requires knowledge and skills, government assistance and legislation, better marketing, and increased competitiveness. This suggests that the reasons for implementing sustainable practices in the rural setting are comparable to those of the tour operators that Hamid and Isa (2020) investigated, who were driven by sustainable marketing, resource conservation, employee development, legal compliance, and customer happiness. It's interesting to notice that procurement strategy, sustainability system, and facility design were not considered to be driving forces in the rural tourism industry.

The results highlight several areas in which the stakeholders in rural tourism need to increase their use of sustainable practices. The areas identified for improvement include acquiring knowledge and skills, enhancing infrastructure and basic facilities, refining marketing strategies and recycling efforts, conserving and preserving the natural environment, and fostering engagement and collaboration in sustainable practices aspects that fall within the social, economic, and environmental categories. The results indicate that more community involvement and participation from rural tourism players, together with joint efforts to implement sustainable practices, are necessary to lead the way in implementing sustainable practices at rural sites because of divergent viewpoints and the ensuing disputes. Therefore, it is imperative to have a deeper comprehension of the advantages of sustainable practices in rural tourism. At the same time, for rural tourism participants to pursue sustainable practices, improving relevant knowledge and skills is crucial. To be clear, there is a notable lack of knowledge about sustainable practices in rural areas. To advance their sustainable practices, the players in rural tourism must employ sustainable practices, knowledge, and skills. For rural tourism businesses, having sufficient infrastructure and basic amenities like internet, electricity, and water is essential. Many respondents said that to draw more visitors to the rural locations, conservation and preservation activities must be strengthened in addition to using innovative marketing techniques. In the same way, more frequent usage of the 3Rs—reduce, reuse, and recycle—at remote locations might improve sustainable practices, to build sustainable rural tourism in Sabah, these are important areas that require more focus.

The techniques proposed in this article enhance the adoption of sustainable practices and bolster the roles and capacities of rural tourism stakeholders in promoting sustainable development. Understanding the concept of sustainable practices is crucial for operators in the rural tourism industry, as it enables them to realize the benefits and foster sustainable rural tourism. To attain a complete approach, sustainable practices necessitate extensive collaboration and engagement among all stakeholders. This acceptance extends to the operators' sustainable activities in the remote location. In order to encourage sustainable practices at the location, operators of rural tourism forge strong partnerships with tour operators and travel agents. To encourage environmentally friendly practices at the site, operators of rural tourism forge close partnerships with tour operators and travel agents. To ensure the effective management and implementation of sustainable practices across social, economic, and environmental aspects in rural areas, it is advisable for the rural tourism organization to establish a committee and develop appropriate criteria. All rural tourism providers essentially need to collaborate with one another at the remote locations and adopt sustainable practices. Establishing a training and development program should be the second step in developing a development strategy for sustainable rural tourist growth, according to the rural tourism association and the destination management organization. The implementation of sustainable rural tourist practices will be supported by the knowledge and technical skills that such a program will provide. Incentives for rural tourist operators should be created to encourage them to become more sustainable, as the sector is still relatively new to the adoption of sustainable practices and ideas. Money-based and non-monetary rewards should be included in the incentives. Thirdly, rural sustainable development needs to be encouraged by big government agencies like the Department of Wildlife and Forestry and the local government. With the Sabah tourism council's assistance, this can be started by the regional rural tourism organization. It is therefore possible to conserve and preserve natural resources. In addition, the growth of rural tourism depends on having sufficient infrastructure and amenities, including road access, electricity, water, and internet for rural tourist destinations. The lack of these facilities currently causes a lot of rural tourism destinations and business operations to be disrupted or closed down. The growth of rural tourism is aided by sustainable practices, but the adoption of these practices by operators of rural tourism may be hindered by significant upfront costs, opposition to changes in sustainability and practices, and the practicality of these practices.

5 Limitations and Suggestions for Future Research Directions

This study highlights the importance of sustainable practices in rural tourism but has certain limitations. One major constraint is its focus on stakeholder perceptions, which may not fully account for broader systemic challenges such as inadequate policies, financial constraints, and infrastructure gaps that hinder sustainable tourism development. The findings are context-specific, making it challenging to generalize them across India's diverse rural regions, each with unique cultural, ecological, and socioeconomic conditions. Additionally, the study lacks a long-term perspective, limiting its ability to assess the sustainability of implemented practices over time. Future research could adopt a more comprehensive approach by combining qualitative and quantitative methods to provide a deeper understanding of these issues. Exploring the role of technology, such as digital tools for tourism management and sustainability tracking, could enhance the efficiency and effectiveness of rural tourism initiatives. Comparative studies across different regions or countries could offer valuable insights into replicable models and best practices. Understanding how travelers' preferences and behaviors, especially post-pandemic, influence rural tourism demand can inform tailored strategies. Expanding the scope to include diverse stakeholder groups, including policymakers, local businesses, and tourists, would help develop more inclusive and adaptable frameworks for promoting sustainability in rural tourism.

6 Implications and Conclusion

This study significantly enhances our understanding of sustainable practices in rural tourism by focusing on rural tourism operators' perspectives, filling a substantial gap in existing literature with empirical evidence. It highlights that sustainable practices in rural tourism extend beyond economic, environmental, and social dimensions, impacting destination sustainability, tourism product quality, operational fairness, and visitor satisfaction. The study identifies economic benefits, social development, environmental protection, skills enhancement, and government support as primary motivations for adopting sustainable practices. Key areas for improvement include education and training, rural infrastructure development, and fostering collaboration among stakeholders. These insights are crucial for rural tourism operators, offering essential guidance on implementing sustainable strategies to secure the long-term viability of rural tourism destinations.

However, the study's limitations, such as its focus on a limited sample from specific districts, suggest the need for broader research to generalize findings across diverse rural tourism contexts. Nonetheless, this research provides valuable insights and practical recommendations for advancing sustainable practices in rural tourism. This review makes a substantial contribution to the existing literature by highlighting the lack of systematic organization of sustainability concerns in rural tourism (RT) literature, despite their extensive discussion. It underscores the timeliness of analyzing RT and sustainability, given the increasing relevance of sustainability in business discussions. The study adds to a growing body of evaluations as discussed by Alfaro Navarro et al. (in the year 2020), Bramwell et al. (in the year 2017), Xu et al. (in the year 2020), and Zhang and Zhang (in the year 2020) that address sustainability challenges in the tourism industry. Furthermore, it reviews stakeholder involvement in RT sustainability, among the first of its kind. This analysis underscores the importance of collaborative action supported by policy and examines the crucial role of entrepreneurs, presenting compelling insights for stakeholders in rural tourism (RT). This information is valuable for policymakers aiming to identify and engage communities capable of organizing effective group initiatives for RT development. Moreover, the review advances discussions on leveraging sustainable RT to achieve rural sustainable development goals. Research indicates that eco-branding and sustainability messaging positively impact tourists (Wang et al., 2018), which can encourage RT business owners to adopt these strategies despite initial hesitation. Additionally, it underscores the critical role of entrepreneurs in identifying and creating new value in rural communities aspiring to develop RT sustainably. Collaborative efforts between communities and policymakers to identify visionary entrepreneurs are crucial for successful sustainability initiatives.

References

Ahmad, M., Akhtar, N., Jabeen, G., Irfan, M., Anser, M. K., Wu, H., & Isik, C. (2021). Intention-based critical factors affecting willingness to adopt novel coronavirus prevention in Pakistan: Implications for future pandemics. *International Journal of Environmental Research and Public Health, 18*(11), 6167–6195.

Alfaro Navarro, J. L., Andrés Martínez, M. E., & Mondéjar Jiménez, J. A. (2020). An approach to measuring sustainable tourism at the local level in Europe. *Current Issues in Tourism, 23*(4), 423–437. https://doi.org/10.1080/13683500.2019.1579174

An, W., & Alarcon, S. (2020). How can rural tourism be sustainable? A systematic review. *Sustainability, 12*(7758), 1–23.

Aydin, E. (2022). Rural tourism researches: A co-citation analysis. *Journal of Tourism Leisure and Hospitality, 4*(1), 30–36.

Bahar, O., & Kozak, M. (2007). Advancing destination competitiveness research: Comparison between tourists and service providers. *Journal of Travel Marketing, 2*, 61–71.

Borto, G. (2002). Rural tourism in Romania - Characteristics and trends. Tourism. *Journal of Department of Geography, Tourism and Hotel Management, Faculty of Sciences, 6*, 111–113.

Bramwell, B., Higham, J., Lane, B., & Miller, G. (2017). Twenty-five years of sustainable tourism and the Journal of Sustainable Tourism: Looking back and moving forward. *Journal of Sustainable Tourism, 25*(1), 1–9.

Chabowski, B. R., Mena, J. A., & Gonzalez-Padron, T. L. (2010). The structure of sustainability research in marketing, 1958-2008: A basis for future research opportunities. *Journal of the Academy of Marketing Science, 39*(1), 55–70.

Coros, M. M., Gica, O. A., Yallop, A. C., & Moisescu, O. I. (2017). Innovative and sustainable tourism strategies: A viable alternative for Romania's economic development. *Worldwide Hospitality and Tourism Themes, 9*, 504–515.

Cucculelli, M., & Goffi, G. (2016). Does sustainability enhance tourism destination competitiveness? Evidence from Italian destinations of excellence. *Journal of Cleaner Production, 11*, 370–382.

Demirović, B. D., Radosavac, A., Cimbaljević, M., Tretiakova, T. N., & Syromiatnikova, Y. A. (2020). Determinants of residents' support for sustainable tourism development: Implications for rural communities. *Sustainability, 2020, 12*(22), 9438–9454. https://doi.org/10.3390/su12229438

Duong, N. H. (2022). Relationship of social sustainability, operational performance and economic performance in sustainable supply chain management. *Global Business and Finance Review, 27*(4), 46–64.

Durana, D. C., Gogana, L. M., Artenea, A., & Duran, V. (2015). The components of sustainable development – A possible approach. *Procedia Economics and Finance, 26*, 806–811.

Elkington, J. (1998). *Cannibals with forks: The triple bottom line of 21st century business*. New Society Publishers.

Getz, D., & Carlson, J. (2000). Characteristics and goals of family- and owner-operated businesses in the rural tourism and hospitality sectors. *Tourism Management, 21*, 547–560.

Hamid, M. A., & Isa, S. M. (2020). Exploring the sustainable tourism practices among tour operators in Malaysia. *Journal of Sustainable Science and Management, 15*(1), 68–80.

Iqbal, Z., & Ahmed, S. (2022). Antecedents and consequences of community-based tourism in border district of Jammu and Kashmir. *International Journal of Tourism Cities, 8*, 569–587. https://doi.org/10.1108/IJTC-05-2020-0108

Ivona, A. (2021). Sustainability of rural tourism and promotion of local development. *Sustainability, 13*, 8854–8861.

Jovanović, V., & Manić, E. (2012). Evaluation of sustainable rural tourism development in Serbia. *Scientific Annals of the Danube Delta Institute, 18*(1), 285–294.

Kantar, S., & Svržnjak, K. (2017). *Development of Sustainable Rural Tourism. Deturope, 9*(1), 26–34.

Kastenholz, E., João Carneiro, M., Peixeira Marques, C., Loureiro, M. C., & S. (2018). The dimensions of rural tourism experience: Impacts on arousal, memory, and satisfaction. *Journal of Travel & Tourism Marketing, 35*(2), 189–201.

Kotler, P. (2011). Reinventing marketing to manage the environmental imperative. *Journal of Marketing, 75*(4), 132–135.

Lane, B. (1994). What is rural tourism? *Journal of Sustainable Tourism, 2*, 7–12.

Liu, C. H., Tzeng, G. H., Lee, M. H., & Lee, P. Y. (2013). Improving metro-airport connection service for tourism development: Using hybrid MCDM models. *Tourism Management Perspectives, 6*, 95–107.

Luštický, M., & Bednářov, M. (2018). Tourism destination competitiveness assessment: Research & planning practice. *Global Business and Finance Review, 23*(3), 49–67.

Niedziółka, I. (2012). Sustainable tourism development. *Regional Formation and Development Studies, 8*(3), 157–166.

Park, D.-B., & Yoon, Y.-S. (2009). Segmentation by motivation in rural tourism: A Korean case study. *Tourism Management, 30*, 99–108.

Podovac, M., & Tončev, M. J. (2016). The importance of sustainable rural tourism development in Serbia. *SINTEZA 2016 International scientific conference on ICT and E-business related research*. ICT in Tourism & Hospitality.

Ritchie, J. R. B., & Crouch, G. I. (2003). *The competitive destination, a sustainable tourism perspective*. Cabi Publishing.

Rosalina, P., Dupre, K., & Wang, Y. (2021). Rural tourism: A systematic literature review on definitions and challenges. *Journal of Hospitality and Tourism Management, 47*, 134–149.

Schmidt, C. M., Cielo, I. D., Wenningkamp, K. R., & Tomio, M. (2016). Collective actions in sustainable rural tourism: A case study of the Western region of Paraná. *Systems Research and Behavioral Science, 33*, 249–258.

Sharpley, R. (2002). Rural tourism and the challenge of tourism diversification: The case of Cyprus. *Tourism Management, 23*, 233–244.

Walker, R. (1985). An introduction to applied qualitative research. In R. Walker (Ed.), *Applied qualitative research*. Gower Publishing.

Wang, W., Wu, J., Wu, M. Y., & Pearce, P. L. (2018). Shaping tourists' green behavior: The hosts' efforts at rural Chinese B&Bs. *Journal of Destination Marketing and Management, 9*, 194–203. https://doi.org/10.1016/j.jdmm.2018.01.006

Xu, F., Nash, N., & Whitmarsh, L. (2020). Big data or small data? A methodological review of sustainable tourism. *Journal of Sustainable Tourism, 28*(2), 144–163. https://doi.org/10.1080/09669582.2019.1631318

Zhang, L., & Zhang, J. (2020). A systematic review on tourism energy consumption, sustainable tourism, and destination development: A behavioral perspective. In *Transport and energy research* (pp. 295–313). https://doi.org/10.1016/B978-0-12-815965-1.00013-2

MIX
Papier aus verantwortungsvollen Quellen
Paper from responsible sources
FSC® C105338

If you have any concerns about our products,
you can contact us on
ProductSafety@springernature.com

In case Publisher is established outside the EU,
the EU authorized representative is:
Springer Nature Customer Service Center GmbH
Europaplatz 3, 69115 Heidelberg, Germany

Printed by Libri Plureos GmbH
in Hamburg, Germany